# Maxima and Minima
# with Applications

# Maxima and Minima with Applications

## Practical Optimization and Duality

**WILFRED KAPLAN**
*Emeritus Professor of Mathematics*
*University of Michigan*

**A Wiley-Interscience Publication**
**JOHN WILEY & SONS, INC.**
New York · Chichester · Weinheim · Brisbane · Singapore · Toronto

This book is printed on acid-free paper.$\odot$

Copyright © 1999 by John Wiley & Sons, Inc. All rights reserved.

Published simultaneously in Canada.

No part of this publication may be reproduced, stored in a retrieval
system or transmitted in any form or by any means, electronic,
mechanical, photocopying, recording, scanning or otherwise, except as
permitted under Sections 107 or 108 of the 1976 United States Copyright
Act, without either the prior written permission of the Publisher, or
authorization through payment of the appropriate per-copy fee to the
Copyright Clearance Center, 222 Rosewood Drive, Danvers, MA
01923, (978) 750-8400, fax (978) 750-4744. Requests to the Publisher for
permission should be addressed to the Permissions Department, John
Wiley & Sons, Inc., 605 Third Avenue, New York, NY 10158-0012,
(212) 850-6011, fax (212) 850-6008, E-Mail: PERMREQ@WILEY.COM.

*Library of Congress Cataloging-in-Publication Data*:

Kaplan, Wilfred, 1915–
    Maxima and minima with applications: practical optimization and
duality / Wilfred Kaplan.
        p.    cm.—(Wiley-Interscience series in discrete mathematics
    and optimization)
    "A Wiley-Interscience publication."
    Includes bibliographical references and index.
    ISBN 0-471-25289-1 (alk. paper)
    1. Maxima and minima.  2. Mathematical optimization.  I. Title.
    II. Series.
    QA306.K36  1999                                    98-7318
    511′.66—dc21

Printed in the United States of America.

10 9 8 7 6 5 4 3 2 1

# Contents

# Preface

This book is intended as a text for an intermediate level course on maxima and minima. Practical applications are very much in the background and many illustrations are given. The mathematical background assumed is that of calculus, preferably advanced. Some knowledge of linear algebra is required; an Appendix summarizes the essential tools of that subject.

A prominent role is played by the concept of convexity. On the one hand, many applications involve finding the minimum (maximum) of a convex (concave) function of one or more variables; on the other hand, the convex and concave functions have many properties that assist greatly in finding their maxima and minima. In particular, they have properties of continuity and differentiability that enhance their value.

The four chapters are of increasing depth and difficulty. In particular, the first chapter requires modest mathematical knowedge, even though it provides insight into basic concepts of maxima and minima. The last chapter makes available profound research of Fenchel and Rockafellar in a presentation of moderate difficulty.

Chapter 1 stresses the geometric aspect of maximum and minimum problems. Ideas familiar in elementary calculus are reviewed and extended. Convexity is introduced and developed to modest extent. There is much emphasis on quadratic functions and quadratic forms, which both illustrate the theory and are an important tool for the general problems. Level sets are introduced, and their structure near extreme points is studied. They lead naturally to ordinary differential equations as a method for computing the extreme points; here the concepts of stability and asymptotic stability are shown to be useful. Norms and distance functions are developed and related to minimum problems.

Chapter 2 considers problems with side conditions: first, those of the form of equations; then those that may also include inequalities. The Lagrange multiplier method is explained, and second-derivative tests are treated fully, along with the concept of index of a critical point. Differential equations are again shown to be a useful tool in locating critical points. The Karush–Kuhn–Tucker necessary conditions for side conditions including inequalities are considered at some length; some corresponding sufficient conditions are also presented.

Chapter 3 is an introduction to optimization, with emphasis on convex problems. More convexity theory is developed. Mathematical programming and duality are discussed and then considered in detail for linear programing and quadratic programming. The Fermat–Weber problem and some generalizations are considered.

Chapter 4 has as its principal goal the development of duality theorems of Fenchel and Rockafellar and illustration of their power. Convexity theory is extended further: in particular, to the fundamental separation theorem and the concept of conjugacy. Minkowski norms are defined and shown to have significant practical applications. A very general form of the Fermat–Weber problem, involving such norms, is considered; a dual problem is established and shown to be applicable to general location problems.

Throughout the four chapters many exercises are provided. They illustrate the concepts presented and, in some cases, provide significant additional theoretical results. Starred problems are more difficult.

The author expresses to his friend and colleague Wei H. Yang his profound appreciation for introducing the author to the field of Optimization and providing many ideas which are incorporated in this book. In particular, he provided much of the exposition in Chapter 3 and the author expresses gratitude for his permission to include this material. The author thanks his colleague Katta G. Murty for valuable advice about finding an initial feasible point for programming problems.

The author also expresses to John Wiley & Sons his appreciation for their fine cooperation during the production of this book and to Aiji K. Pipho his thanks for the effort she made to produce the many illustrations.

WILFRED KAPLAN

*Ann Arbor, Michigan*
*February 1998*

# 1

# Maxima and Minima in Analytic Geometry

## 1.1 MAXIMA AND MINIMA; CASE OF FUNCTIONS OF ONE VARIABLE

Throughout this book we consider functions $f(x)$, $f(x, y)$, and, in general, real-valued functions of one or more real variables. One could also consider real-valued functions defined on general sets. For example, one may have a function defined on a set of functions; such a function occurs in elementary calculus, namely the *integral* of a function:

$$I = \int_a^b g(x)dx. \tag{1.10}$$

The value of $I$ depends on the choice of $g$ so $I$ is a function defined on a set of functions. One refers to such a function as a *functional*.

On occasion we consider functions defined on *finite* sets: for example, the function

$$f(n) = \frac{n}{n^2 + 1} \quad \text{for} \quad n = 1, 2, \ldots, 10. \tag{1.11}$$

We can give one definition of maximum and minimum to cover all these cases. The function $f$ defined on a set $E$ has maximum $M$ (*finite*) on $E$ if $f$ has the value $M$ for one member of the set $E$ and $f$ has value at most $M$ for all members of $E$; we call a member of $E$ at which $f$ has its maximum, $M$, a *maximizer* of $f$. When $f$ has a maximum on $E$, then $f$ must have at least one maximizer, but there may well be more than one maximizer. It can also occur that $f$ has no maximum at all.

The definitions of *minimum* and *minimizer* are similar: $f$ has minimum $m$ on $E$ if $f$ has value $m$ for one member of $E$ (a minimizer) and $f$ has value at least equal to $m$ for all members of $E$. We use the word *extremum* to cover the two cases of maximum and minimum; thus we say that $f$ has an extremum at a

1

certain member of $E$ if that member is a maximizer or minimizer of the function.

We also refer to the maximum and minimum just defined as *global maximum* and *global minimum*, respectively. It is preferable to insert the word "global" when one is also considering "local" maxima and minima, to be defined shortly.

***Example 1.*** The function $f(x) = \ln x$ for $0 < x \le 1$ has global maximum 0, with maximizer 1; it has no global minimum.

***Example 2.*** The function $f(x) = \sin x$ for $0 \le x \le 3\pi$ has global maximum 1 with maximizers $\pi/2$ and $5\pi/2$; it has global minimum $-1$ with minimizer $3\pi/2$.

***Example 3.*** The function of (1.11) has maximum $\frac{1}{2}$ with maximizer 1 and minimum $10/101$ with minimizer 10. (See Problem 1.1(e).)

Differential calculus provides powerful tools for finding maxima and minima of functions of one or more real variables. We here consider only the functions $f(x)$ of one variable. If such a function is defined over a closed interval $[a, b]$ (all $x$ such that $a \le x \le b$) and $f$ is continuous on the interval, then it is known that $f$ has a global maximum $M$ and global minimum $m$ on the interval. This is intuitively obvious from a graph (Fig. 1.1); a proof is given in advanced calculus (see Kaplan, 1991, p. 178). Now let the function $f$ be dif-

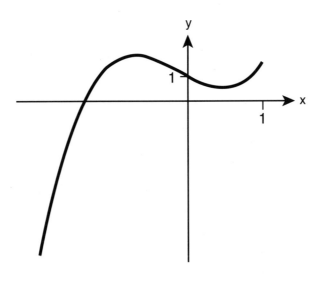

**Figure 1.1.** Graph of Example 4.

ferentiable on the interval (i.e., $f$ has a derivative $f'(x)$ at every $x$ of the interval, where $f'(x)$ is interpreted as a right-hand derivative at $a$ and a left-hand derivative at $b$). If $f(x_0) = M$ or $f(x_0) = m$ at a point $x_0$ *interior* to the interval (i.e., $a < x_0 < b$), then necessarily $f'(x_0) = 0$. This very important rule is proved in elementary calculus (see Kaplan, 1991, p. 149). A point $x$ at which $f'(x) = 0$ is called a *critical* point of $f$. This result leads one to the standard procedure for finding global maxima and minima of a function $f(x)$: find all the critical points of $f$ and evaluate $f$ at each; the global maximum (or minimum) is one of these values or else it is $f(a)$ or $f(b)$.

***Example 4.*** $f(x) = 3x^3 + x^2 - 3x + 2$ for $-2 \le x \le 1$. The function is graphed in Fig. 1.1. The critical points are found to be the two numbers $(-2 \pm \sqrt{112})/18$, or 0.4768 and $-0.6991$. Both lie in the given interval $[-2, 1]$. One finds (Problem 1.1(d))

$$f(0.4768) = 1.1221, \quad f(-0.6991) = 3.5610, \quad f(-2) = -12, \quad f(1) = 3.$$

Hence $M = 3.5610$ with maximizer $-0.6991$ and $m = -12$ with minimizer $-2$.

For our function $f(x)$ on $[a, b]$, we now say that $f$ has a *local maximum* at a point $x_0$ of the interval if $f(x) \le f(x_0)$ for all $x$ on $[a, b]$ that are sufficiently close to $x_0$; that is, for which $|x - x_0| < \delta$ for some positive number $\delta$. Equivalently we can state that $f$ has a local maximum at $x_0$ if, on some smaller open interval (or half-open interval if $x_0$ is an endpoint of $[a, b]$) containing $x_0$ and contained in $[a, b]$, $f$ has a global maximum $f(x_0)$, with maximizer $x_0$. A local minimum is defined similarly.

From the previous results we see that a local extremum of $f$ can occur only at a critical point interior to the interval $[a, b]$ or at an endpoint of the interval. Thus to find all local maxima and minima, we again find all the critical points and examine the values at these points and at $a$ and $b$. The points just described (in all practical cases, a finite set) divide the interval $[a, b]$ into a finite number of subintervals, inside each of which the derivative $f'(x)$ is never 0, so the function $f$ is either steadily increasing or steadily decreasing. From this property we can locate the local maxima and minima easily.

***Example 5.*** Let $a = 0$ and $b = 10$, and let critical points of $f$ occur only at $x = 2, 4, 7, 8, 9$; if $f(0) = 3$, $f(2) = 5$, $f(4) = -1$, $f(7) = -2$, $f(8) = 6$, $f(9) = 3$, $f(10) = 2$, then there is a local minimum at $x = 0$, a local maximum at $x = 2$, a local minimum at $x = 7$, a local maximum at $x = 8$, and a local minimum at $x = 10$. (See Problem 1.4.)

For a critical point $x_0$ inside $[a, b]$, one can also obtain useful information from higher derivatives of $f(x)$. If $f''(x_0) < 0$, then the critical point provides a local maximum; if $f''(x_0) > 0$, then the critical point provides a local minimum. These rules are proved in elementary calculus. The sign of the second derivative is related to the *concavity* of the graph of $f$: where it is positive, the graph is concave upward; where it is negative, the graph is concave downward (see Section 1.2). At a point where the second derivative is 0, an inflection point may occur, that is, a point where the graph crosses the tangent line. The second derivative tests can be illustrated by Example 4: at the critical point $-0.6991$ the second derivative $18x + 2$ is negative, and there is a local maximum; at the critical point $0.4768$ it is positive, and there is a local minimum.

For an endpoint the first derivative (a one-sided derivative, as described above) alone may suffice to determine whether there is an extremum: If $f'(a) > 0$, then there is a local minimum at $a$; if $f'(a) < 0$, then there is a local maximum at $a$; at $b$, there are similar results, with reversal of the two inequalities. These rules follow from the relationship between the sign of the derivative and the increasing or decreasing nature of the function. An illustration is provided by Example 4 (Fig. 1.1), for which the derivative is positive at both endpoints; hence there is a local minimum at $-2$ and a local maximum at 1.

Finally we observe that if $f$ has its global minimum or maximum at a point, then it also has a local minimum or maximum at the point; if the point happens to be the endpoint $a$, then in the case of a global maximum necessarily $f'(a) \leq 0$ and in the case of a global minimum necessarily $f'(a) \geq 0$. These results follow from those of the preceding paragraph.

**Example 6.** *An example from mechanics.* Let a particle of mass $m$ move on the $x$-axis subject to a force derived from a potential $V(x)$; that is, the force is $-V'(x)$. It is shown in mechanics that conservation of energy then holds; that is, for each particular motion of the particle,

$$\tfrac{1}{2}mv^2 + V(x) = c, \tag{1.12}$$

where $v = dx/dt$, the velocity of the particle, and $c$ is a constant, the total energy of the particle. (See Problem 1.5(b).) (a) If $x_0$ is a critical point of $V$, then a possible "motion" of the particle is to remain at rest at $x_0$; one calls $x_0$ an *equilibrium point*. (See Problem 1.5(c).) (b) If $V = kx^2$, where $k$ is a positive constant and hence $V$ has a local minimum at $x_0 = 0$, then the equilibrium at 0 is *stable*; that is, a motion starting close to $x = 0$ with small speed $|v|$ remains close to 0 for all time $t$. (See Problem 1.5(d).) (c) If in (b) $k$ is a negative constant and hence $V$ has a local maximum at $x_0 = 0$, then the equilibrium at 0 is *unstable*: that is, one cannot ensure that the motion remains close to $x_0 = 0$ by making the initial position close to 0 and the initial speed sufficiently small. (See Problem 1.5(e).) One shows that the assertions in (b) and (c) apply quite generally to local minima and maxima of the potential energy (see Section 1.10).

## PROBLEMS

**1.1.** Find the global maximum and global minimum, if either exists, for each of the following functions:

(a) $y = 1/x$ for $0 < x \leq 1$.

(b) $y = \sec x$ for $0 \leq x < \pi/2$.

(c) $y = x^3 - 3x$ for $0 \leq x < \infty$.

(d) The function of Example 4 in Section 1.1.

(e) The function of (1.11).

(f) The function $f(n) = n^2/2^n$ for $n = 1, 2, \ldots$.

(g) The length of a line segment joining two points on different sides of the pentagon in the $xy$-plane whose vertices are the points $(0, 0)$, $(1, 0), (3, 1), (1, 3), (0, 1)$.

(h) The area of a circular area which covers all five points of part (g).

**1.2.** Find the critical points of each function and determine whether the function has a local maximum or local minimum at the point:

(a) $y = x^3 + 6x^2 - 15x + 2$ for $-\infty < x < \infty$.

(b) $y = x^2 e^x$ for $-\infty < x < \infty$.

(c) $y = 3\cos 2x + 4\sin 2x$ for $0 \leq x \leq 2\pi$.

(d) $y = (x^2 - 1)^{-2}$ for all $x$ except $\pm 1$.

**1.3.** Determine whether the function has a local maximum or minimum at an endpoint of the interval of definition:

(a) $y = x^2 - 7x + 12$ for $0 \leq x \leq 5$.

(b) $y = e^{2x} + 3e^{-5x} + 2e^{7x}$ for $1 \leq x \leq 4$.

**1.4.** Verify the assertions made in Example 5 in Section 1.1.

**1.5.** Consider the mechanics problem of Example 6 in Section 1.1.

(a) Show that the motion is governed by the differential equation

$$m\frac{dv}{dt} \equiv m\frac{d^2x}{dt^2} = -V'(x).$$

(b) Multiply both sides of the differential equation by $v = dx/dt$ and integrate to obtain (1.12).

(c) Show that if $V'(x_0) = 0$, then the particle can remain at rest at $x_0$ (equilibrium state).

(d) Let $V(x) = kx^2$, where $k$ is a positive constant and hence $V$ has a local minimum at $x = 0$. Obtain an expression for the solution of the second-order differential equation for $x$ as function of $t$ in terms of the initial values $x_0$ and $v_0$ of $x$ and $v$. From this expression show that the equilibrium position $x = 0$ is stable.

**(e)** Show that if **(d)** is changed by making $k$ a negative constant (and so $V$ has a local maximum at $x = 0$), then the equilibrium is unstable.

## 1.2  CONVEXITY

We will often use the phrase *lies below* to compare the graphs of two functions. For example, the graph of $f(x)$ lies below the graph of $g(x)$ if $f(x) \leq g(x)$ wherever both functions are defined. We say that the graph of $f$ is *strictly below* that of $g$ if $f(x) < g(x)$ wherever both functions are defined. We use the word "above" in a similar way for the opposite inequalities. We also write, for example: $f$ is above $g$ at $x = c$ to mean that $f(c) \geq g(c)$, and $f$ is strictly above $g$ at $x = c$ to mean that $f(c) > g(c)$.

Now let $f(x)$ be defined on the closed interval $[a, b]$. This function is said to be *convex* if its graph lies below the chord joining each two points of the graph, as in Fig. 1.2; here the chord is the graph of a linear function defined on an interval $[x', x'']$. This geometrical definition of convexity is equivalent to the inequality

$$f((1 - \alpha)x' + \alpha x'') \leq (1 - \alpha)f(x') + \alpha f(x'')  \tag{1.20}$$

for each choice of $x'$, $x''$ on $[a, b]$ and each $\alpha$ on the interval $[0, 1]$. To see this, we first observe that for $\alpha = 0$ both sides of (1.20) reduce to $f(x')$ and for $\alpha = 1$

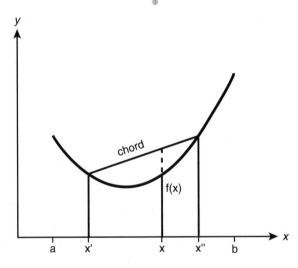

**Figure 1.2.** Convex function.

both sides reduce to $f(x'')$. For $\alpha = \frac{1}{2}$ the left side of (1.20) is the value of $f$ at the midpoint of the interval $[x', x'']$, while the right side is the value of $y$ at the midpoint of the chord joining the endpoints of the graph over the interval $[x', x'']$; because of the inequality, the value of $f$ at the midpoint is at most equal to the value of $y$ at the midpoint of the chord—that is, the graph of $f$ is below the chord at the midpoint. A similar reasoning applies to each point on $[a, b]$. The left side is the value of $f$ at the point dividing $[a, b]$ in the ratio $\alpha : (1 - \alpha)$ and the right side is the value of $y$ at the point of the chord dividing it in the same ratio.

We have defined convexity only for a function defined on a closed interval. However, the same definition can be used for a function defined on other types of interval: for example, $a < x < b$ (open), $a \leq x < b$ (half-open), $a \leq x < \infty$ (infinite, half-open), $-\infty < x < \infty$ (infinite, open). (Occasionally we abbreviate such intervals as $(a, b)$, $[a, b)$, $[a, \infty)$, $(-\infty, \infty)$, respectively.) In each case the graph of the function is required to be below the chord joining each pair of points on the graph of $f$. By graphing, we see that the functions $y = x^2$ and $y = x^4$ are convex on the interval $(-\infty, \infty)$. Also the function $y = x$ is convex on this interval, as is every linear function, since the graph coincides with the chord on every closed interval. The function $y = |x|$ is also convex on the infinite interval; the graph coincides with the chord in some cases, is below the chord in all cases (Fig. 1.3).

We have made no mention of continuity and, in fact, a convex function can have discontinuities (see below). If $f$ is continuous on $[a, b]$ and has continuous derivatives through the second order, then $f$ *is convex if and only if* $f''(x) \geq 0$ *for*

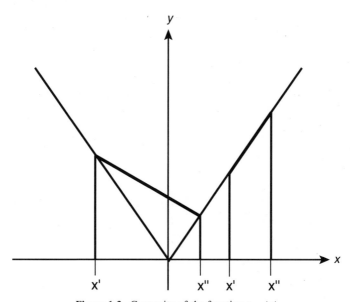

**Figure 1.3.** Convexity of the function $y = |x|$.

*all x in the open interval* (*a, b*). The proof of this theorem is left as an exercise (Problem 1.9(a) and (c)).

A convex function is said to be *strictly convex* if its graph is always *strictly below* the chord minus its endpoints. Thus in the inequality (1.20) the sign $\leq$ is replaced by $<$ for $0 < \alpha < 1$. The function $y = x^2$ is strictly convex over the whole $x$-axis, but the function $y = |x|$ is convex, not strictly, on every interval. As in the preceding paragraph, for the case of continuous second derivatives one has a rule: *If $f''(x) > 0$ on an interval, then $f$ is strictly convex on that interval* (see Problem 1.9b). In this case there is no converse; for example, the function $y = x^4$ is strictly convex over the whole $x$-axis, but its second derivative is $0$ at $x = 0$.

A function $f(x)$ is said to be *concave* if its graph always lies *above* the chord; that is, if in (1.20) $\leq$ is replaced by $\geq$. Equivalently we can define a concave function as one whose negative is convex. Strict concavity is defined by analogy with strict convexity. Generally, results about concave functions are established by referring to the convex functions obtained by reversing signs.

We observe that the strictly convex functions correspond to the functions whose graphs are concave upward and that the strictly concave functions correspond to the functions whose graphs are concave downward, as mentioned near the end of Section 1.1.

***Example 1.*** *Example of a discontinuous convex function.* The function $f(x)$ on the interval [0, 1] such that $f(0) = 1$ and $f(x) = x$ for $0 < x \leq 1$ is discontinuous at the endpoint 0 but is convex, as one sees from its graph (Fig. 1.4).

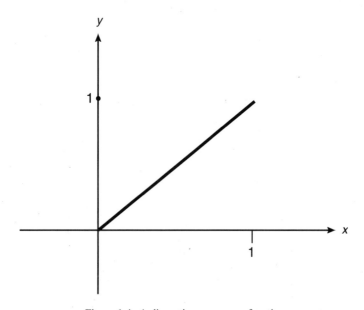

**Figure 1.4.** A discontinuous convex function.

However, *a convex function is necessarily continuous at each* **interior** *point of the interval over which it is defined*. This one sees easily by geometric reasoning. As in Fig. 1.5, let the convex function $f$ be defined on the interval $[a, b]$, and let $a < p < q < b$, determining corresponding points $A$, $P$, $Q$, $B$ on the graph of $f$. Then $Q$ must lie below the chord $PB$, by convexity. But $Q$ must also lie above the chord $AP$ extended, for otherwise $P$ would lie strictly above the chord $AQ$. Thus $Q$ is "trapped" in a sector with vertex at $P$, and as $q \to p$ from the right, $Q$ must have $P$ as limit. This shows that $f$ is continuous to the right at $p$. A similar argument shows that $f$ is continuous to the left at $p$.

A similar argument shows that *at each interior point the convex function $f$ has derivatives from the right and from the left*. To prove this, we refer to Fig. 1.5 and observe that the slope of the chord $PQ$ is less than or equal to that of the chord $PB$; in general, as $x$ decreases from $q$ toward $p$, $Q$ moves along the curve toward $P$, and the slope is decreasing (or nonincreasing). Also the slope of the chord $PQ$, for all choices of $Q$ on the arc $PB$, is at least equal to that of the chord $AP$. Hence, as $x$ decreases from $q$ toward $p$, the slope has a finite limit. This proves existence of a derivative to the right at $p$. A similar reasoning shows existence of a derivative to the left at $p$. In general, these two derivatives need not be equal, as is illustrated by the function of Fig. 1.3.

## 1.3 CONVEXITY AND MAXIMA AND MINIMA

A convex function $f(x)$ may have no local minima or maxima, as is shown by the function $y = e^{-x}$ on the whole $x$-axis. However, *if it has a local minimum, say at $x_0$, then $f(x_0) = m$, the global minimum of the function*. Indeed, if $f(x_1) < m$ for some $x_1 \neq x_0$ on the interval on which the function is given, then by convexity

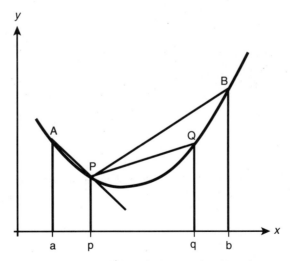

**Figure 1.5.** Proof of continuity at an interior point.

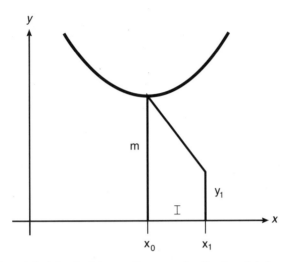

**Figure 1.6.** A local minimum of a convex function is a global one.

the graph of $f$ on the interval $I$ with endpoints $x_0$, $x_1$ lies below the line segment joining $(x_0, y_0)$ and $(x_1, y_1)$, where $y_0 = f(x_0)$, $y_1 = f(x_1)$. (See Fig. 1.6.) Since $y_1 < m$, we conclude that $f(x) < m$ for all $x$ other than $x_0$ on the interval $I$. This contradicts the assumption that there is a local minimum at $x_0$. The corresponding assertion about local *maxima* of a convex function is not correct (see Problem 1.10(a)).

*Every critical point $x_0$ of a convex function must be a minimizer.* Indeed, if $f(x_1) < f(x_0)$ for some $x_1$ on the interval of definition of $f$ and, for example, $x_1 > x_0$, then for $x_0 + \Delta x$ on the interval $(x_0, x_1)$ the corresponding $\Delta y$ satisfies $\Delta y/\Delta x \leq k < 0$, where $k$ is the slope of the chord (see Fig. 1.6). Hence $\Delta y/\Delta x$ cannot have limit 0 as $\Delta x \to 0$. This proves the assertion. One should observe that the assertion and proof apply to the case when the critical point occurs at an endpoint of the interval where the convex function is defined.

The result just established gives another way of showing that a critical point $x_0$ at which the second derivative (assumed continuous) is positive must be a local minimizer. From $f''(x_0) > 0$ and continuity we conclude that $f''(x)$ is positive in some interval to which $x_0$ is interior (or having $x_0$ as endpoint). Thus $f$ must be convex (even strictly) in such an interval, and by the previous result $x_0$ is a minimizer in the interval; that is, $x_0$ is a local minimizer.

A typical function $f(x)$ may have several local minima (and maxima). However, if $f$ is convex and $f$ has local minima at several values of $x$, then $f$ *must be constant over an interval containing all these values.* By the result proved at the beginning of this section, $f$ must equal its global minimum $m$ at each of the local minima. If there are two local minima, say at $x_0$ and $x_1$, then $f(x_0) = f(x_1) = m$, so the corresponding chord is horizontal. By convexity, the graph lies below the chord, but if it actually falls strictly below the chord, $m$ would not be the global minimum. Hence the graph of $f$ coincides with the chord. From this we conclude that $f$ must be constant over an interval, each point of which provides a local minimum of $f$. An example is suggested in Fig. 1.7.

If $f$ is *strictly convex*, then there can be *at most one* local minimum, so an example like that of Fig. 1.7 cannot occur. Indeed, strict convexity requires that the graph of $f$ lies *strictly below* the chord minus its endpoints; thus $f$ cannot be constant over an interval.

In general, a local minimum $x_0$ of a function is said to be a *strong local minimum* if, for all $x$ sufficiently close to $x_0$ but not equal to $x_0$, the values of the function are strictly greater than $f(x_0)$. A strong local maximum is defined similarly, with "greater than" replaced by "less than." A local maximum or minimum that is not strong is said to be *weak*. These concepts are illustrated in Fig. 1.8.

The function of Fig. 1.7 is an example of a function with local minima none of which is a strong local minimum. As remarked above, such a graph cannot arise for a strictly convex function and, by the same reasoning, every local minimum (or maximum) of a strictly convex function is a strong one. In particular, if $f''(x)$ is continuous and $f''(x_0) > 0$ at a critical point $x_0$ of $f$, then $f$ must be strictly convex in some interval containing $x_0$, so there is a strong local minimum at $x_0$.

As in Section 1.1, we can use the first derivative to test for local extrema at the endpoints of the interval $[a, b]$ of definition of function $f$. If, for example, $f'(a) > 0$, then there is a strong local minimum at $a$.

### Functions Strictly Convex on the Whole x-Axis

If $f(x)$ has a continuous second derivative and $f''(x) > 0$ for all $x$ and hence $f$ is strictly convex, then three cases arise: (a) $f$ has a critical point $x_0$, the minimizer of $f$; (b) $f'(x) > 0$ for all $x$; (c) $f'(x) < 0$ for all $x$. In case (a), $f'(x) > 0$ for $x > x_0$ and $f(x) \to \infty$ as $x \to \infty$, $f'(x) < 0$ for $x < x_0$ and $f(x) \to \infty$ as

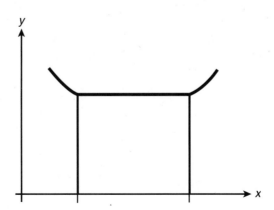

**Figure 1.7.** Convex function with many local minima.

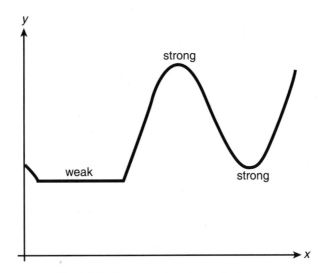

**Figure 1.8.** Strong local maxima and minima.

$x \to -\infty$; in case (b), $f(x) \to \infty$ as $x \to \infty$ but not as $x \to -\infty$; in case (c), $f(x) \to \infty$ as $x \to -\infty$ but not as $x \to \infty$. The three cases are illustrated by the functions $x^2$, $e^x$, and $e^{-x}$ respectively. (See Problem 1.14.)

### Convexity versus Strict Convexity

We observe that the examples of convex functions which are not strictly convex all have graphs including a *linear* portion. This property is typical: *A convex function whose graph includes no line segment must be strictly convex.* (See Problem 1.16.)

### PROBLEMS

**1.6.** For each of the following functions, verify graphically that it is convex for $-\infty < x < \infty$, and state whether the function is strictly convex:

    **(a)** $y = x^2$.

    **(b)** $y = x^4$.

    **(c)** $y = |x|$.

    **(d)** $y = |x + 1| + |x - 1|$.

**1.7.** As in Example 1 in Section 1.2, let $f(x) = x$ for $0 < x \le 1$ and $f(0) = 1$. Show that $f(x)$ is convex on the interval $[0, 1]$ but is not continuous on the interval.

**1.8.** Show that each of the following functions is strictly convex:

(a) $y = 1/(1 - x^2)$ on $(-1, 1)$.

(b) $y = x^6 - 15x^2$ on $[2, 3]$.

(c) Every solution of the differential equation $y'' = 1 + y'^2$.

(d) Every function $y = f(x)$ whose graph has positive curvature $\kappa$ (defined as $y''(1 + y'^2)^{-3/2}$).

**1.9.** Let $f(x)$ be defined, with continuous first and second derivatives on the interval $[a, b]$.

(a) Prove: If $f''(x) \geq 0$ on $[a, b]$, then $f$ is convex. [Hint: Let $a \leq x_1 < x_2 \leq b$, and suppose that there is an $x_3$ between $x_1$ and $x_2$ such that $f(x_3)$ is greater than the value of $y$ at $x_3$ on the chord through $(x_1, f(x_1))$ and $(x_2, f(x_2))$. Let the chord have slope $m$. Then by the mean value theorem $f'(x_4) > m > f'(x_5)$ for some $x_4, x_5$, with $a < x_4 < x_3 < x_5 < b$. Hence by the mean value theorem $f''(x_6) < 0$ for some $x_6$ between $x_4$ and $x_5$, contrary to the assumption that $f''(x) \geq 0$. See Fig. 1.9.]

(b) Prove: If $f''(x) > 0$ on $[a, b]$, then $f$ is strictly convex. [Hint: One must show that the graph of $f$ generally lies strictly below the chord minus its endpoints. Assume that the graph has one point, not an endpoint, on the chord and imitate the reasoning of part (a).]

(c) Prove: If $f$ is convex, then $f''(x) \geq 0$ on $[a, b]$. [Hint: If $f''(x_0) < 0$ for one $x_0$ on the interval, then by continuity $f''(x) < 0$ in some interval containing $x_0$. By analogy with (b), this implies that $f$ is *strictly concave* on that interval, so the graph would generally have

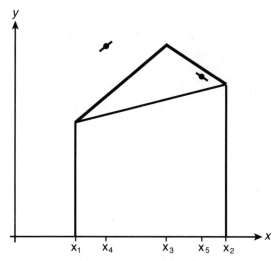

**Figure 1.9.** Proof of part (a) in Problem 1.9.

to be *strictly above* the chord minus its endpoints. This contradicts the assumed convexity of $f$.]

REMARK. The results of this problem remain valid if one assumes that $f$ is continuous on $[a, b]$, that $f'(x)$ and $f''(x)$ are continuous on $(a, b)$, and that the conditions on $f''$ hold on $(a, b)$. In fact, the same proofs are valid.

**1.10. (a)** Show by example that a convex function can have a local maximum that is not its global maximum.

    **(b)** Show by example that a function $f$ can have $f''(x_0) = 0$ at a critical point but not have a local minimum or maximum at the point.

**1.11.** Show that the function of Problem 1.6(**d**) is an example of a convex function with many local minima.

**1.12.** Let $y = f(x)$ be defined by the following conditions: $y = 2 - x$ for $x < 1$; $y = 1$ for $1 \leq x \leq 3$; $y = 4 - x$ for $3 < x \leq 4$; and $y = (4 - x)^2$ for $x > 4$. Graph and locate all local minima, stating which are strong local minima.

**1.13.** Determine whether there is a strong local minimum or maximum at an endpoint of the interval of definition:

    **(a)** $y = \sin^2 x$ for $0 \leq x \leq 3\pi/2$.

    **(b)** $y = 5x^4 + 2x^3 + 3x^2 + 7x + 2$ for $1 \leq x \leq 3$.

**1.14.** Let $f$ be strictly convex on the whole $x$-axis as near the end of Section 1.3.

    **(a)** Justify the discussion of the three cases given.

    **(b)** Show by examples that in case (b) as $x \to -\infty$, $f$ may have a finite limit or may have limit $-\infty$.

    **(c)** (Difficult) Show that every strictly convex function $f(x)$ on $(-\infty, \infty)$ (with no assumption about differentiability) having a minimizer $x_0$ is strictly increasing for $x \geq x_0$ and strictly decreasing for $x \leq x_0$ and has limit $+\infty$ as $x \to \pm\infty$.

**1.15.** Let the function $y = f(x)$ be defined and continuous, with continuous derivative, on an interval.

    **(a)** Show that if $f$ is convex, then for each pair of points $a$, $b$ of the interval, one has

$$f(b) - f(a) \geq f'(a)(b - a). \qquad (*)$$

[Hint: First assume that $a < b$. Since the graph of $f$ lies below each chord, one has for each $x$ such that $a < x < b$

$$f(x) \leq f(a) + (x - a)\frac{f(b) - f(a)}{b - a}.$$

From this relation obtain an inequality for the difference quotient $[f(x) - f(a)]/[x - a]$ and pass to the limit. If $b < a$, proceed similarly and observe that the direction of the inequality is reversed twice.]

**(b)** Prove a converse: If the inequality $(*)$ of **(a)** is valid for every pair of points $a$, $b$ on the interval, then $f$ is convex. [Hint: The conclusion of **(a)** can be interpreted as asserting that the graph of $f$ lies above each tangent line to the graph. Show that if $f$ fails to be convex and thus at some point the graph is strictly above the chord, then there is a tangent line to the graph having a point of the graph strictly below it.]

**(c)** Show that if $f$ is convex, then for all $a$, $b$ on the interval, $a < b$ implies $f'(a) \leq f'(b)$. [Hint: Apply the result $(*)$ of **(a)**, and this result with $a$, $b$ interchanged.]

**(d)** Prove a converse of **(c)**: If, for all $a$, $b$ on the interval, $a < b$ implies $f'(a) \leq f'(b)$, then $f$ is convex. [Hint: Apply the mean value theorem of differential calculus to show that the result $(*)$ of **(a)** holds for $a < b$ and also for $b < a$.]

**(e)** Prove: If in **(d)** $a < b$ implies $f'(a) < f'(b)$, then $f$ is strictly convex.

**(f)** Prove: If $f''(x) > 0$ except at a finite number of points, at which $f''(x) = 0$, then $f$ is strictly convex. [Hint: Show that **(e)** applies.]

**1.16.** Let $f(x)$ be convex for $a \leq x \leq b$, and let the graph of $f$ meet the chord joining $(a, f(a))$ to $(b, f(b))$ at a point that is not an endpoint of the chord. Show that $f$ is a linear function for $a \leq x \leq b$. Deduce the rule: *A convex function whose graph contains no line segment is strictly convex.*

## 1.4   PROBLEMS IN TWO DIMENSIONS

A function $f(x, y)$ of two variables is typically defined in an open region or a closed region (analogous to the open and closed intervals for functions of one variable). An open region is illustrated by a set consisting of the points inside a closed curve, such as a circle or ellipse (see Fig. 1.10). If one includes the boundary points—that is, the points on the closed curve—then one obtains a closed region.

For completeness, we give general definitions and refer to Chapter 2 of Kaplan (1991) for a full discussion. The $\delta$-neighborhood of a point $P$ in the plane is the set formed of all points whose distance from $P$ is less than $\delta$; here $\delta$ is a positive number. An open set in the plane is one such that for each point $P$ of the set, all points within some $\delta$-neighborhood of $P$ are in the set. A closed set is either the entire $xy$-plane or a set such that the points not in the set form an open set. An open region is an open set such that each pair of points of the set can be joined by a broken line (polygonal path) in the set. A boundary point of a set is a point such that each $\delta$-neighborhood of the point contains points of the

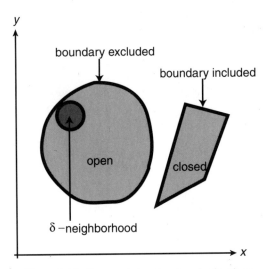

**Figure 1.10.** Open and closed regions in the plane.

set and points not in the set; all boundary points together form the boundary of the set. The boundary of every set (if nonempty) is closed, and the union of every set and its boundary is a closed set. A point of a set that is not a boundary point is an interior point of the set; the interior of the set $E$ is the set formed of all the interior points. A closed region is the set formed of an open region and its boundary. A set is bounded if it can be enclosed in a circle of sufficiently large (finite) radius. These definitions generalize to space of three or higher dimension and also apply in the one-dimensional case, that is, the $x$-axis.

We state two rules concerning a function of two variables:

A function $f(x, y)$ defined and continuous on a bounded closed set $E$ has a global maximum $M$ and global minimum $m$ on $E$.

If the function $f$ is defined on the set $E$, with maximizer or minimizer at an interior point $(x_0, y_0)$ of $E$, and if $f$ is differentiable at $(x_0, y_0)$, then the partial derivatives of $f$ must be 0 at the point, so the point is a *critical point* of $f$.

These rules are established in elementary or advanced calculus; see Kaplan (1991, pp. 178, 151). Accordingly, just as for one variable, one seeks the global maximum or minimum of $f$ on a bounded closed region $E$ by finding the critical points of $f$ interior to $E$ and comparing the values of $f$ at these points with the values of $f$ on the boundary of $E$.

***Example 1.*** Find the global extrema of the function $z = xy - x + y$ for $-2 \leq x \leq 2, -2 \leq y \leq 2$. Here the set $E$ is a bounded closed region, a square. One finds that there is one critical point interior to $E$, at $(1, -1)$, where $z = -3$.

Along each boundary line segment, the function becomes a nonconstant linear function of $x$ or $y$ and its extrema occur at the ends of the segments. Hence we need only examine the values of $z$ at $(-1, 1)$ and the four corners of the square. One finds that $M = 4$, with maximizers $(2, 2)$ and $(-2, -2)$, and that $m = -4$, with minimizer $(2, -2)$. (See Problem 1.17.)

A function $f(x, y)$, defined on set $E$, has a *local maximum* at a point $(x_0, y_0)$ of $E$ if, at all points of $E$ in some $\delta$-neighborhood of $(x_0, y_0)$, the values of $f$ are all at most equal to $f(x_0, y_0)$. A local minimum is defined similarly, and strong local extrema are defined as in Section 1.3. If $E$ is an open or closed region and $f$ is differentiable on the interior of $E$, then each local extremum of $f$ at an interior point must occur at a critical point of $f$.

Many extremum problems for functions $f(x, y)$ of two variables can be reduced to such problems for functions of one variable.

***Example 2.*** Find the global maximum and minimum of the function $z = x^4 + y^4$ on the ellipse $x^2 + 2y^2 = 1$. We verify that on the ellipse $x$ can take values between $-1$ and $1$, and that for each such $x$, $y$ has two values (which coincide for $x = \pm 1$). If we solve the ellipse equation for $y^2$, we can then express $z$ as a function of $x$ alone:

$$z = x^4 + \frac{(x^2 - 1)^2}{4} = \frac{5}{4}x^4 - \frac{1}{2}x^2 + \frac{1}{4}, \quad -1 \le x \le 1.$$

By the methods of Section 1.1, we find that this function has global maximum 1, with maximizers $x = \pm 1$, and has global minimum $1/5$, with minimizers $\pm 1/\sqrt{5}$ (Problem 1.18). Thus for the given function, $M = 1$, with maximizers $(\pm 1, 0)$ and $m = 1/5$, with minimizers $(\pm 1/\sqrt{5}, \pm \sqrt{2/5})$ (four points).

***Example 3.*** Find the global maximum and minimum of a quadratic function $Q(x, y) = ax^2 + 2bxy + cy^2$ on the circle $x^2 + y^2 = 1$. This problem yields important information about the geometry of second-degree equations in the plane. We discuss it here and leave to Problem 1.19 the verification of some of the assertions made.

Before considering the example, we remark that in analytic geometry it is shown that by rotating the axes through angle $\alpha$ about the origin to obtain new coordinates $x'$, $y'$, the quadratic function $Q$ takes a similar form:

$$Q = a'x'^2 + 2b'x'y' + c'y'^2. \tag{1.40}$$

We discuss the process further below (see especially Problem 1.19).

The problem of Example 3 can be reduced to one for for functions of one variable by using a parametrization of the circle:

$$x = \cos t, \quad y = \sin t, \quad 0 \le t \le 2\pi.$$

The function $Q$ becomes a function

$$g(t) = a\cos^2 t + 2b\cos t \sin t + c\sin^2 t, \quad 0 \le t \le 2\pi.$$

One finds that $g(t)$ has critical points for four values of $t$: $t = \alpha$, $t = \alpha + \pi/2$, $t = \alpha + \pi, t = \alpha + 3\pi/2$, where

$$\alpha = \left(\frac{1}{2}\right) \arctan\left[\frac{2b}{a-c}\right]. \tag{1.41}$$

We observe that the four values of $t$ determine four points on the circle $x^2 + y^2 = 1$ whose directions from the origin are at right angles, as in Fig. 1.11. If we choose new coordinates $(x', y')$ by rotating axes about the origin through the angle $\alpha$ given by (1.41), then the four directions are simply those of the new axes and the extrema of $Q$ occur only at the points where the new axes meet the circle. Hence in the new coordinates the given function $Q$ must take the special form $a'x'^2 + c'y'^2$, and we are considering this on the circle $x'^2 + y'^2 = 1$. Now the extrema must occur at the points $(\pm 1, 0)$ and $(0, \pm 1)$ in the new coordinates; the values of the function at these four points are $a'$ and $c'$, and the relations between these two numbers determine which points are maximizers and which are minimizers.

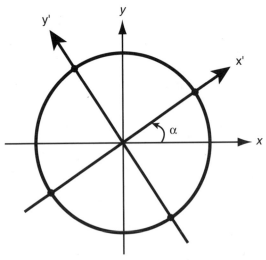

**Figure 1.11.** Extrema of a quadratic function on the circle $x^2 + y^2 = 1$.

REMARK. The case $b = 0$, $a = c$ requires special treatment.

We have incidentally shown that a rotation of axes through the special angle $\alpha$ given by (1.41) reduces the function $Q$ to the form $a'x'^2 + c'y'^2$ and thus in (1.40) $b' = 0$. In Problem 1.19(d) it is shown that after a rotation of axes through an *arbitrary* angle $\alpha$, the function $Q$ takes the form (1.40) and that $a' + c' = a + c$, $a'c' - b'^2 = ac - b^2$. For the case where $b' = 0$, we have $a' + c' = a + c$ and $a'c' = ac - b^2$. In the case where $Q$ has a positive minimum on the circle, $a'$ and $c'$ are both positive, so $a + c$ and $ac - b^2$ are both positive (and hence so also are $a$ and $c$). Conversely, when $a + c$ and $ac - b^2$ are both positive, $a'$ and $c'$ are also positive, so $Q$ has a positive minimum on the circle. We thus conclude that *Q has a positive minimum on the circle precisely when a, c and ac* $- b^2$ *are all positive.* (See Problem 1.19(e).) Similarly, one can show that $Q$ *has minimum* 0 *on the circle precisely when* $a + c \geq 0$ *and* $ac - b^2 = 0$ (Problem 1.19(f)).

REMARK. The technique of rotating axes is used in analytic geometry to study the graph of a second-degree equation such as the equation

$$ax^2 + 2bxy + cy^2 = 1$$

or the general equation

$$ax^2 + 2bxy + cy^2 + dx + ey + f = 0.$$

By rotating axes through angle $\alpha$ as in (1.41), one eliminates the term in $xy$ and has an easier equation to graph. One finds that the graph is a conic section (ellipse, hyperbola, or parabola) that may be degenerate.

## Convexity

A set in the $xy$-plane is termed *convex* if the set contains the line segment joining each pair of points of the set. Figure 1.12 suggests typical convex sets, as well as one nonconvex set. In general, each $\delta$-neighborhood is convex as is each open or closed elliptical region or rectangular region. The whole $xy$-plane is convex, and each straight line or line segment, with or without endpoints, is a convex set.

A function $f(x, y)$ defined on a convex set $E$ is said to be a convex function if it is convex as a function of one variable along each line segment in $E$; that is, *for every such line segment, its graph along the segment lies below the chord,* as in Fig. 1.13. If we parametrize the line segment by equations

$$x = (1 - t)x_1 + tx_2, \quad y = (1 - t)y_1 + ty_2, \quad 0 \leq t \leq 1, \tag{1.42}$$

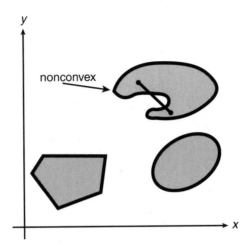

**Figure 1.12.** Convex sets in the plane.

then the condition for convexity becomes

$$f((1 - t)x_1 + tx_2, \ (1 - t)y_1 + ty_2) \leq (1 - t)f(x_1, \ y_1) + tf(x_2, \ y_2) \qquad (1.43)$$

for all $t$ in the interval $[0, 1]$ and each pair of points $(x_1, \ y_1)$, $(x_2, \ y_2)$ in the set $E$.

The convex function $f$ is termed *strictly convex* if in (1.43) $\leq$ is replaced by $<$ for $0 < t < 1$. A function $f$ is termed *concave* or *strictly concave* if its negative is convex or strictly convex, respectively.

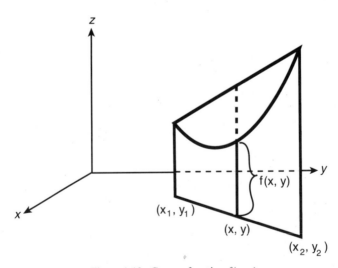

**Figure 1.13.** Convex function $f(x, y)$.

For functions of one variable, the second derivative was helpful in testing for convexity. A function $f(x, y)$ of two variables has generally three different second partial derivatives. From these we can form the quadratic expression

$$Q = f_{xx}u^2 + 2f_{xy}uv + f_{yy}v^2. \tag{1.44}$$

The value of $Q$ depends on the point $(x, y)$ at which the partial derivatives of $f$ are evaluated and on the choices of $u$ and $v$.

We now have the rules: *If $f(x, y)$ has continuous derivatives through the second order in an open convex set $E$, then $f$ is convex on $E$ precisely when, for each point of $E$, the expression $Q$ is nonnegative for $u^2 + v^2 = 1$. If in each case the function $Q$ is positive for $u^2 + v^2 = 1$, then $f$ is strictly convex.*

The rules are an immediate consequence of the rules for functions of one variable. Along a line segment (1.42) the function $f(x, y)$ becomes a function $g(t)$ whose second derivative is

$$f_{xx}(x_2 - x_1)^2 + 2f_{xy}(x_2 - x_1)(y_2 - y_1) + f_{yy}(y_2 - y_1)^2. \tag{1.45}$$

This expression follows from the chain rules of calculus (Problem 1.21). We can write $u = (x_2 - x_1)/d$, $v = (y_2 - y_1)/d$, where $d$ is the distance between $(x_1, y_1)$ and $(x_2, y_2)$ and hence $u^2 + v^2 = 1$. The function of (1.45) becomes $d^2 Q(u, v)$, where $Q$ is as in Example 3 above, with $a = f_{xx}$, $b = f_{xy}$, $c = f_{yy}$ evaluated at a chosen point of the line segment. Now $Q(u, v)$ is nonnegative for $u^2 + v^2 = 1$ precisely when the second derivative is nonnegative, or precisely when $g(t)$ is convex. If at each point of the line segment, $Q$ is positive for $u^2 + v^2 = 1$, then the second derivative is positive at every point and $g$ is strictly convex. If $Q$ is nonnegative for $u^2 + v^2 = 1$ at every point of $E$, then $f$ is convex along every line in $E$ and is hence a convex function in $E$; if "nonnegative" is replaced by "positive," then $f$ is strictly convex in $E$.

From the results of Example 3, we now obtain specific tests for convexity and strict convexity: Let

$$a = f_{xx}, \quad b = f_{xy}, \quad c = f_{yy}.$$

Then *$f$ is convex on $E$ precisely when at each point of $E$ either $a$, $c$, and $ac - b^2$ are all positive or else $a + c \geq 0$ and $ac - b^2 = 0$; if $a$, $c$, and $ac - b^2$ are all positive at each point of $E$, then $f$ is strictly convex on $E$.*

REMARK. For the quadratic function $Q$ the case $a' > 0$, $c' < 0$ is of interest. From the discussion above, the case is characterized by the condition $ac - b^2 < 0$. The function $Q$ now has a positive maximum and a negative minimum on the circle. An example is the function $z = xy$ whose graph is a *saddle surface* (see Kaplan, 1991, p. 157).

## Critical Points of a Convex Function

Let $f(x, y)$ be convex and differentiable in the open convex set $E$. Then *each critical point* $(x_1, y_1)$ *of* $f$ *provides a global minimum of* $f$. Indeed, let $(x_2, y_2)$ be a second point of $E$ and form the line segment (1.42) joining the two points. As above, along the segment $f$ becomes a function $g(t)$, with $t = 0$ at $(x_1, y_1)$. By the chain rule

$$g'(t) = f_x(x_2 - x_1) + f_y(y_2 - y_1), \tag{1.46}$$

where $f_x$ and $f_y$ are evaluated at the point $(x, y)$ given by (1.42). Hence $g'(0) = 0$. But the convexity of $f$ implies that $g(t)$ is convex for $0 \le t \le 1$. Thus as in Section 1.3, $g$ has a global minimum at $t = 0$ so that $g(1) \ge g(0)$, or $f(x_2, y_2) \ge f(x_1, y_1)$.

From this result we deduce the following much used test for a local minimum of a function $f$ with continuous first and second partial derivatives in an open region $D$. Let $a, b, c$ be defined as above. *If at a critical point* $(x_0, y_0)$ *of* $f$ *in* $D$ *the functions* $a, c, ac - b^2$ *are all positive, then* $f$ *has a strong local minimum at* $(x_0, y_0)$. By continuity the three functions remain positive in the convex set $E$ formed by a $\delta$-neighborhood of $(x_0, y_0)$, where $\delta$ is chosen so that $E$ is contained in $D$. Hence $f$ is strictly convex in $E$, and as in the preceding paragraph, $f$ has a global minimum on $E$ at $(x_0, y_0)$. Since $f$ is strictly convex on $E$, this is a strong local minimum.

From the two results just found, we derive corresponding results for maxima: Under the proper continuity assumptions, a critical point of a concave function provides a global maximum; if at a critical point $a, c,$ and $b^2 - ac$ are all negative, then that critical point provides a strong local maximum.

Other results of Sections 1.2 and 1.3 extend to functions of two variables: (1) Each local minimum of a convex function is a global minimum (see Problem 1.23(a)). (2) If a convex function has several minimizers, then these minimizers form a convex set on which $f$ is constant (see Problem 1.23(b)). (3) A strictly convex function has at most one minimizer (Problem 1.23(c)). (4) A convex function is continuous at each interior point of the set on which it is defined (Problem 1.27). (5) A convex function $f$ defined on an open region has a "directional derivative" in each direction at each point $(x_0, y_0)$ of the region; that is, for each $(\alpha, \beta)$ not $(0, 0)$, the function of $t$,

$$f(x_0 + \alpha t, y_0 + \beta t)$$

has a right-hand derivative and a left-hand derivative at $t = 0$. This follows from the result for functions of one variable proved at the end of Section 1.2. (6) A convex function whose graph includes no line segment must be strictly convex (see Problem 1.16).

## Saddle Points

If at a critical point of $f$ one has $ac - b^2 < 0$, then the corresponding quadratic function $Q$ takes on both positive and negative values on the unit circle. One verifies that $f$ can have neither a maximum nor a minimum at the critical point. One says that $f$ has a "saddle point" at the critical point. This case is illustrated by the function $f(x, y) = x^2 - y^2$. (See the end of Section 1.9.)

## Positive Definite Quadratic Functions

A quadratic function $Q(u, v) = au^2 + 2buv + cv^2$ is said to be positive definite if $Q > 0$ for all $(u, v)$ except $(0, 0)$. If $Q$ is positive definite, then in particular $Q > 0$ for $u^2 + v^2 = 1$. Conversely, if $Q > 0$ for $u^2 + v^2 = 1$, then $Q$ is positive definite, for if $(u_1, v_1)$ is not $(0, 0)$, then as above one can write $u_1 = ku, v_1 = kv$, where $k \neq 0$ and $u^2 + v^2 = 1$, and so $Q(u_1, v_1) = k^2 Q(u, v) > 0$.

One calls the quadratic function $Q$ *positive semidefinite* if $Q \geq 0$ for all $(u, v)$. By similar reasoning, we see that $Q$ is positive semidefinite if and only if $Q(u, v) \geq 0$ for $u^2 + v^2 = 1$.

We can thus restate the tests for convexity given above as follows:

> If $f(x, y)$ has continuous derivatives through the second order in an open convex set $E$, then $f$ is convex on $E$ precisely when, for each point of $E$, the quadratic function $Q$ defined by (1.44) is positive semidefinite. If in each case the function $Q$ is positive definite, then $f$ is strictly convex.

There is a similar discussion for *concave* functions, with "positive" replaced by "negative" and reversal of the inequality signs.

## PROBLEMS

**1.17.** Fill in all the details for Example 1 in Section 1.4.

**1.18.** Fill in all the details for Example 2 in Section 1.4.

**1.19.** Fill in all the details for Example 3 in Section 1.4:
   (a) Verify that the parametrization reduces the problem to one for the stated function $g(t)$.
   (b) First exclude the case $b = 0$, $a = c$ and find the four critical points given. Then show how they determine four points dividing the circle into four equal arcs, as in Fig. 1.11.
   (c) A rotation of axes through angle $\alpha$ is given by equations

$$x = x' \cos \alpha - y' \sin \alpha, \quad y = x' \sin \alpha + y' \cos \alpha.$$

(See part (**h**).) Show that when the function $Q$ is expressed in terms of the new coordinates $x'$, $y'$ by these equations, with $\alpha$ as in (1.41), it has the form stated and hence has extrema at the points $(\pm 1, 0)$, $(0, \pm 1)$ in the new coordinates, at which $Q$ has the values $a'$ and $c'$.

(**d**) Show that for rotation through an arbitrary angle $\alpha$, $Q$ becomes $a'x'^2 + 2b'x'y' + c'y'^2$, and express $a'$, $b'$, $c'$ in terms of $a$, $b$, and $c$. Show from these formulas that $a' + c' = a + c$ and $a'c' - b'^2 = ac - b^2$.

(**e**) Show that if the rotation is chosen as in (**c**) and thus $b' = 0$, one has $a' + c' = a + c$ and $a'c' = ac - b^2$. From (**c**) we know that $Q$ has a positive minimum on the circle precisely when $a'$ and $c'$ are both positive. Conclude that this occurs precisely when $a$, $c$, and $ac - b^2$ are all positive.

(**f**) Reason as in (**e**) to show that $Q$ has global minimum 0 on the circle precisely when $a + c \geq 0$ and $ac - b^2 = 0$.

(**g**) Show that in the excluded case $b = 0$, $a = c$, $Q$ is constant on the circle. Show that the conditions given in (**e**) and (**f**) for a positive minimum or a zero minimum remain valid.

(**h**) Derive the formula for change of coordinates by rotating axes, as given in part (**c**). [Hint: Use polar coordinates so $x = r\cos\theta$, $y = r\sin\theta$, $x' = r\cos(\theta - \alpha)$, $y' = r\sin(\theta - \alpha)$. Now write $x = r\cos[(\theta - \alpha) + \alpha]$ and expand to obtain the formula for $x$. Proceed similarly for $y$.]

**1.20.** Each of the following sets of inequalities describes a set $E$ in the $xy$-plane. Determine whether $E$ is convex:

(**a**) $x > 0, y > 0$.

(**b**) $xy > 1$.

(**c**) $3x^2 + 2y^2 \leq 1$.

(**d**) $|x|^{1/3} + |y|^{1/3} \leq 1$.

(**e**) $x \geq 0, x \leq 0, y > 1, y < 3$.

**1.21.** Let $g(t) = f[x(t), y(t)]$, where the functions $x(t)$, $y(t)$ are given by (1.42). Show by the chain rules of calculus that the second derivative of $g(t)$ is given by (1.45).

**1.22.** Show that the function has a strong local minimum at the point $(0, 0)$:

(**a**) $z = 1 + x^2 + y^2$.

(**b**) $z = (x^2 + y^2)/(x^2 + y^2 + 1)$.

(**c**) $z = \exp(x^2 + y^2)$.

**1.23.** Let $f(x, y)$ be defined on the convex set $E$, and let $f$ be a convex function. Prove the following statements:

(**a**) Each local minimum of $f$ is a global minimum.

**(b)** The set of all minimizers of $f$ forms a convex subset of $E$.

**(c)** If $f$ is strictly convex, then $f$ has at most one minimizer.

**1.24.** Find all minimizers of the function

$$f(x, y) = |x + y - 1| + |x| + |y|$$

and check the result by Problem 1.23(**b**).

**1.25.** Let a particle $P$ of mass $m$ move in the $xy$-plane subject to the attractive force of three springs, obeying Hooke's law with the same spring constant $k$, attached at the points $(0, 2)$, $(\sqrt{3}, \pm 1)$; thus the total potential energy is $V = k(r_1^2 + r_2^2 + r_3^2)$, where the $r_i$ are the distances from $P$ to the three points. Show that $V$ is strictly convex and has a unique minimizer. [As shown in Section 1.10, one can conclude that the minimizer is a stable equilibrium position for the particle.]

**1.26.** Let $E$ be a closed convex region in the $xy$-plane, and let the graph of a continuous function $y = f(x)$, $a \le x \le b$, form part of the boundary of $E$.

**(a)** Show that $E$ lies wholly on one side of the graph of $f$: that is, either $y \ge f(x)$ for all $x$ in $[a, b]$ for all $(x, y)$ in $E$ or $y \le f(x)$ for all $x$ in $[a, b]$ for all $(x, y)$ in $E$.

**(b)** Show that if in part (**a**) the $\ge$ sign holds, then $f$ is a convex function.

**(c)** Under the assumption of part (**b**), let $f$ have a continuous derivative. Show that $E$ lies above the tangent line at each point of the graph of $f$.

**(d)** Under the assumptions of part (**c**), if $E$ lies above a straight line through a point $(x_0, f(x_0))$, where $a < x_0 < b$, then that straight line is the tangent line to the graph of $f$ at the point.

**1.27.** (Difficult) Prove that a convex function of $(x, y)$ is continuous at each interior point of its set $E$ of definition. [Hint: Extend the argument given at the end of Section 1.2. This argument shows that the graph of a convex function of one variable lies *above* each line which is an extension of a chord. For simplicity let the origin be a point of the graph of $z = f(x, y)$ such that $O : (0, 0)$ is interior to $E$ and prove continuity of $f$ at $(0, 0)$. First consider the behavior of $f$ in the first quadrant of the $xy$-plane. Choose $a > 0$ so that the $2a$-neighborhood of $(0, 0)$ is contained in $E$ and hence so is the the triangle with vertices $O$, $Q : (a, 0)$, $R : (0, a)$. For each point $(x, y, f(x, y))$ for which $(x, y)$ is on the segment $QR$, join the point to the origin by a line segment (a chord of the graph of $f$). By the continuity of convex functions of one variable, $f$ is continuous along $QR$, and hence the surface formed by the chords drawn is the graph of a continuous function $g(x, y)$. By convexity $f(x, y) \le g(x, y)$ in the triangular region. Now carry out the same process in the third quadrant, with the aid of the vertices $Q':(-a, 0)$ and $R':(0, -a)$, to obtain a surface $z = g'(x, y)$. Extend the chords thereby formed to positive $x$ and $y$, thereby extending the graph of $g'$ to cover the triangle in the first quadrant. By the remark above on extending chords, conclude that $g'(x, y) \le f(x, y) \le g(x, y)$ in the triangular region in the first quadrant. A similar procedure in the other quadrants leads to a similar double inequality in the corresponding triangular regions. Thus in the square region with vertices $R, Q, R', Q'$ the function $f$ is "trapped" between two continuous functions with value 0 at the origin, and $f$ is continuous at the origin.]

## 1.5 SOME GEOMETRIC EXTREMUM PROBLEMS

We mention here, without proof, some famous maximum and minimum problems of two-dimensional geometry.

**Four Vertex Theorem.** Let $C$ be a simple closed curve in the $xy$-plane; that is, $C$ is a curve with parametric equations

$$x = f(t), \quad y = g(t), \quad 0 \le t \le T, \tag{1.50}$$

where $f$ and $g$ are continuous and $f(0) = f(T)$, $g(0) = g(T)$, but otherwise, the values of $(f(t), g(t))$ for two different values of $t$ are never the same (Fig. 1.14). We assume further that $f$ and $g$ have continuous derivatives through the second order and that $f'(t)^2 + g'(t)^2$ is positive throughout and so the vector $\mathbf{v} = (f'(t), g'(t))$ is a nonzero tangent vector to the path (1.50) (which can be interpreted as the velocity vector if $t$ is time). Then arc length $s$ along the path is

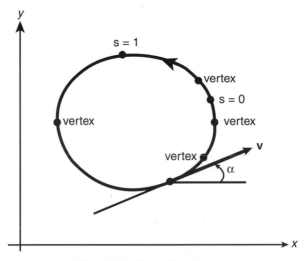

**Figure 1.14.** Four vertex theorem.

defined by $ds/dt = |\mathbf{v}|$, and one can use $s$ as parameter instead of $t$. We assume that the path is directed counterclockwise, with increasing $s$, as in Fig. 1.14, and thus the angle $\alpha$ between $\mathbf{v}$ and the positive $x$-direction can be chosen to vary continuously with $s$ and to increase from 0 to $2\pi$ in one circuit of $C$. The *curvature $\kappa$* of the path then equals $d\alpha/ds$.

We now make the additional assumption that $C$ is a *convex* curve. The condition of convexity can be defined in various equivalent ways: that $\kappa = d\alpha/ds \geq 0$ throughout, or that the region bounded by $C$ is a convex region, or that at each point $P$ of $C$ the whole curve $C$ lies on one side of the tangent line at $P$. (See the discussion at the end of Section 4.4.)

The Four Vertex Theorem can now be stated as follows: *For such a convex simple closed curve $C$ the curvature has at least four extrema on $C$.* Each extremum of the curvature is termed a *vertex* of the curve. The theorem is illustrated by an ellipse, for which there are precisely two maxima and two minima at the ends of the axes. For a circle, the curvature is constant and every point is an extremum. For a proof see, for example, Chern (1967, pp. 23–25) or Struik (1950, pp. 48–49).

## Isoperimetric Inequality

This again concerns a simple closed curve $C$. One now assumes that the length $L$ of $C$ is given, and one tries to choose $C$ to maximize the area $A$ enclosed. (Here we are dealing with a *functional*, as mentioned at the beginning of this chapter, for the area depends on the choice of the pair of functions $f$, $g$ that determine $C$.)

The solution is very simple: The maximum area is achieved if and only if $C$ is a circle, in which case $A = L^2/(4\pi)$. Thus in general

$$4\pi A \leq L^2. \tag{1.51}$$

The statement (1.51) is the *isoperimetric inequality*.

For most problems concerning functionals, one is dealing with a problem of the *calculus of variations*, and the problem considered here can in fact be treated in the context of the calculus of variations (see Akhiezer, 1962, pp. 200–204). However, there are various quite simple derivations of the result stated above, not involving the calculus of variations. Two such proofs are given in Chern (1967, pp. 25–29).

## Maximum Problem Concerning Plane Lattices

By a plane lattice in the $xy$-plane, we mean the set of all points having coordinates $(ma_1 + na_2, mb_1 + nb_2)$, where $a_1$, $b_1$, $a_2$, $b_2$ are given numbers such that $a_1b_2 - a_2b_1 \neq 0$ and $m$, $n$ take on all integer values 0, $\pm 1$, $\pm 2$, .... If $a_1 = b_2$ and $a_2 = -b_1$, we get a square lattice. In general, each lattice can be thought of as dividing the plane into congruent parallelograms whose vertices are the points of the lattice. The way this is done is not unique; however, one shows that for every way of doing it, the area of each parallelogram is the same number. When that area is 1, the lattice is called a *unit lattice*. One is illustrated in Fig. 1.15. For each such unit lattice we denote by $c$ the minimum distance between lattice points (see Fig. 1.15).

Now our maximum problem is as follows: *Choose a unit lattice, if possible, to maximize the minimum distance c.* The problem has a solution: Namely, the maximum value of $c$ is

$$C = \sqrt{\frac{2}{\sqrt{3}}}, \tag{1.52}$$

and this value is achieved for

$$(a_1, b_1) = \left(\frac{1}{C}, 0\right), \qquad (a_2, b_2) = \left(\frac{1}{2C}, \frac{\sqrt{3}}{2C}\right). \tag{1.53}$$

The corresponding lattice can be formed of parallelograms each of which is formed of two equilateral triangles, as in Fig. 1.16.

From this maximizing lattice one derives a useful "packing of circles" in the plane. One constructs about each lattice point the circle whose radius is $c/2$. Then no two circles overlap, but tangencies occur. This packing can be interpreted as the densest packing of circles in the plane; the ratio of the area of each circle to that of one of the parallelograms has its maximum value, namely $\pi/(2\sqrt{(3)})$. (See Problem 1.34.) For a detailed discussion of this topic, see Hilbert and Cohn-Vossen (1952, pp. 32–37).

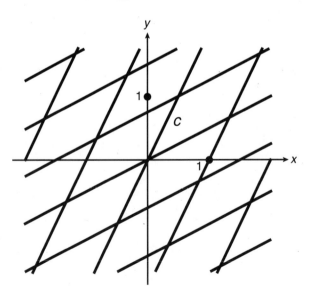

**Figure 1.15.** Unit lattice.

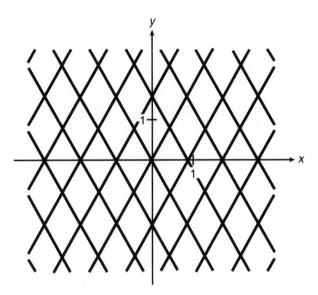

**Figure 1.16.** Best lattice.

## Four Color Problem

One asks how many colors are sufficient to assign a different color to each country in such a way that countries with a common boundary line have different colors. This question is made mathematically precise by assuming that each country is a closed region and that the whole surface of the earth is partitioned into a finite set of such regions, as in a typical map. The problem is to minimize the number of colors used. Experimentation led to the conjecture that *four* was the minimum, but proving this turned out to be extraordinarily difficult. A proof was finally achieved in 1976 by mathematicians at the University of Illinois but only with the aid of a digital computer. See Jensen and Toft (1995) for details.

## Two Minimum Problems with No Solution

The book of Hilbert and Cohn-Vossen (1952, p. 214) mentions two minimum problems that might appear to have a solution, though they have none. The first problem is that of joining two given points $A$ and $B$ in the plane by a curve of minimum length which meets the line segment $AB$ at $A$ at right angles (Fig. 1.17). As the figure suggests, one can make the length of the curve as close to that of the line segment as desired but cannot achieve this minimum value.

The second problem is to move a rod in the plane continuously in such a way that at the end of the motion the rod has turned through angle $\pi$ and that the area of the region formed of all positions of the points of the rod is

Figure 1.17. A minimum length problem with no solution.

minimized. As explained in Hilbert and Cohn-Vossen (1952, pp. 214, 280–281), the area can be made as close to 0 as desired, so that again there is no solution.

## PROBLEMS

**1.28.** It is shown in the calculus that, for a plane curve given in polar coordinates by an equation $r = f(\theta)$, the curvature $d\alpha/ds$ can be evaluated as

$$\kappa = \frac{-rr'' + r^2 + 2r'^2}{(r'^2 + r^2)^{3/2}}. \qquad (\dagger)$$

(For example, see Kaplan and Lewis 1970, p. 440.)

(a) Apply this formula to the curve $r = 3 - \cos\theta$ to show that the curve is a closed convex curve.

(b) Verify the Four Vertex Theorem graphically for the curve of part (a).

(c) Show by calculus that $\kappa$ has four extrema on the curve.

(d) Derive the formula ($\dagger$).

**1.29.** Suggest by graphs convex curves with more than four vertices.

**1.30.** Verify the isoperimetric inequality (1.51) for each of the following curves $C$:

(a) An equilateral triangle.

(b) A square.

(c) A regular hexagon.

(d) An isosceles triangle. [Hint: Let the triangle have vertices $(x, 0)$, $(-x, 0)$, $(0, h)$, where $x$ and $h$ are positive. Show that for $h$ fixed and $0 < x < \infty$, the ratio $L^2/A$ is minimized when the triangle is equilateral and apply the isoperimetric inequality for equilateral triangles.]

(e) An arbitrary triangle. [Hint: Let the vertices be $(c, 0)$, $(-c, 0)$, and $(x, h)$, where $c$ and $h$ are fixed and positive. Show that the ratio $L^2/A$ is minimized when $x = 0$ and hence the triangle is isosceles and apply the result of (d).]

*1.31. (A project) Find a proof of the isoperimetric inequality with the aid of the article by Chern (1967).

1.32. Draw a plane lattice, and then verify experimentally that different ways of using the lattice to divide the plane into congruent parallelograms lead to parallograms all having the same area.

1.33. Graph the plane lattice for each of the given choices of $a_1$, $a_2$, $b_1$, $b_2$; verify that it is a unit lattice and find the number $c$ for each:

(a) $a_1 = 3, a_2 = 7, b_1 = 2, b_2 = 5$.

(b) $a_1 = \sqrt{2}, a_2 = 3/\sqrt{2}, b_1 = \sqrt{2}, b_2 = 2\sqrt{2}$.

1.34. Verify the graph in Fig. 1.16 of the maximizing lattice, using (1.53), and verify that the minimum distance is given by (1.52).

1.35. Verify experimentally that four colors suffice to color a map.

1.36. Consider the two minimum problems mentioned at the end of Section 1.5.

(a) Explain in detail why the first one has no solution.

(b) Use graphical experiments to indicate why the second problem has no solution.

## 1.6   GEOMETRY OF *n*-DIMENSIONAL SPACE

The concepts of geometry in space of dimension $n$, where $n$ is a positive integer, are most easily introduced with the aid of linear algebra: vectors, matrices and linear operations. We assume that the reader is familiar with these tools of linear algebra. They are summarized in Appendix A.

By analogy with 2-dimensional space (the plane) and 3-dimensional space, we define *n*-dimensional space $\mathcal{R}^n$, for each positive integer $n$, as the set of all $n$-tuples $(x_1, \ldots, x_n)$ of real numbers. Each *n*-tuple defines a *point* of $\mathcal{R}^n$, and the $x_i$ are the *coordinates* of the point. The point $(0, \ldots, 0)$ is the *origin* of $\mathcal{R}^n$, often denoted by $O$. We often denote points by capital letters: $P, Q, \ldots$ . The *distance d* between two points $(u_1, \ldots, u_n)$, $(v_1, \ldots, v_n)$ is given by the formula

$$d = \sqrt{(u_1 - v_1)^2 + \cdots + (u_n - v_n)^2}. \tag{1.60}$$

As in linear algebra, we can interpret each such *n*-tuple as a row vector or a column vector. We choose to interpret it as a column vector and write accordingly:

$$\mathbf{x} = (x_1, \ldots, x_n)^t,$$

where $t$ denotes the transpose. We can think of the vector $\mathbf{x}$ as a directed line segment from the origin to the point $P$ corresponding to the $n$-tuple, as in Fig. 1.18 for $n = 3$.

It is helpful to freely change point of view and regard our $n$-tuples as points or vectors (column vectors) as convenient. For example, the distance between $\mathbf{u}$ and $\mathbf{v}$, as in (1.60), can be written as

$$d = \sqrt{(\mathbf{u} - \mathbf{v})^t(\mathbf{u} - \mathbf{v})} = \|\mathbf{u} - \mathbf{v}\|. \tag{1.60'}$$

Here we are using the standard inner product $\mathbf{w}^t\mathbf{z}$ of two column vectors of $\mathcal{R}^n$ and the corresponding norm:

$$\mathbf{w}^t\mathbf{z} = w_1 z_1 + \cdots + w_n z_n, \quad \|\mathbf{w}\| = \sqrt{\mathbf{w}^t\mathbf{w}}. \tag{1.61}$$

In linear algebra it is shown that the inner product satisfies the *Schwarz inequality*:

$$|\mathbf{w}^t\mathbf{z}| \le \|\mathbf{w}\| \, \|\mathbf{z}\|. \tag{1.61'}$$

(See Kaplan, 1991, p. 49.)

REMARK. The distance formula (1.60) and the corresponding vector norm of (1.61) generalize the familiar formulas of Euclidean geometry (and are based on the Pythagorean theorem, discovered about 2500 years ago). Hence we term our $n$-dimensional space with this distance rule *Euclidean n*-dimensional space. Below we introduce other distances and norms. We may also introduce other coordinate systems in $\mathcal{R}^n$. Those that do not change the distance formula (1.60) are called *Cartesian coordinates*.

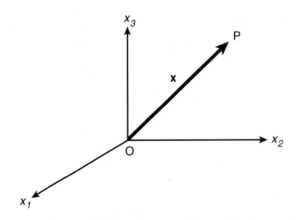

**Figure 1.18.** Vector corresponding to point $P$.

A *straight line* in our *n*-dimensional space $\mathcal{R}^n$ can be given in vector form as the set of all points

$$\mathbf{x} = \mathbf{a} + t\mathbf{b}, \quad -\infty < t < \infty, \tag{1.62}$$

where $\mathbf{a}$ and $\mathbf{b}$ are two vectors of $\mathcal{R}^n$ and $\mathbf{b}$ is not the zero vector $\mathbf{0}$. Two such lines

$$\mathbf{x} = \mathbf{a}_1 + t\mathbf{b}_1, \quad \mathbf{x} = \mathbf{a}_2 + t\mathbf{b}_2,$$

are said to be *parallel* if $\mathbf{b}_1$ and $\mathbf{b}_2$ are linearly dependent; this includes the case when the lines coincide (Problem 1.39). The lines are termed *perpendicular* if the vectors $\mathbf{b}_1$, $\mathbf{b}_2$ are orthogonal: That is, their inner product is 0. In general, the angle between the lines can be defined as the number $\theta$ with

$$\cos \theta = \pm \frac{\mathbf{b}_1^t \mathbf{b}_2}{\|\mathbf{b}_1\| \|\mathbf{b}_2\|}, \quad 0 \le \theta \le \frac{\pi}{2}. \tag{1.63}$$

If one omits the plus or minus sign and chooses $\theta$ between 0 and $\pi$, inclusive, then one terms $\theta$ the angle between the *directed* lines, directed by the vectors $\mathbf{b}_1$, $\mathbf{b}_2$, respectively; this $\theta$ is the angle between the vectors $\mathbf{b}_1$, $\mathbf{b}_2$.

A *line segment* is defined by (1.62), with $t$ restricted to a closed interval $[t_1, t_2]$. A *plane* in $\mathcal{R}^n$ ($n \ge 2$) is given in vector form in terms of two parameters $u$ and $v$ as the set of all points

$$\mathbf{x} = \mathbf{a} + u\mathbf{b} + v\mathbf{c}, \quad -\infty < u < \infty, \quad -\infty < v < \infty, \tag{1.64}$$

where $\mathbf{b}$ and $\mathbf{c}$ are linearly independent. More generally, a *linear variety* of dimension $k$ in $\mathcal{R}^n$, where $k$ is an integer, $0 \le k \le n$, is the set of all points

$$\mathbf{x} = \mathbf{a} + u_1\mathbf{b}_1 + \cdots + u_k\mathbf{b}_k, \quad -\infty < u_i < \infty, \quad i = 1, \ldots, k, \tag{1.65}$$

where the vectors $\mathbf{b}_1, \ldots, \mathbf{b}_k$ are linearly independent. For $k = 0$ the linear variety reduces to the point $\mathbf{a}$; for $k = n$ it coincides with the whole *n*-dimensional space $\mathcal{R}^n$.

From linear algebra we see that in (1.65) we could allow the vectors $\mathbf{b}_1, \ldots, \mathbf{b}_k$ to be linearly dependent. Then their linear combinations form a *subspace* $W$ of the vector space $\mathcal{R}^n$, and one obtains the same set of points $\mathbf{x}$ in (1.65) if one replaces these linear combinations by the linear combinations of an arbitrary *basis* of the subspace $W$. The dimension of the linear variety is then that of the subspace $W$. In particular, one can always choose an *orthonormal* basis for $W$.

To pursue this idea further, we assume that the vectors $\mathbf{b}_1, \ldots, \mathbf{b}_k$ already form an orthonormal system (and so they are necessarily linearly independent). Then we can choose the point $\mathbf{a}$ as a new origin in $\mathcal{R}^n$ and consider the $k$-tuple

$(u_1, \ldots, u_k)$ as the coordinates of the point $\mathbf{x}$ of the variety with respect to new axes having the directions of $\mathbf{b}_1, \ldots, \mathbf{b}_k$ respectively. This is illustrated in Fig. 1.19 for the case $n = 3$, $k = 2$; that is, of a plane in 3-dimensional space. One then verifies that the points of the variety can be considered as the points of $\mathcal{R}^k$, with coordinates $(u_1, \ldots, u_k)$; the distance formula remains the usual one in terms of these coordinates. We can say that *each linear variety of dimension $k$ in $\mathcal{R}^n$ is congruent to $\mathcal{R}^k$.* In particular, all planes in 3-dimensional space are the same as the plane of plane geometry.

Linear varieties appear naturally as the sets of solutions of linear equations. Let $A$ be an $m \times n$ matrix and $\mathbf{y}$ a column vector of $\mathcal{R}^m$. Then the equation

$$A\mathbf{x} = \mathbf{y} \tag{1.66}$$

is equivalent to $m$ equations in the $n$ unknowns $x_1, \ldots, x_n$. The equation (1.66) may have no solutions. However, if it has solutions, they are given by a vector equation of the form of (1.65) with linearly independent $\mathbf{b}_1, \ldots, \mathbf{b}_k$. Here $k$ is the *nullity* of the matrix $A$; $k = n - r$, where $r$ is the rank of $A$, the maximum number of linearly independent columns (or rows) of $A$. We give a simple example.

***Example 1.***   The equations are

$$x_1 + x_2 = 1, \quad x_2 + x_3 = 2, \quad x_3 + x_4 = 3.$$

One can write the solutions as $x_1 = t$, $x_2 = 1 - t$, $x_3 = 1 + t$, $x_4 = 2 - t$ or as the vector equation

$$\mathbf{x} = (0, \ 1, \ 1, \ 2)^t + t(1, \ -1, \ 1, \ -1)^t.$$

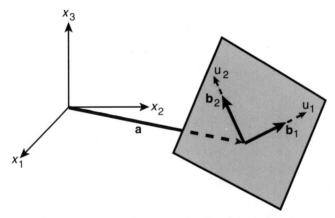

**Figure 1.19.** Geometry of a plane in space.

The matrix $A$ has rank 3 and nullity 1. The linear variety has dimension 1; that is, it is a straight line in $\mathcal{R}^4$, as in (1.62). (See Problem 1.38.)

## Whole Space $\mathcal{R}^n$ as a Linear Variety

We can represent the whole *n*-dimensional space as the set of points

$$\mathbf{x} = x_1\mathbf{e}_1 + \cdots + x_n\mathbf{e}_n,$$

where the vectors $\mathbf{e}_1 = (1, 0, \ldots, 0)^t, \ldots, \mathbf{e}_n = (0, \ldots, 0, 1)^t$ form the *standard basis* of $\mathcal{R}^n$. The $x_i$, which are the standard coordinates of $\mathbf{x}$, are now the parameters, like the $u_i$ in (1.65). The *n* vectors of the standard basis, which are the columns of the identity matrix $I$, form an orthonormal system.

If instead of the standard basis one uses another orthonormal system of *n* vectors, then they form the columns of an *orthogonal* matrix $B$. The equation

$$\mathbf{x} = B(x'_1, \ldots, x'_n)^t = (x_1, \ldots, x_n)^t \tag{1.67}$$

then introduces the new coordinates of $\mathbf{x}$, with respect to the new orthonormal basis, and also expresses the old coordinates in terms of the new ones. Since $B$ is an orthogonal matrix, $B^{-1} = B^t$, and hence we derive from (1.67) the equation

$$(x'_1, \ldots, x'_n)^t = B^t(x_1, \ldots, x_n)^t, \tag{1.67'}$$

which expresses the new coordinates in terms of the old. We can interpret the change of coordinates as a *rotation of axes* in $\mathcal{R}^n$ (provided the orthogonal matrix $B$ has determinant $+1$ rather than $-1$). The process is illustrated in Problem 1.19, parts (c) and (h), where

$$B = \begin{bmatrix} \cos\alpha & -\sin\alpha \\ \sin\alpha & \cos\alpha \end{bmatrix}.$$

For Euclidean geometry we allow only changes of coordinates which preserve the distance formula (1.60). One verifies that the distance formula is indeed correct for the change of coordinates (1.67) (see Problem 1.41). The most general linear change of coordinates in $\mathcal{R}^n$ which preserves the distance formula has the form

$$(x_1, \ldots, x_n)^t = \mathbf{a} + B(x'_1, \ldots, x'_n)^t, \tag{1.68}$$

where $B$ is an orthogonal matrix. (See Birkhoff and MacLane, 1977, p. 225.) As explained in this reference, the equation (1.68) is usually termed an *affine transformation*, rather than a linear one, because of the constant term $\mathbf{a}$.

**Extremum Problems on Sets in $\mathcal{R}^n$**

The definitions concerning sets in the plane given at the beginning of Section 1.4 generalize at once to our $n$-dimensional space $\mathcal{R}^n$: A $\delta$-neighborhood of point $\mathbf{x}$ in $\mathcal{R}^n$ is the set of all $\mathbf{y}$ in $\mathcal{R}^n$ such that $\|\mathbf{x} - \mathbf{y}\| < \delta$. An open set in $\mathcal{R}^n$ is one such that each point of the set has a $\delta$-neighborhood contained in the set. A closed set is one whose complement (the set of points not in the set) is open or empty (i.e., $\mathcal{R}^n$ is a closed set). An open region is an open set such that each two points of the set can be joined by a broken line (formed of a finite number of line segments) contained in the set. The definitions given in Section 1.4 for boundary point, boundary of a set, interior point, interior of a set, closed region can all be repeated. A set of $\mathcal{R}^n$ is bounded if, for some $R > 0$, $\|\mathbf{x}\| < R$ for all points $\mathbf{x}$ of the set; that is, all points lie within distance $R$ of the origin.

It is at times convenient to use the term *neighborhood of a point* to mean an open set containing the point. A function $f(x_1, \ldots, x_n)$ can be written as $f(\mathbf{x})$, where $\mathbf{x}$ is in $\mathcal{R}^n$. Continuity and differentiability of such a function are defined in the calculus. If the function is defined on a closed bounded set $E$ in $\mathcal{R}^n$ and the function is continuous on $E$, then the function has a global maximum $M$ and a global minimum $m$ on $E$. (See Kaplan, 1991, p. 178.) If $f$ is also differentiable on $E$ and a maximizer or minimizer is an interior point of $E$, then it is a critical point of $f$—that is, a point at which $\partial f / \partial x_i = 0$ for $i = 1, \ldots, n$, or equivalently, a point at which the *gradient* of $f$ is $\mathbf{0}$. The gradient is regarded as a column vector:

$$\operatorname{grad} f = \nabla f = \left( \frac{\partial f}{\partial x_1}, \ldots, \frac{\partial f}{x_n} \right)^t.$$

Local extrema and strong local extrema are defined as in Sections 1.3 and 1.4. If such a point has a $\delta$-neighborhood in which $f$ is defined and differentiable, then it is a critical point of $f$.

## 1.7   CONVEX FUNCTIONS OF $n$ VARIABLES

The ideas of Section 1.4 carry over at once to $\mathcal{R}^n$. A set $E$ in $\mathcal{R}^n$ is convex if $E$ contains the line segment joining each pair of points of the set. A function $f(\mathbf{x})$ defined on a convex set $E$ is convex if

$$f((1 - t)\mathbf{x}_1 + t\mathbf{x}_2) \leq (1 - t)f(\mathbf{x}_1) + tf(\mathbf{x}_2) \tag{1.70}$$

for all $t$ in the interval $[0, 1]$ and each pair of points $\mathbf{x}_1, \mathbf{x}_2$ in $E$. This is equivalent to the statement that as a function of one variable, the function $f$ is convex along each line or line segment in $E$. Strict convexity, concavity, and strict concavity of functions are defined as in Section 1.4.

We observe that each linear variety in $\mathcal{R}^n$ is a convex set. Let the linear variety (1.65) be given, and let $x_1$, $x_2$ be points in it, say

$$x_1 = a + u_1 b_1 + \cdots + u_k b_k, \quad x_2 = a + v_1 b_1 + \cdots + v_k b_k.$$

Then for $0 \leq t \leq 1$,

$$(1 - t)x_1 + tx_2 = a + w_1 b_1 + \cdots + w_k b_k,$$

where $w_i = (1 - t)u_i + tv_i$ for $i = 1, \ldots, k$.

For a function $f$ of $n$ variables defined in a convex open region $E$, we obtain a condition for convexity with the aid of second partial derivatives (assumed continuous in $E$). Let the line segment

$$x = (1 - t)x_1 + tx_2, \quad 0 \leq t \leq 1, \tag{1.71}$$

be contained in $E$, and let $u = x_2 - x_1$. Let $H(x)$ be the matrix $\partial^2 f/(\partial x_i \partial x_j)$, formed of the second partial derivatives of $f$, evaluated at the point $x$ of $E$; $H$ is called the *Hessian matrix* of the function $f$. We observe that $H$ is a *symmetric* matrix (Problem 1.43). Then by the chain rules of calculus, we find that as a function of $t$ along the segment, the function $f$ has second derivative at each $t$ equal to

$$Q(u) = \sum f_{x_i x_j} u_i u_j = u^t H(x) u, \tag{1.72}$$

where $x$ is the point corresponding to the chosen $t$ as determined by (1.71); the sum is over all $i$ and $j$ from 1 to $n$. (See Problem 1.44.) Thus at each such point the second derivative is the value of a quadratic function (also called quadratic *form*)

$$Q(u_1, \ldots, u_n) = \sum h_{ij} u_i u_j, \tag{1.73}$$

where the $h_{ij}$ are the entries in the Hessian matrix of $f$ at the point. We call a quadratic form *positive definite* if its values are greater than 0 for $u \neq 0$; we call it *positive semidefinite* if its values are all nonnegative. (See Section 1.4.)

We now conclude: *The function $f$ is convex on $E$ if and only if at each point of $E$ the quadratic form $Q$ is positive semidefinite; $f$ is strictly convex on $E$ if at each point $Q$ is positive definite.*

We also verify that the quadratic form is positive definite precisely when its values are all greater than 0 for $\|u\| = 1$ and positive semidefinite precisely when its values are all nonnegative for $\|u\| = 1$ (Problem 1.46). Thus the tests for convexity can be expressed in terms of the values of $Q$ on the *unit sphere*: $\|u\| = 1$. For concave functions we have an analogous discussion, with the word "positive" replaced by the word "negative" throughout.

One also terms a symmetric matrix $H$ positive definite, and so on, according as the corresponding quadratic form $Q = \mathbf{u}^t H\mathbf{u}$ is positive definite, and so on. We remark that a positive definite symmetric matrix $H$ is necessarily invertible, for $H\mathbf{u} = \mathbf{0}$ implies that $\mathbf{u}^t H\mathbf{u} = 0$, and hence $\mathbf{u} = \mathbf{0}$. The properties of quadratic forms are of such importance that we devote the following section to them.

## 1.8   QUADRATIC FORMS

We consider a quadratic form in $n$ variables:

$$Q(u_1, \ldots, u_n) = \sum h_{ij} u_i u_j = \mathbf{u}^t H\mathbf{u}, \tag{1.80}$$

where $H = (h_{ij})$ is a symmetric matrix with constant entries. An example of such a quadratic form is

$$Q(x_1, \ x_2) = 5x_1^2 + 3x_1 x_2 + 3x_2 x_1 + 7x_2^2.$$

Here the two middle terms could be combined to yield $6x_1 x_2$, but we do not do so in order to preserve the symmetry of the matrix $H$. In general, a quadratic form can be written in precisely one way in terms of a symmetric matrix $H$.

We now give a geometric reasoning to show that *new Cartesian coordinates can be chosen in $\mathcal{R}^n$ (as in (1.67)) so that, in terms of the new coordinates $v_1, \ldots, v_n$, the quadratic form takes the form*

$$Q(v_1, \ldots, v_n) = \lambda_1 v_1^2 + \cdots + \lambda_n v_n^2, \tag{1.81}$$

*where $\lambda_1, \ldots, \lambda_n$ are constants, which are the eigenvalues of the matrix H.*

To show this, we observe that the function $Q(\mathbf{u})$ is continuous on the unit sphere $\|\mathbf{u}\| = 1$ (a bounded closed set) and hence has a global maximum $\lambda_1$ at some point $\mathbf{u}_1$ on the unit sphere. We now choose new coordinates in $\mathcal{R}^n$, with the same origin, so that in these coordinates $v_1, \ldots, v_n$ the maximum occurs at the point $(1, \ldots, 0)$. As in Section 1.6, the change from the old coordinates to the new coordinates is given by an equation

$$\mathbf{u} = B\mathbf{v}, \tag{1.82}$$

where $B$ is an orthogonal matrix. In the new coordinates the quadratic form can be written as

$$Q_1(\mathbf{v}) = \mathbf{v}^t B^t H B\mathbf{v} = \mathbf{v}^t H_1 \mathbf{v}, \tag{1.83}$$

where $H_1 = B^t H B$ is again a symmetric matrix. Now we consider the values of the quadratic form on the portion of the unit sphere in the coordinate plane in which all coordinates are 0 except for $v_1$ and $v_2$. In this coordinate plane the quadratic form is a quadratic form in two variables, with maximum on the circle $v_1^2 + v_2^2 = 1$ occurring at the point $(1, 0)$ and maximum value $M = \lambda_1$. As in Section 1.4 the quadratic form in two variables can have no term in $v_1 v_2$, and the coefficient of $v_1^2$ must be $\lambda_1$. If we apply this reasoning to the other coordinate planes in which only $v_1$ and some $v_j$ may differ from 0, we conclude that the only term involving $v_1$ in the quadratic form (1.83) is the term $\lambda_1 v_1^2$. The remaining terms form a quadratic form in $v_2, \ldots, v_n$. We consider this quadratic form in the portion of the unit sphere for which $v_1 = 0$, that is, for which

$$v_2^2 + \cdots + v_n^2 = 1.$$

This is simply the unit sphere for the $(n-1)$-dimensional space with $v_2, \ldots, v_n$ as coordinates. By choosing new axes in this space, as before, we ensure that the corresponding quadratic form in the $n-1$ variables has only one term involving $v_2$, namely one of form $\lambda_2 v_2^2$. Here $\lambda_2$ is the maximum of the original quadratic form on the portion of the original unit sphere for which $v_1 = 0$. This cannot exceed the global maximum on the whole sphere, and so $\lambda_2 \leq \lambda_1$.

Proceeding in this way, continually choosing new coordinates, we eventually eliminate all the terms in products such as $v_i v_j$ with $i \neq j$. Thus we conclude that new coordinates can be chosen through one orthogonal transformation as in (1.82) so that the quadratic form takes the form (1.81), and incidentally,

$$\lambda_1 \geq \lambda_2 \geq \lambda_3 \geq \ldots .$$

From (1.83) we see that, after this transformation, the new coefficient matrix $H_1$ must be the diagonal matrix

$$\mathrm{diag}(\lambda_1, \ldots, \lambda_n).$$

Thus $H_1$ has eigenvalues $\lambda_1, \ldots, \lambda_n$. By (1.83) (where $B$ is orthogonal), $H_1 = B^t H B$ so that these are the eigenvalues of $H$ also. The corresponding eigenvectors of $H_1$ are the standard basis vectors $\mathbf{e}_j$ of $\mathcal{R}^n$, $(j = 1, \ldots, n)$, and thus the eigenvectors of $H$ are the vectors $B\mathbf{e}_j$ (Problem 1.47). Since the standard basis vectors form an orthonormal system, so do the eigenvectors of $H$.

REMARK. As a corollary of this result on quadratic forms, we deduce a result about symmetric matrices: *Every symmetric matrix can be reduced to diagonal form by an orthogonal transformation (1.82); it has a set of eigenvectors forming a complete orthonormal system.*

We can now obtain criteria for positive definiteness: *A quadratic form (1.80) is positive definite precisely when all eigenvalues of the matrix H are positive; it is positive semidefinite (but not positive definite) precisely when all eigenvalues are nonnegative and at least one eigenvalue is 0.*

Another test for positive definiteness can be given in terms of the determinants $\det(H_j)$, where $H_j$ is obtained from $H$ by deleting all rows and columns after the $j$th. The form is positive definite if and only if all these determinants $(j = 1, \ldots, n)$ are positive. (See Strang, 1976, ch. 6.)

As a final remark, we observe that $\lambda_1$, the largest eigenvalue of $H$, was found as the global maximum of the quadratic form $Q$ on the unit sphere. Hence it is also the global maximum of the function

$$g(\mathbf{u}) = \frac{Q(\mathbf{u})}{\|\mathbf{u}\|^2} \qquad (1.84)$$

for $\mathbf{u} \neq 0$. Indeed, we verify that the right side of (1.84) can be written as $Q(\mathbf{v})$, where $\mathbf{v} = \mathbf{u}/\|\mathbf{u}\|$, so $\|\mathbf{v}\| = 1$, and every $\mathbf{v}$ on the unit sphere is related to some nonzero vector $\mathbf{u}$ in this way. By similar reasoning we conclude that $\lambda_n$, the *smallest* eigenvalue of $H$, is the global minimum of $Q$ on the unit sphere or the global minimum of the function $g(\mathbf{u})$ of (1.84) for $\mathbf{u} \neq 0$. Therefore for all $\mathbf{u}$ ($0$ is not an exception) we can write

$$\lambda_n\|\mathbf{u}\|^2 \leq Q(\mathbf{u}) \leq \lambda_1\|\mathbf{u}\|^2. \qquad (1.85)$$

In general, we call the quadratic form $Q$ *nonsingular* if the corresponding symmetric matrix $H$ is nonsingular, that is, invertible. Thus $Q$ is nonsingular precisely when none of the eigenvalues $\lambda_j$ of $H$ is 0. A nonsingular quadratic form can be positive definite (all $\lambda_j$ positive) or negative definite (all $\lambda_j$ negative) or *indefinite*, when some $\lambda_j$ are positive, some are negative. In the last case the nonsingular matrix $H$ is also called indefinite.

### Gradient of a Quadratic Form

From (1.80), for $i = 1, \ldots, n$,

$$\frac{\partial Q}{\partial u_i} = 2 \sum_{j=1}^{n} h_{ij} u_j,$$

since there are two terms $h_{ij}u_i u_j$ and $h_{ij}u_j u_i$ for $i \neq j$ and one term $h_{ii}u_i^2$. Therefore

$$\nabla Q = 2H\mathbf{u}. \qquad (1.86)$$

## Gradient of a Quadratic Function

A quadratic function in $\mathcal{R}^n$ is a function of form

$$f(\mathbf{x}) = \mathbf{x}^t A \mathbf{x} + \mathbf{b}^t \mathbf{x} + c, \tag{1.87}$$

where $A$ is a symmetric matrix, $\mathbf{b}$ is a constant vector of $\mathcal{R}^n$, and $c$ is a constant scalar. From (1.86) the gradient of $f$ is

$$\nabla f = 2A\mathbf{x} + \mathbf{b}.$$

Hence the critical points of $f$ are the solutions of the linear equation

$$A\mathbf{x} = -\tfrac{1}{2}\mathbf{b}.$$

If $A$ is nonsingular, there is a unique critical point:

$$\mathbf{x} = \mathbf{x}^* = -\tfrac{1}{2}A^{-1}\mathbf{b}.$$

This holds, in particular, if $A$ is positive definite. In that case, $2A$ is the (constant) Hessian matrix of $f$, $f$ is strictly convex, and $\mathbf{x}^*$ is the unique minimizer of $f$.

## PROBLEMS

**1.37.** In $\mathcal{R}^4$ let the four points $P$: (3, 5, 1, 7), $Q$: (1, −2, 0, 2), $R$: (4, 4, 9, 1), and $S$: (0, 3, 0, 6) be given.

(a) Show that $P$, $Q$, $R$ do not lie on a line, and find the area of the triangle of which they are the vertices.

(b) Find equations for the line through $P$ and $Q$ and for the line through $R$ and $S$, and determine whether they are parallel.

(c) Find points dividing the line segment from $P$ to $Q$ into three equal parts.

(d) Find the dimension of the subspace $W$ formed of the linear combinations of the vectors from the origin to the four points.

(e) Find an orthonormal basis for the subspace $W$ of part (d).

**1.38.** Verify the details of Example 1 in Section 1.6.

**1.39.** Let two straight lines in $\mathcal{R}^n$ be given: $\mathbf{x} = \mathbf{a}_1 + t\mathbf{b}_1$ and $\mathbf{x} = \mathbf{a}_2 + t\mathbf{b}_2$. Determine the conditions on $\mathbf{a}_1, \mathbf{a}_2, \mathbf{b}_1, \mathbf{b}_2$ such that the lines (a) coincide, (b) intersect in a point, (c) are parallel but do not coincide, (d) are skew.

**1.40.** Let $A$ be the matrix

$$\begin{bmatrix} 2 & 2 & 3 & -1 & 4 \\ 1 & -1 & 2 & 1 & -1 \\ 2 & 3 & -1 & 3 & -2 \end{bmatrix}.$$

(a) Find the dimension of the null space of $A$.

(b) Represent the linear variety $A\mathbf{x} = (5, 1, 3)^t$ by parametric equations as in (1.65).

**1.41.** In (1.65) let the vectors $\mathbf{b}_j$ form an orthonormal system. Show that the square of the distance between the points $\mathbf{a} + u_1\mathbf{b}_1 + \cdots + u_k\mathbf{b}_k$ and $\mathbf{a} + v_1\mathbf{b}_1 + \cdots + v_k\mathbf{b}_k$ equals $(u_1 - v_1)^2 + \cdots + (u_k - v_k)^2$.

**1.42.** Find the Hessian matrix of each function:

(a) $f(x_1, x_2) = x_1^2 \cos(3x_1 - 2x_2)$.

(b) $f(x_1, x_2, x_3) = x_1^3 + 4x_1^2x_2 + 6x_1^2x_3 + 5x_1x_2^2 + x_1x_3^2 + x_2^3 + x_3^3$.

**1.43.** Explain why the Hessian matrix is symmetric (for a function with continuous second partial derivatives).

**1.44.** Let $f(\mathbf{x})$ have continuous second partial derivatives in the convex open region $E$ and let the line segment (1.71) be contained in $E$. Show that the composite function

$$g(t) = f[(1 - t)\mathbf{x}_1 + t\mathbf{x}_2]$$

has second derivative $g''(t)$ equal to the expression given by (1.72), where $\mathbf{u} = \mathbf{x}_2 - \mathbf{x}_1$.

**1.45.** Verify the result of Problem 1.44 for the special case $f(x_1, x_2, x_3) = x_1^2(x_2 + 3x_3)$ and line segment $\mathbf{x} = (1 - t)(2, 4, 7)^t + t(3, 0, 5)^t, 0 \le t \le 1$.

**1.46.** Let a quadratic form $Q(\mathbf{u})$ be given as in (1.73).

(a) Show that for each scalar $c$, $Q(c\mathbf{u}) = c^2 Q(\mathbf{u})$.

(b) Use the result of part (a) to show that the quadratic form is positive definite precisely when its values are all positive for $\|\mathbf{u}\| = 1$.

(c) Use the result of part (a) to show that the quadratic form is positive semidefinite precisely when its values are all nonnegative for $\|\mathbf{u}\| = 1$.

**1.47.** Let the $n \times n$ matrices $B, H, H_1$ be related by the equation $H_1 = B^t H B$ and let $B$ be an orthogonal matrix, let $H_1$ be diag$(\lambda_1, \ldots, \lambda_n)$. Show that each $\lambda_j$ is an eigenvalue of $H$, with associated eigenvector $B\mathbf{e}_j$.

**1.48.** Let the quadratic form

$$Q(x, y, z) = 11x^2 + 8xy + 9y^2 + 8yz + 7z^2$$

be given, with corresponding coefficent matrix

$$H = \begin{bmatrix} 11 & 4 & 0 \\ 4 & 9 & 4 \\ 0 & 4 & 7 \end{bmatrix}.$$

(a) Verify (with the aid of a computer program) that $H$ has eigenvalues 15, 9, 3, with corresponding eigenvectors

$$(\tfrac{1}{3})(2, \ 2, \ 1)^t, \quad (\tfrac{1}{3})(-2, \ 1, \ 2)^t, \quad (\tfrac{1}{3})(-1, \ 2, \ -2)^t$$

and that these form an orthonormal system.

(b) From the result of part (a), the global maximum of $Q$ on the sphere $x^2 + y^2 + z^2 = 1$ should be 15. Test this numerically by evaluating $Q$ at a number of points on the sphere. To obtain such points, one can use spherical coordinates to write $x = \sin\phi\cos\theta$, $y = \sin\phi\sin\theta$, $z = \cos\phi$ and select values of $\phi$ between $-\pi/2$ and $\pi/2$ and values of $\theta$ between 0 and $2\pi$.

**1.49.** Let $A$ be an $m \times n$ matrix.

(a) Show that the matrix $B = A^t A$ is a positive semidefinite symmetric matrix. Show that if $A$ has rank $n$, then $B$ is positive definite.

(b) *The 2-norm of $A$.* This quantity, denoted by $\|A\|_2$, is defined as the maximum of $\|A\mathbf{x}\|$ on the sphere $\|\mathbf{x}\| = 1$ in $\mathcal{R}^n$. Show that the square of this norm is the largest eigenvalue of the matrix $B$ of part (a).

(c) Let $A$ be an invertible matrix. Show that $\|A^{-1}\|_2$ equals $1/\mu$, where $\mu$ is the minimum of $\|A\mathbf{x}\|$ on the sphere $\|\mathbf{x}\| = 1$ in $\mathcal{R}^n$, and hence its square is the smallest eigenvalue of the (positive definite) matrix $B$ of part (a).

(d) Show that if $A$ is invertible and $\mathbf{u}_1, \ldots, \mathbf{u}_n$ is an orthonormal system of eigenvectors of the matrix $B$ of part (a), then the vectors $\mathbf{v}_i = A\mathbf{u}_i$ $(i = 1, \ldots, n)$ form an orthogonal system.

**1.50.** *Geometry of a linear mapping.* Let a linear mapping be given by a vector equation $\mathbf{y} = A\mathbf{x}$, where $A$ is an invertible $n \times n$ matrix. In this problem we consider only the case $n = 2$ and interpret the mapping as assigning to each point $\mathbf{x}$ in the $x_1 x_2$-plane a point $\mathbf{y}$ in the $y_1 y_2$-plane. Since $A$ is invertible, the correspondence between the two planes is one-to-one.

(a) Show that each straight line in the $x_1 x_2$ plane (given as in (1.62)) corresponds to a straight line in the $y_1 y_2$ plane; in particular, give the values of $\mathbf{a}$ and $\mathbf{b}$ for the line in the $y_1 y_2$-plane.

(b) Show that each circle $\|\mathbf{x}\| = c$ with center at the origin and radius $c > 0$ in the $x_1 x_2$-plane corresponds to an ellipse with center at the origin in the $y_1 y_2$-plane, and give the equation of the ellipse. Under what conditions is the ellipse also a circle? [Hint: $\|\mathbf{x}\| = c$ is equivalent to $\|D\mathbf{y}\|^2 = c^2$, where $D = A^{-1}$.]

(c) Take $c = 1$. Show that that the major and minor axes of the ellipse equal $2a$ and $2b$, respectively, where $a^2$ is the larger eigenvalue of the symmetric matrix $A^t A$ and $b^2$ is the smaller eigenvalue (or $b = a$ when the eigenvalues are equal). Conclude from Problem 1.49 that $a = \|A\|_2$. [Hint: At the ends of the major axis, $\|\mathbf{y}\|^2 = \|A\mathbf{x}\|^2$ is maximized for $\|\mathbf{x}\| = 1$.]

(d) Let axes be rotated in the $x_1 x_2$-plane so that the new basis vectors $\mathbf{e}'_1$, $\mathbf{e}'_2$ are eigenvectors of $A^t A$ associated with the eigenvalues $a^2$, $b^2$ respectively, and let axes be rotated in the $y_1 y_2$-plane so that the ellipse is in standard position, with major axis along the $y'_1$-axis. (See Fig. 1.20.) Show that in the new coordinates, the linear mapping has matrix $\operatorname{diag}(a, b)$.

(e) Conclude from the results of part (d) that $2 \times 2$ orthogonal matrixes $Q_1$ and $Q_2$ can be found such that

$$A = Q_1 \Sigma Q_2, \qquad (*)$$

where $\Sigma = \operatorname{diag}(a, b)$ and $a \geq b > 0$. The equation $(*)$, with the conditions stated, is called the *singular value decomposition of the matrix A*. The numbers $a$, $b$ are called the *singular values* of $A$.

**1.51. (a)⟹(e).** Extend the results of Problem 1.50 to the case $n = 3$.

**1.52. (a)⟹(e).** Extend the results of Problem 1.50 to arbitrary $n$. In (e) one now has $\Sigma = \operatorname{diag}(\sigma_1, \dots, \sigma_n)$, where

$$\sigma_1 \geq \sigma_2 \geq \cdots \geq \sigma_n > 0.$$

REMARK. The singular value decomposition $(*)$ of Problem 1.50 can be extended to an arbitrary $m \times n$ matrix $A$, where $\Sigma$ is now formed of a diagonal

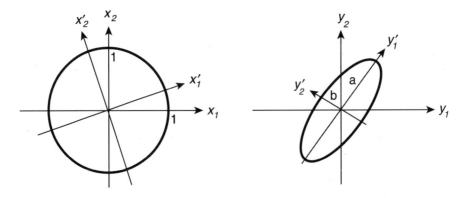

**Figure 1.20.** Invertible linear mapping for $n = 2$.

matrix of order $r$, with decreasing nonzero entries as in Problem 1.52, to which $m - r$ rows of zeros and $n - r$ columns of zeros have been adjoined; here $r$ is the rank of $A$. (See Strang, 1976, p. 135, for a proof.)

**1.53.** *Condition number of matrix A.* This is defined, for an invertible matrix $A$, as

$$\kappa(A) = \|A\|_2 \|A^{-1}\|_2.$$

Use the results of Problems 1.49 to 1.52 to show that

$$\kappa(A) = \frac{\sigma_1}{\sigma_n};$$

thus it is the ratio of the largest singular value to the smallest. For the case $n = 2$ it is the ratio of the major axis of the ellipse of Fig. 1.20 to the minor axis. The condition number is of much importance in the numerical solution of linear equations. If the matrix $A$ has condition number close to 1, then typically it is easy to solve the equation $A\mathbf{x} = \mathbf{b}$ with high accuracy, but if the condition number is much larger than 1, then solution with high accuracy becomes difficult. (See Isaacson and Keller, 1966, pp. 37–49.)

**1.54.** Let $Q(\mathbf{u})$ be a positive definite quadratic form in $\mathcal{R}^n$. Show that the set $E$ of all $\mathbf{u}$ such that $Q(\mathbf{u}) \leq 1$ is a convex set. [Hint: Note that $Q(\mathbf{u})$ is a convex function in $\mathcal{R}^n$ and is hence also convex as a function of the usual parameter along each line segment joining two points of $E$. Therefore $Q$ takes its maximum value along such a segment at one or both endpoints (why?). Therefore the line segment must lie in $E$.]

**1.55.** (Generalization of Problem 1.15, following Section 1.3, to $\mathcal{R}^n$) Let $f(\mathbf{x})$ be defined and have continuous first partial derivatives on a convex open region $E$ in $\mathcal{R}^n$. Show that $f$ is a convex function if and only if, for every $\mathbf{a}$ and $\mathbf{b}$ in $E$,

$$f(\mathbf{b}) - f(\mathbf{a}) \geq (\mathbf{b} - \mathbf{a})^t \nabla f(\mathbf{a}).$$

[Hint: Apply the results of Problem 1.15.]

## 1.9 CONVEXITY AND EXTREMA, LEVEL SETS AND SUBLEVEL SETS

The results of Section 1.4 extend at once to convex functions in $\mathcal{R}^n$. Thus, if $f$ is a convex function defined on a convex open set $E$ contained in $\mathcal{R}^n$ and $f$ is differentiable on $E$, then each critical point of $f$ provides a global minimum of $f$. If $f$ has continuous second partial derivatives, which are the entries in its Hessian matrix $H$, then convexity follows from the property that the corresponding quadratic form $Q(\mathbf{u})$ as in (1.80) is positive semidefinite at each point of $E$. If $Q$ is positive definite at every point, then $f$ is strictly convex. If $f$ has continuous second partial derivatives in an open region and at a critical point of $f$ the quadratic form $Q$ is positive definite, then $f$ is strictly convex in a $\delta$-neighborhood of the critical point, and hence it has a strong local minimum at the point.

The results stated at the end of Section 1.4 also extend without change: Each local minimum of a convex function is a global minimum; if a convex function has several minimizers, then the minimizers form a convex set on which the function is constant; a strictly convex function has at most one minimizer; at each interior point a convex function is continuous and has a "directional derivative" in each direction; and a convex function whose graph includes no line segment is strictly convex. As in Section 1.4 all of these results have their counterparts for concave functions, with minima replaced by maxima (and positive definiteness replaced by negative definiteness with its obvious change of sign).

### Level Sets and Sublevel Sets

The level sets of a function are the sets on which the function takes a chosen constant value. By a *sublevel set* of a function $f$, we mean a set on which the function $f$ has values *less than or equal to* a chosen constant value. Thus, for the function $f(x, y) = x^2 + y^2$, the level sets are the curves $x^2 + y^2 = c$, for varying choices of $c$; hence they are circles when $c$ is positive, for $c = 0$ the level set is a point, and for $c < 0$ the level set is empty. For the same function the sublevel sets are defined by the inequality $x^2 + y^2 \leq c$ for varying choices of $c$. Hence for $c$ positive, each one is a circle plus interior; for $c$ negative, the sublevel set is empty.

Occasionally we will refer to the set on which $f$ has values *less than $c$*. We call such a set an *o-sublevel set* of $f$. If $f$ is defined and continuous in an open region $E$ in $\mathcal{R}^n$, then each nonempty o-sublevel set of $f$ is an open set; furthermore each boundary point in $E$ of a nonempty o-sublevel set $f(\mathbf{x}) < c$ is a point of the level set $f(\mathbf{x}) = c$ (Problem 1.59).

REMARK. In some books on convexity, the term "level set" is used to mean the set on which the function has values less than or equal to a given number. We have preferred the old meaning, as defined above. A preferable notation for level sets uses the symbol $\{\ldots \mid \ldots\}$ for "the set of all ... such that...." Thus $\{\mathbf{x} \mid f(\mathbf{x}) = c\}$ denotes a level set, and $\{\mathbf{x} \mid f(\mathbf{x}) \leq c\}$ a sublevel set. However, we will often use the shorter notation of the preceding paragraphs.

The sublevel sets of an arbitrary function $f$ have a property that will be important in our discussion: *If $c_1 < c_2$, and the sublevel set $f(\mathbf{x}) \leq c_1$ is nonempty, then this set is contained in the sublevel set $f(\mathbf{x}) \leq c_2$* (Problem 1.60).

*If $f$ is a convex function, defined on the convex set $E$ in $\mathcal{R}^n$, then each nonempty sublevel set $f \leq c$ of $f$ is convex, as is each nonempty o-sublevel set*

$f < c$. The proof of the assertion is left as an exercise (Problem 1.61). An example is provided by the function $x^2 + y^2$ in the $xy$-plane.

We explore the level sets for the case in which $f$ is strictly convex on an open convex set $E$ in $\mathcal{R}^n$ and $f$ has continuous partial derivatives through the second order in $E$. Let $f$ have a critical point at $\mathbf{x}_0$ in $E$. Then this point provides a global minimum $m$ of $f$ and is the only critical point of $f$. The level set on which $f$ has the value $m$ is thus the single point $\mathbf{x}_0$. For each $c > m$ in the range of $f$, the level set is defined by a single equation in $n$ variables: $f(x_1, \ldots, x_n) = c$, and $f$ has no critical points on the set. It follows from calculus (the implicit function theorem; see Section 2.1) that the set is formed of one or more smooth hypersurfaces. Thus for $n = 2$ the set is formed of smooth curves in the plane, as in the example $x^2 + y^2$ with $m = 0$. The configuration of the example is typical of the general case for $n = 2$, for which the level sets near the minimizer form a pattern of "concentric" ovals as suggested in Fig. 1.21. For $n = 3$ the ovals are replaced by spherelike surfaces, such as ellipsoids, and a similar description applies for $n > 3$. We describe the configuration as a *center*.

We justify our assertions about the configuration for the case $n = 2$. Similar reasoning applies for larger $n$. We choose a $\delta_0$-neighborhood of the minimizer $\mathbf{x}_0$ contained in $E$ and choose a positive $\delta$ less than $\delta_0$. The circle $C$ of radius $\delta$ about $\mathbf{x}_0$ is then a bounded closed set contained in E, and so our strictly convex function $f$ has a minimum $m_\delta$ on $C$. Since $m$ is the global minimum of $f$, necessarily $m_\delta > m$. From this inequality it follows that *each sublevel set $f \le c$ for $m < c < m_\delta$ must lie inside $C$*; such a sublevel set is nonempty, since $f(\mathbf{x}_0) = m < c$, and it cannot meet $C$, since $f(\mathbf{x}) \ge m_\delta$ on $C$. The sublevel set cannot contain a point outside of $C$, since it is convex, and one cannot join a point outside of $C$ to $\mathbf{x}_0$ without crossing $C$. Thus the sublevel set lies inside $C$.

Now we consider a line segment $\mathbf{x} = \mathbf{x}_0 + r\mathbf{u}$, where $\mathbf{u}$ is a unit vector, joining $\mathbf{x}_0$ to a point of $C$. Along this segment $f$ becomes a strictly convex function of $r$ with unique minimizer $r = 0$ and thus $f$ is strictly increasing in $r$. Accordingly the function $f$ attains the value $c$ at a unique point on the segment, at which $r$ has a positive value less than $\delta$. In terms of polar coordinates with origin at $\mathbf{x}_0$, the value found determines a function $r = g(\theta), 0 \le \theta \le 2\pi$, whose

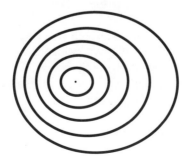

**Figure 1.21.** Level sets of $f$ near a minimizer.

graph is the level curve $f = c$. As noted above, by the implicit function theorem (Section 2.1), the level curve is a smooth curve, so $g$ is continuous (and differentiable). This shows that each level curve $f = c$ for $c$ sufficiently close to $m$ is an oval, as asserted. As $c$ increases, the oval expands and one has the center configuration of Fig. 1.21.

REMARK. The assumption of differentiability of the strictly convex function $f$ is not essential here. The level curve $f = c$ is the boundary of a convex region, and as shown in Section 4.2, in a neighborhood of each point, the curve is the graph of a *continuous* function. Hence the assertion about expanding ovals remains correct. This reasoning extends to higher dimensions.

### Case of Functions Convex in $\mathcal{R}^n$

Let the convex set $E$ of the preceding discussion be the whole space $\mathcal{R}^n$. Then *all* the level sets $f = c$, where $c > m$, have the same structure as that found near the minimizer: For every such value of $c$, the level set is a spherelike surface or hypersurface, representable in spherical coordinates by an equation such as $\rho = F(\phi, \theta)$ for $n = 3$. Thus the family of level surfaces again has a center at $\mathbf{x}_0$. Since $c$ can be arbitrarily large (Problem 1.62), the strictly convex function must approach $+\infty$ as the norm $\|\mathbf{x}\|$ approaches $\infty$. (See Section 1.12.)

### Saddle Points

If at a critical point of function $f$ the corresponding quadratic form $Q$ is indefinite, then the function $f$ has neither maximum nor minimum, but has a saddle point, as at the end of Section 1.4. To study this in detail, we assume for simplicity that the critical point is the origin of $\mathcal{R}^n$ and consider the behavior of the function $f$ along a line $\mathbf{x} = t\mathbf{u}$ through the origin, where $\mathbf{u}$ is a unit vector. As in Section 1.7, $f$ becomes a function of $t$ alone with a critical point for $t = 0$ and having second derivative at $t = 0$ equal to

$$\mathbf{u}^t H(\mathbf{0})\mathbf{u} = Q(\mathbf{u}).$$

If now $\mathbf{u}$ is an eigenvector of $H(\mathbf{0})$ with associated eigenvalue $\lambda$, then this second derivative reduces to $\lambda$. If $\lambda > 0$, we conclude that $f$, as a function of $t$ along the line, has a strong local minimum; if $\lambda < 0$, $f$ has a strong local maximum. With $Q$ indefinite, both cases occur, so $f$ can have neither local maximum nor local minimum at the origin.

The level sets of $f$ near the point can then be expected to resemble those of the quadratic function $Q$. For $n = 2$ these level sets are hyperbolas, as illustrated in Fig. 1.22.

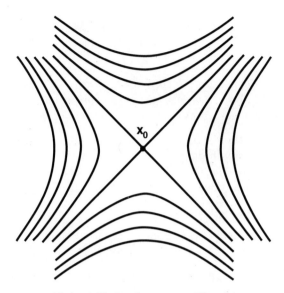

**Figure 1.22.** Level sets at a saddle point.

## PROBLEMS

**1.56.** Verify that each function is strictly convex:

   (a) $x^2 + 2y^2 + 3z^2 + xy + xz + yz$ in $\mathcal{R}^3$.

   (b) $(x + y)^2 + (y + 2z)^2 + (z + 3w)^2 + (w + 4x)^2$ in $\mathcal{R}^4$.

**1.57.** Show that each function is strictly convex in $\mathcal{R}^2$ and sketch the corresponding level sets:

   (a) $3x^2 + 5y^2$.

   (b) $x^4 + x^2y^2 + y^4$ [Hint: Let $Q$ be as in (1.44) and verify that $a + c \geq 0$, $ac - b^2 \geq 0$, so $f$ is convex. Now apply the result of Problem 1.16.]

**1.58.** Sketch each of the following sublevel sets:

   (a) $3x^2 + 5y^2 \leq 1$.

   (b) $3x^2 + 5y^2 \leq 5$.

   (c) $|x|^{3/2} + |y|^{3/2} \leq 1$.

**\*1.59.** Let $f(\mathbf{x})$ be continuous in an open region $D$ of $\mathcal{R}^n$. Let $E$ be a nonempty o-sublevel set: $f(\mathbf{x}) < c$ of the function $f$.

   (a) Show that $E$ is an open set.

   (b) Let $\mathbf{x}_0$ be a point of $D$ that is a boundary point of $E$. Show that $f(\mathbf{x}_0) = c$.

**1.60.** Let $f(\mathbf{x})$ be defined on a set in $\mathcal{R}^n$ and have a nonempty sublevel set $f(\mathbf{x}) \leq c_1$. Let $c_1 < c_2$. Show that this sublevel set is contained in the sublevel set $f(\mathbf{x}) \leq c_2$.

**1.61.** Let $f(\mathbf{x})$ be defined on the convex set $E$ in $\mathcal{R}^n$, and let $f$ be a convex function. **(a)** Show that each nonempty sublevel set of $f$ is convex. **(b)** Show that each nonempty o-sublevel set $f(\mathbf{x}) < c$ is also convex.

**1.62.** As near the end of Section 1.9, let the function $f$ be strictly convex in all of $\mathcal{R}^n$, with minimum $m$ and minimizer $\mathbf{x}_0$. Show that for each $c > m$ each ray $\mathbf{x} = \mathbf{x}_0 + \rho\mathbf{u}$, $0 \leq \rho < \infty$, meets the level set $f = c$ at exactly one point. [Hint: Consider the whole line instead of just the ray so $\rho$ goes from $-\infty$ to $+\infty$. By definition of strict convexity, $f$ when restricted to this line, is a strictly convex function of one variable, $\rho$. As in Problem 1.14(c), this function of $\rho$ is monotone strictly increasing for positive $\rho$, with limit $+\infty$ as $\rho \to +\infty$; the function is also continuous.]

**1.63.** Let $z = f(\mathbf{x})$ be a convex function defined on a convex set $E$ in $\mathcal{R}^n$. One denotes by $[f]$ the set of all $(\mathbf{x}, z)$ of the $(n+1)$-dimensional space of coordinates $(x_1, \ldots, x_n, z)$ such that $\mathbf{x} \in E$ and $z \geq f(\mathbf{x})$. One calls this set the *epigraph* of $f$. Show that the epigraph is a convex set. (See Section 4.4.)

## 1.10  STABILITY

We consider a system of ordinary differential equations:

$$\frac{dx_i}{dt} = X_i(x_1, \ldots, x_N), \quad i = 1, \ldots, N. \tag{1.100}$$

We assume that the functions $X_i(\mathbf{x})$ are all defined and have continuous first partial derivatives in an open region $E$ of $\mathcal{R}^N$. It follows that for each set of initial conditions $\mathbf{x} = \mathbf{x}_0$ (in $E$) for $t = t_0$, there is a unique solution $\mathbf{x} = \mathbf{x}(t)$ of (1.100) satisfying these conditions. The solution is defined in an open interval containing $t_0$, and that interval includes all $t > t_0$ provided the solution does not approach the boundary of $E$ as $t \to t_1$ for some finite $t_1$ greater than $t_0$. Furthermore the solution depends continuously on the choice of initial values $\mathbf{x}_0, t_0$. (See Kaplan, 1958, ch. 12, for this theory.)

### Equilibrium Solution: Stability

If $\mathbf{X}(\mathbf{x}_0) = \mathbf{0}$ for some point $\mathbf{x}_0$ of $E$, then $\mathbf{x} \equiv \mathbf{x}_0$ for $-\infty < t < \infty$ is a solution of (1.100), called an *equilibrium solution*. The solution "path" here reduces to a single point. This solution is said to be *stable* if, for each $\varepsilon > 0$, there is a $\delta > 0$ such that each solution with initial $t$ equal to 0 and initial $\mathbf{x}$ at distance at most $\delta$

from $\mathbf{x}_0$ is defined for all positive $t$ and remains at distance at most $\varepsilon$ for all positive $t$. The choice of initial $t$ is irrelevant here, since the differential equations are unchanged by the replacement of $t$ by $t - t_0$.

## Stability of Vibrations

In many physical systems the behavior is described by a system of ordinary differential equations having the form

$$m_i \frac{d^2 x_i}{dt^2} = -\frac{\partial V}{\partial x_i}, \quad i = 1, \ldots, n. \tag{1.101}$$

Here in the typical application in mechanics the $m_i$ are *masses*, the $x_i$ are position coordinates, $t$ is time, and $V$ is potential energy. We will assume that the $m_i$ are all positive, that $V$ is a function $V(\mathbf{x})$ of the $n$ $x_i$ which has continuous partial derivatives through the second order in an open region $D$ of $\mathcal{R}^n$, and that $V$ has a single critical point; for simplicity, we assume that the critical point is the origin of $\mathcal{R}^n$.

We can write the system (1.101) as follows:

$$\frac{dx_i}{dt} = v_i, \quad m_i \frac{dv_i}{dt} = -\frac{\partial V}{\partial x_i}, \quad i = 1, \ldots, n. \tag{1.101'}$$

This system is a special case of the system (1.100) with $N = 2n$ and $x_{n+1} = v_1, \ldots, x_N = v_n$. The open region $E$ now consists of all points of $\mathcal{R}^N$ for which the first $n$ coordinates specify a point of $D$ and the last $n$ coordinates specify an arbitrary point of $\mathcal{R}^n$. One can think of the points of $E$ as formed of pairs $(\mathbf{x}, \mathbf{v})$; in mechanics these are termed *phases* of the mechanical system and $E$ is the *phase space*.

Since $V$ has a critical point at the origin of $\mathcal{R}^n$, the right sides of (1.101') are all 0 when $\mathbf{x} = \mathbf{0}$ and $\mathbf{v} = \mathbf{0}$; that is, at the origin of $\mathcal{R}^N$. Therefore there is a corresponding equilibrium solution for which $\mathbf{x} \equiv \mathbf{0}$ and $\mathbf{v} \equiv \mathbf{0}$ for all $t$. We now show that this equilibrium solution is stable if at $\mathbf{x} = \mathbf{0}$ the eigenvalues of the Hessian matrix of $V$ are all positive, and hence at $\mathbf{x} = \mathbf{0}$ the corresponding quadratic form

$$Q = \mathbf{u}^t H \mathbf{u} \tag{1.102}$$

is positive definite.

In order to prove our assertion, we first note that by continuity the quadratic form $Q$ is positive definite in a $\delta$-neighborhood of the origin, for some positive $\delta$, so $V$ is strictly convex in this neighborhood and has its unique minimum at the origin. For convenience we take this minimum to be 0.

We next multiply the $i$th equation in (1.101) by $dx_i/dt$ and add all $n$ resulting equations. By calculus (Problem 1.64(a)), we see that the left side of the summed equation is the derivative, with respect to $t$, of the function

$$\sum_{i=1}^{n} \frac{1}{2} m_i \left( \frac{dx_i}{dt} \right)^2 \tag{1.103}$$

which we denote by $T$; in the mechanical application it is the total *kinetic energy* of the system. The right side of the summed equation is simply $-dV/dt$; that is, it is the derivative, with respect to $t$, of the composite function $-V[\mathbf{x}(t)]$ (Problem 1.64(b)). Accordingly we conclude that

$$T + V = \text{const} = c \tag{1.104}$$

for each solution of (1.101). The equation (1.104) expresses the *conservation of energy* for the physical system: kinetic energy plus potential energy remains constant for each allowed motion.

Now the function $T$ is also a quadratic form in the variables $v_1 = dx_1/dt, \ldots, v_n = dx_n/dt$, and since all the $m_i$ are positive, it is also a strictly convex function. The sum on the left of (1.104) can be considered as a function of $N = 2n$ variables, and it is strictly convex in the convex set of $\mathcal{R}^N$ for which $\mathbf{x}$ is restricted to the $\delta$-neighborhood of $\mathbf{0}$ chosen above and $\mathbf{v}$ is arbitrary. (See Problem 1.64(c).) The function has its global minimum $m = 0$ at the origin of $\mathcal{R}^N$.

The law of conservation of energy (1.104) now asserts that each solution of (1.101) is represented by a path in $\mathcal{R}^{2n}$ along which each pair $(\mathbf{x}, \mathbf{v})$ remains on one level set of the function $T + V$. As near the end of Section 1.9, for $c$ sufficiently small and positive, these level sets are "concentric" hypersurfaces, all contained in one $\delta_0$-neighborhood of the origin so one has a *center* configuration. This shows that each solution of the differential equations, with sufficiently small $\mathbf{x}$ and $\mathbf{v}$ for $t = 0$, must remain in the $\delta_0$-neighborhood for all $t$ (positive and negative). Accordingly stability is established.

***Example 1.*** *Simple pendulum.* Here there is only one differential equation:

$$mL \frac{d^2\theta}{dt^2} = -mg \sin \theta,$$

where $L$ is the length of the pendulum (Fig. 1.23). We take $V = mg(1 - \cos \theta)$ and so $V$ is strictly convex for $|\theta| < \pi/2$ and $V$ has its global minimum of 0 at $\theta = 0$. The quadratic form $Q$ here reduces to $(mg/2)\theta^2$ at the minimizer, and this is clearly positive definite. We write $\omega$ for $d\theta/dt$, since here we are dealing with an angular velocity. The conservation of energy is then expressed by the equation

$$\frac{1}{2} mL\omega^2 + mg(1 - \cos \theta) = c.$$

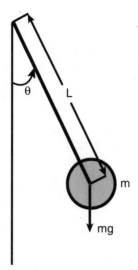

**Figure 1.23.** Simple pendulum.

The function on the left is a strictly convex function of $(\theta, \omega)$ for $|\theta| < \pi/2$, $-\infty < \omega < \infty$, a convex set in $\mathcal{R}^2$, and its level sets are shown in Fig. 1.24. Its global minimum $m$ is 0, and for $c$ sufficiently small and positive, the level sets are ovals enclosing the minimizer, which is the origin. From this figure the stability is clear. It should be observed that because we are dealing with a 1-dimensional problem, the ovals in Fig. 1.24 give the solutions of the differential equation: They are curves parametrized by the functions $\theta = \theta(t)$, $\omega = d\theta/dt = \omega(t)$; in

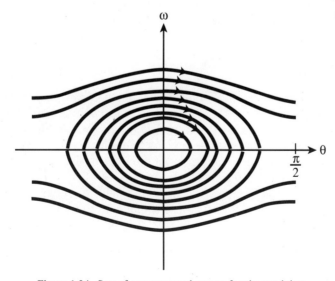

**Figure 1.24.** Sets of constant total energy for the pendulum.

the figure we can see the direction in which they are traced with increasing $t$; however, the figure alone does not give the time values; they can be obtained by integration (Problem 1.66).

The general stability criterion established above applies to *nonlinear* systems of the form considered, such as the simple pendulum of Example 1. It is common practice to establish stability by first proving it for the "linearized" system, obtained from (1.101) by replacing $V$ by the terms of at most second degree in its Taylor series expansion at the equilibrium point, and then showing that stability of the linearized system implies it for the nonlinear system. (For example, see Coddington and Levinson, 1955, pp. 327–329.) The linearized system replaces $V$ by $(\frac{1}{2})Q(\mathbf{x})$, where $Q$ is as above, with Hessian matrix evaluated at the origin. As in Section 1.8, one can change coordinates in $\mathcal{R}^n$ to make $Q$ have the form

$$\lambda_1 x_1^2 + \cdots + \lambda_n x_n^2,$$

where here all $\lambda_i$ are positive. Then the differential equations (1.101) have right-hand sides $-\lambda_i x_i$ and are easily solved linear equations. The solutions are sinusoidal and the stability follows. (See Problem 1.68.) However, proving that stability of the linearized system implies that of the given nonlinear system is more difficult.

REMARK 1. The discussion of the system (1.101) can be applied to an arbitrary system (1.100) for which there is a function $F(\mathbf{x})$, analogous to the total energy $T + V$, which remains constant along each solution of (1.100). Such a function is called a *first integral* of the system of differential equations. If (1.100) has an equilibrium solution $\mathbf{x} \equiv \mathbf{x}_0$ and $F$ is strictly convex in a $\delta$-neighborhood of the point $\mathbf{x}_0$, which is a minimizer of $F$, then the same proof shows that the equilibrium solution is stable.

### Asymptotic Stability

For a real pendulum, the oscillations about the equilibrium point not only remain small; they also die out as time increases, because of friction, which causes a *loss* of energy. In general, we say that an equilibrium solution of (1.100) is *asymptotically stable* if it is stable, and for solutions starting close enough (in all initial conditions) to the equilibrium solution, the deviation from the equilibrium solution approaches $\mathbf{0}$ as $t \to \infty$.

Precisely this type of stability occurs for a physical system described by (1.101) if we adjoin dissipative terms, corresponding to friction, to obtain the system

$$m_i \frac{d^2 x_i}{dt^2} = -\frac{\partial V}{\partial x_i} - h_i \frac{dx_i}{dt}, \quad i = 1, \ldots, n. \tag{1.105}$$

We make the same assumptions as above about the potential energy $V$ and assume that each $h_i$ is positive. As for the frictionless system (1.101), we consider the system in the form of (1.100):

$$\frac{dx_i}{dt} = v_i, \quad m_i \frac{dv_i}{dt} = -\frac{\partial V}{\partial x_i} - h_i v_i, \quad i = 1, \ldots, n. \tag{1.105'}$$

The proof of asymptotic stability involves reasoning similar to that for establishing stability of the equilibrium solution. We write $K$ for the total energy $T + V$. Then a calculation like that leading to the conservation of energy law (1.104) (Problem 1.69) shows that along each solution of (1.105)

$$\frac{dK}{dt} = -\sum_{i=1}^{n} h_i v_i^2. \tag{1.106}$$

The right side of (1.106) is negative except when $\mathbf{v} = \mathbf{0}$. Hence it is a law of *dissipation of energy*.

As above the function $K = T + V$ is a function defined on a convex set $E$ in $\mathcal{R}^{2n}$, and it is strictly convex. The dissipation law (1.106) asserts that along each solution of (1.105), this function is a decreasing function of $t$, except where $\mathbf{v} = \mathbf{0}$. From this fact alone the stability of the equilibrium solution follows. The proof of stability given above can be repeated, with the modification that a solution starting on a level set $K = c$, for $c$ positive and sufficiently close to 0, must remain in the corresponding sublevel set $K \leq c$ as $t$ increases.

It is plausible that on each solution whose initial conditions (values of $\mathbf{x}$ and $\mathbf{v}$ for $t = 0$) determine a point of $E$, the function $K$ approaches its minimum value, 0, as $t \to \infty$. This would mean that on the solution $\mathbf{x}$ and $\mathbf{v}$ approach $\mathbf{0}$ as $t \to \infty$, and the asymptotic stability would follow.

To justify the conclusion, we assume that for some solution the function $K$ does not approach 0. By the dissipation equation $K$ cannot increase as $t$ increases. By stability we know that the solution determines a path inside the convex set $E$ and is hence defined for all $t \geq 0$. Since the function $K$ cannot increase, it must approach a limiting value $K_0$ as $t \to \infty$, and we are assuming $K_0 > 0$. Thus in $\mathcal{R}^{2n}$ the solution is approaching the level set $K = K_0$ of the convex function $K$. Now at each point $P : (\mathbf{x}, \mathbf{v})$ of this level set for which $\mathbf{v} \neq \mathbf{0}$, the dissipation equation implies that a solution through the point must *enter* the convex o-sublevel set $K < K_0$. By continuity, the same statement applies to a solution passing sufficiently close to $P$. For a point $P_1$ on the level set $K = K_0$ for which $\mathbf{v} = \mathbf{0}$, the differential equations (1.105') imply that the derivative of $\mathbf{v} = d\mathbf{x}/dt$ is not zero, for the only critical point of $V$ is at $\mathbf{x} = \mathbf{0}$. Hence, as $t$ increases, the solution through $P_1$ moves at once to points at which $\mathbf{v} \neq \mathbf{0}$. This implies that $K$ must in fact decrease as $t$ increases from its value at $P_1$, and again the path enters the o-sublevel set $K < K_0$. By continuity, the same assertion applies to all points $P$ sufficiently close to $P_1$. Now, if, as above, we have a solution on which $K$ remains strictly above $K_0$ but on which $K$ approaches $K_0$

as $t \to \infty$, then by the Weierstrass-Bolzano theorem (Kaplan, 1991, p. 177), there is a sequence $t_k, k = 1, 2, \ldots$, such that, as $k \to \infty, t_k \to \infty$ and the corresponding points $\mathbf{x}(t_k), \mathbf{v}(t_k)$ on the solution considered have as limit a point $P$ on the level set $K = K_0$. This conclusion violates the assertion that for each point sufficiently close to $P$ the solution through the point must enter the o-sublevel set $K < K_0$ as $t \to \infty$. Thus the asymptotic stability is proved.

REMARK 2. The proof just given shows that one can establish asymptotic stability of the equilibrium solution of (1.100) under the conditions stated in Remark 1 above, with the condition that the function $F$ be a first integral of the differential equations (1.100) replaced by the following: $dF/dt \le 0$ along each solution in the $\delta$-neighborhood described and the function $F$ is identically constant on *no solution* of (1.100).

## 1.11   GLOBAL ASYMPTOTIC STABILITY, APPLICATION TO FINDING MINIMIZER

We consider a system (1.100) for which the functions $X_i$ satisfy the hypotheses stated at the beginning of Section 1.10 in the whole space $\mathcal{R}^N$. Then the equilibrium solution is said to be globally asymptotically stable if *every solution* approaches the equilibrium state as $t \to \infty$.

Precisely this kind of stability occurs for a system (1.100) under the conditions stated in Remark 2 provided that the $\delta$-neighborhood is replaced by the whole space $\mathcal{R}^N$. Indeed, as remarked at the end of Section 1.9, the level sets of $F$ have a center at $\mathbf{x}_0$. The argument given to show asymptotic stability can be repeated, for an *arbitrary* initial choice of $\mathbf{x}$ in $\mathcal{R}^N$ and so the global asymptotic stability follows. In particular, the equilibrium solution of (1.105) is globally asymptotically stable, under the hypotheses stated above, when the potential energy function $V$ is strictly convex in all of $\mathcal{R}^n$.

These results suggest that differential equations can be used as a computational tool for finding the minimizer of a given strictly convex function $F$. If, for example, $F$ is strictly convex in all of $R^N$, $F$ has continuous first partial derivatives and $F$ has a minimizer $\mathbf{x}_0$, then this point must be a critical point of $F$. Hence it is an equilibrium point for the differential equations:

$$\frac{dx_i}{dt} = -\frac{\partial F}{\partial x_i}, \quad i = 1, \ldots, N, \tag{1.110}$$

or in vector form,

$$\frac{d\mathbf{x}}{dt} = -\nabla F. \tag{1.110'}$$

As shown in calculus (see Kaplan, 1991, pp. 134–137), the vector $-\nabla F$ points in the direction of *maximum rate of decrease* of the function $F$. It therefore follows from (1.110′) that along each solution of the differential equations (other than the equilibrium solution), the function $F$ decreases as $t$ increases. Thus as above, the equilibrium solution is globally asymptotically stable. Accordingly, if we select an arbitrary initial point $\mathbf{x}_1$ and find the solution of the system (1.110) that has this value for $t = 0$, then as $t \to \infty$, the solution will approach the value $\mathbf{x}_0$, which is the minimizer sought.

***Example 1.*** Let $F(x, y) = x^2 + y^2 + e^{-x-y}$. Here one can verifies that $F$ is strictly convex in $\mathcal{R}^2$ and that it has a critical point $(k, \; k)$ where $k = 0.28357\ldots$ is the solution of the equation $e^{-2x} - 2x = 0$. (See Problem 1.71.) The equations (1.110) are as follows:

$$\frac{dx}{dt} = -2x + e^{-x-y}, \quad \frac{dy}{dt} = -2y + e^{-x-y}.$$

By MATLAB we find the solution with $x = 1$ and $y = 1$ for $t = 0$. It is graphed in Fig. 1.25. For $t = 10$, we have $x = y = 0.28335\ldots$. The accuracy can be improved by using the point $(x, \; y)$ found as a new starting point and decreasing the tolerance in the MATLAB program.

***Example 2.*** Let

$$F(x, y) = 3x + 2y + x^2 + y^2 + x^4 + x^2y^2 + y^4.$$

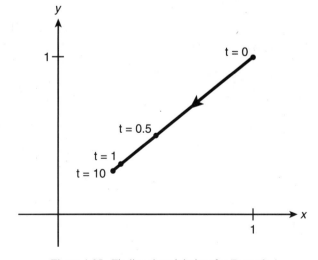

**Figure 1.25.** Finding the minimizer for Example 1.

Again one verifies that the function is strictly convex in $\mathcal{R}^2$. One can show by the methods of Section 1.12 that $F$ has a unique critical point (see Problem 1.84). We form the equations (1.110) for this function and seek the minimizer by simply choosing an initial point and computing the solution by MATLAB. The example illustrates a difficulty that can arise when one attempts to compute the solution. If the initial point has large values of $x$ or $y$ or both, the right-hand sides of the differential equations may be extremely large, and the MATLAB program may report that there is a singularity and so the solution cannot be computed up to the chosen value of $t$; precisely this happens if one starts at the point $(4, -3)$. To rescue the solution, one can modify the differential equations by dividing both right-hand sides by a positive function $s(x, y)$ which is large where $\|\nabla F\|$ is large and is close to 1 otherwise. Specifically, we divide the right-hand members by the function

$$\max(1, \|\nabla F\|).$$

If this is done, we find that there is no difficulty in solving the differential equation even for an initial point which turns about to be very far from the minimizer, which is found to be $(-0.6843, -0.5053)$. If we start at the point $(50, 49)$, then one application of MATLAB with tolerance 0.001 and $t$ going from 0 to 10 takes one to $(42.8, 42.0)$. With this point as initial point we repeat the process and obtain successively the points $(34.9, 34.9)$, $(27.9, 27.9)$, $(20.9, 20.9)$, $(13.9, 13.9)$, $(6.9, 6.9)$, $(-0.247, -0.028)$, and $(-0.6843, -0.5053)$.

The procedure illustrated can be used in other cases when the conditions ensure that the solutions of the differential equations (1.110) with appropriate initial values do have the minimizer as limit as $t \to \infty$. For example, if $F$ is strictly convex in a spherical region $E: \|\mathbf{x}\| \le k$ and $F$ has the spherical boundary of $E$ as a level set, then $F$ has a unique minimizer in $E$ and it can be found from an arbitrary initial point in $E$. (See Problem 1.73.)

## PROBLEMS

**1.64.** Multiply the $i$th equation in (1.101) by $dx_i/dt$, and add the resulting equations.

    **(a)** Show that the left side of the resulting equation is the derivative, with respect to $t$, of the total kinetic energy $T$ given by (1.103).

    **(b)** Show that the right side of the equation is the derivative, with respect to $t$, of minus the potential energy $V$.

(c) Show that under the hypotheses made for the potential energy $V$, the total energy $T + V$ is a strictly convex function of $2n$ variables

$$x_1, \ldots, x_n, v_1, \ldots, v_n$$

in the convex subset of $\mathcal{R}^{2n}$ for which $\mathbf{x}$ is in the chosen $\delta$-neighborhood of the origin and the $v_i$ are arbitrary.

**1.65.** When $n = 1$ in (1.101), the energy equation (1.104) is a first-order differential equation for $x = x_1$ as a function of $t$. Solve the equation for $dx/dt$, and then separate variables to obtain the solutions in implicit form in terms of an integral.

**1.66.** (a) Apply the procedure of Problem 1.65 to the pendulum equation of Example 1 in Section 1.10.

(b) Take $L = 100$ cm, $g = 980$ cm/sec$^2$ and use a numerical process to find $\theta$ at $t = 0.1$, 0.2, 0.3 for the solution such that $\theta = 0$ and $\omega = 1$ rad/sec for $t = 0$.

(c) Find the maximum value of $\theta$ (amplitude of the oscillation).

(d) Use a numerical process to find the period.

**1.67.** Find the energy integral and sketch the level sets of the total energy in the phase space (here a plane) for the system (1.101) with $n = 1$, $m = 1$, and $V = x_1^2 + (\frac{1}{4})x_1^4$ (nonlinear spring).

**1.68.** Show that the equilibrium solution of (1.101) is stable if $V$ is a positive definite quadratic form

$$\lambda_1 x_1^2 + \cdots + \lambda_n x_n^2.$$

**1.69.** Follow the steps described in the text to derive the law of dissipation of energy (1.106) for the system (1.105) under the hypotheses stated.

**1.70.** Verify the asymptotic stability of the equilibrium solution $\theta = 0$ of the damped pendulum equation

$$\frac{d^2\theta}{dt^2} = -\sin\theta - 2\frac{d\theta}{dt}$$

by using software such as MATLAB to find solutions with $\theta = \pi/4$ and various choices of $d\theta/dt$ for $t = 0$.

**1.71.** Consider Example 1 in Section 1.11.

(a) Verify that the function $F$ is strictly convex in $\mathcal{R}^2$.

(b) Use software such as MATLAB to find the solution with $x = 1$ and $y = 1$ for $t = 0$, and verify that it approaches the minimizer of $F$ given.

**1.72.** Consider Example 2 in Section 1.11.

(a) Form the differential equations (1.110) for this case, and use software such as MATLAB to try to obtain the solution for $0 \le t \le 10$ starting at $(10, 10)$ for $t = 0$.

(b) Modify the differential equations of part (a), as suggested in the text, by dividing the right-hand members by $s(x, y) = \max(1, \|\nabla F\|)$, and again seek the particular solution specified in part (a). Explain the effect on the solutions of dividing by $s(x, y)$.

*1.73. Let $F$ have continuous second partial derivatives in an open region of $\mathcal{R}^n$ containing a spherical closed region $E: \|\mathbf{x}\| \le k$ (with $k > 0$), and let $F$ be strictly convex on $E$, so $F$ has a unique minimizer $\mathbf{x}_0$ in $E$. Let the spherical boundary of $E: \|\mathbf{x}\| = k$ be a level set of $F$. Show that the differential equations (1.110) for this $F$ have the equilibrium solution $\mathbf{x} = \mathbf{x}_0$ and that this solution is asymptotically stable. Show that every solution of the differential equations starting at a point of $E$ approaches $\mathbf{x}_0$ as $t \to \infty$.

## 1.12  EXTREMA OF FUNCTIONS ON UNBOUNDED CLOSED SETS

We know that a function which is defined and continuous on a bounded closed set $E$ has a global maximum and a global minimum. If the set $E$ is closed but unbounded, the assertion is no longer true. For example, the function $y = x$ is continuous for $x \ge 0$ but has no global maximum. However, it has a global minimum, $m = 0$. This simple example suggests a definition and a general rule.

Let the function $f(\mathbf{x})$ be defined and continuous on an unbounded set $E$ in $\mathcal{R}^n$. We say that the function $f$ has *limit* $\infty$ *for* $\|\mathbf{x}\| \to \infty$ if for each number $k$, there is a positive $r_0$, such that $f(\mathbf{x}) > k$ for $\mathbf{x}$ in $E$ and $\|\mathbf{x}\| = r > r_0$. For $n = 2$ an example of such a function is given by the quadratic function $f(x, y) = 3x^2 + y^2$ with $E$ the set $x^2 + y^2 \ge 1$. Here $f > k$ for

$$ r = \sqrt{x^2 + y^2} > r_0 = \sqrt{|k|}. $$

We remark that if $f$, defined on the unbounded set $E$, has limit $\infty$ as $\|\mathbf{x}\| \to \infty$, then $f$ has the same property on each unbounded subset of $E$. In particular, if $f$ has the property on the whole space $\mathcal{R}^n$, then it has the property on every unbounded set $E$ in $\mathcal{R}^n$.

*If $E$ is closed and $f$ has limit $\infty$ for $\|\mathbf{x}\| \to \infty$, then $f$ has a global minimum on $E$.* To show this, we let $\mathbf{x}_0$ be a point of $E$, and let $m_0 = f(\mathbf{x}_0)$. We now choose $r_0$, greater than $\|\mathbf{x}_0\|$ and so large that $f(\mathbf{x}) > m_0$ for $r > r_0$. The set of points of $E$ for which $\|\mathbf{x}\| \le r_0$ form a bounded closed set $E_0$ (Problem 1.76). Hence $f$ has a global minimum $m$ on $E_0$. Since $\mathbf{x}_0$ is in $E_0$, necessarily $m \le m_0$. Since $f(\mathbf{x}) > m_0$ for the points of $E$ outside of $E_0$, $m$ must be the global minimum of $f$ on $E$.

For our function $3x^2 + y^2$ on the set $x^2 + y^2 \geq 1$, the global minimum is 1, with minimizer $(0, 1)$.

REMARK. If $f$, defined on the unbounded set $E$, has limit $\infty$ as $\|x\| \to \infty$, then each nonempty level set (or sublevel set) of $f$ is bounded; for the sets $\{f(x) = k\}$ and $\{f(x) \leq k\}$ are contained in the set $\{\|x\| \leq r_0\}$ for $r_0$ chosen as above. If $f$ is a polynomial of degree $m$ in $x_1, \ldots, x_n$, then one can write

$$f(x) = g_m(x) + \cdots + g_0(x),$$

where $g_i(x)$ is formed of the terms of degree $i$ in $f$. Let $h_i$ be the global minimum of $g_i$ on the unit sphere $\|x\| = \|u\| = 1$. If $h_m > 0$, then $f$ has limit $\infty$ as $\|x\| \to \infty$. Indeed, one can write $x = ru$, where $r \geq 0$ and $u$ is a unit vector. Then

$$f(x) = r^m g_m(u) + \cdots + r^0 g_0(u),$$

so for $r \geq 0$,

$$f(x) \geq h_m r^m + \cdots + h_0$$

and the polynomial on the right has limit $\infty$ as $r = \|x\| \to \infty$.

## Application to Distance Function

In $\mathcal{R}^n$ the distance of a general point from a fixed point $x_0$ is a function $f(x) = \|x - x_0\|$ defined and continuous in $\mathcal{R}^n$. (See Problem 1.77.) If $E$ is a closed set in $\mathcal{R}^n$ that is bounded, then $f(x)$ has a global minimum (and global maximum) on $E$. We call the global minimum the *shortest distance from $x_0$ to $E$* or simply the *distance from $x_0$ to $E$*. If, however, $E$ is closed but unbounded, then there is still a point of $E$ closest to $x_0$, for we verify that the function $f$ has limit $\infty$ as $\|x\| \to \infty$ (Problem 1.78). Hence we can always speak of the distance from a point $\|x_0\|$ to a closed set $E$. (See Fig. 1.26 for examples.)

We remark that the minimizer for the distance function is generally not unique (see the set $E_1$ in Fig. 1.26). However, *if $E$ is a convex set, it is unique.* We show this most easily by observing that a minimizer of the distance function is also a minimizer of its square:

$$\|x - x_0\|^2 = (x - x_0)^t (x - x_0).$$

The Hessian matrix of this function (a polynomial in the $x_i$) is simply twice the identity matrix! Thus the corresponding quadratic form is positive definite at every point. (See Problem 1.79.) Accordingly the squared distance function is a strictly convex function in $\mathcal{R}^n$ and has at most one minimizer on each convex set.

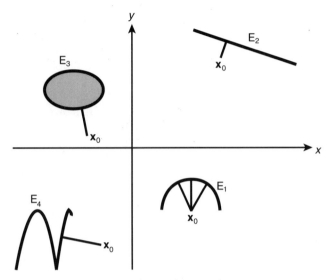

**Figure 1.26.** Distance from a point to a set.

## Functions Strictly Convex in $\mathcal{R}^n$

Let $f$ be such a function. At the end of Section 1.9, we saw that if $f$ is strictly convex and has a global minimum $m$, with minimizer $\mathbf{x}_0$, then for every $c > m$ the level set $f = c$ is a spherelike set enclosing $\mathbf{x}_0$, and so the level sets form a center configuration. In particular, each level set $f = c$ can be represented in the form $\rho = F(u_1, \ldots, u_n)$, where $\mathbf{u}$ is a variable point on the sphere of radius 1 with center $\mathbf{x}_0$ and $F$ is continuous and positive; $\rho$ is a polar coordinate, the distance from $\mathbf{x}_0$. By continuity, $F$ has a maximum value, say $k$. Thus on the level set chosen $\rho = \|\mathbf{x} - \mathbf{x}_0\| \leq k$. Since the distance function has limit $\infty$ as $\|\mathbf{x}\| \to \infty$, we can choose $r_0$ so large that the level set chosen is contained in the set $\|\mathbf{x}\| \leq r_0$. It follows that $f(\mathbf{x}) > c$ for $\|\mathbf{x}\| > r_0$. Since $c$ can be chosen arbitrarily large, we conclude that $f$ has limit $\infty$ as $\|\mathbf{x}\| \to \infty$. Thus *every strictly convex function in $\mathcal{R}^n$ which has a minimum has limit $\infty$ as $\|\mathbf{x}\| \to \infty$.* Conversely, if $f$ is a strictly convex function in $\mathcal{R}^n$ and $f$ has limit $\infty$ as $\|\mathbf{x}\| \to \infty$, then $f$ has a minimum, since $f$ is continuous and $\mathcal{R}^n$ itself is an unbounded closed set in $\mathcal{R}^n$.

We observe that the condition that $f$ have a minimum is crucial for the preceding paragraph. For example, the function $e^{-x}$ in $\mathcal{R}^1$ is strictly convex but has no minimum and does not have limit $\infty$ as $|x| \to \infty$.

## Case of Quadratic Functions

Let

$$f(\mathbf{x}) = \mathbf{x}^t A \mathbf{x} + \mathbf{b}^t \mathbf{x} + c, \tag{1.120}$$

where $A$ is symmetric (not the $O$ matrix), $\mathbf{b}$ is a constant vector of $\mathcal{R}^n$ and $c$ is a constant scalar. We call $f$ a quadratic function. We now assume that $A$ is positive definite. Then, since $2A$ is the Hessian matrix of $f$, $f$ is strictly convex in $\mathcal{R}^n$. Moreover $f$ has a critical point. From (1.120), as at the end of Section 1.8, we find that

$$\nabla f = 2A\mathbf{x} + \mathbf{b}.$$

As noted at the end of Section 1.7, a positive definite matrix is invertible. Hence the equation $\nabla f = \mathbf{0}$ has a unique solution

$$\mathbf{x}^* = -(\tfrac{1}{2})A^{-1}\mathbf{b},$$

which is the unique critical point of $f$ and is the minimizer of $f$. Therefore $f$ has limit $\infty$ as $\|\mathbf{x}\| \to \infty$ in $\mathcal{R}^n$. Accordingly $f$ has limit $\infty$ as $\|\mathbf{x}\| \to \infty$ on each unbounded set in $\mathcal{R}^n$. One can prove this result in other ways. For example, from (1.85) we conclude that

$$\mathbf{x}^t A\mathbf{x} \geq \lambda_n \|\mathbf{x}\|^2,$$

where $\lambda_n$ (necessarily positive) is the smallest eigenvalue of $A$, and by the Schwarz inequality of Section 1.6,

$$\mathbf{b}^t \mathbf{x} \geq -\|\mathbf{b}\|\,\|\mathbf{x}\|.$$

Therefore from (1.120),

$$f(\mathbf{x}) \geq \lambda_n \|\mathbf{x}\|^2 - \|\mathbf{b}\|\,\|\mathbf{x}\| - |c|.$$

The quadratic function of $\|\mathbf{x}\|$ on the right has limit $\infty$ as $\|\mathbf{x}\| \to \infty$, so the same statement applies to $f$. Another proof is given in Remark 3 in Section 1.15.

## 1.13 SHORTEST DISTANCE FROM A LINEAR VARIETY

As in Section 1.6 a linear variety can be given in two ways: by parametric equations (1.64) or by a set of linear equations (1.66) in the coordinates $x_i$. Each linear variety $E$ (of dimension greater than 0) in $\mathcal{R}^n$ is an unbounded closed convex set (Problem 1.81). (Convexity was proved in Section 1.7.) Hence as in Section 1.12 the distance function $f(\mathbf{x}) = \|\mathbf{x} - \mathbf{x}_0\|$ has a unique minimizer $\mathbf{x}^*$ in $E$, and $\|\mathbf{x}^* - \mathbf{x}_0\|$ is the shortest distance from $\mathbf{x}_0$ to $E$.

Let the linear variety $E$ be given by parametric equations as in (1.65):

$$\mathbf{x} = \mathbf{a} + u_1\mathbf{b}_1 + \cdots + u_k\mathbf{b}_k = B\mathbf{u} + \mathbf{a}, \tag{1.130}$$

where the vectors $\mathbf{b}_1, \ldots, \mathbf{b}_k$ are a basis of a subspace $W$ of $\mathcal{R}^n$ and the corresponding matrix $B$ is an $n \times k$ matrix of rank $k$. To find the point of $E$ nearest to $\mathbf{x}_0$, we can seek the minimum of the *squared* distance function $\|\mathbf{x} - \mathbf{x}_0\|^2$ for $\mathbf{x}$ in $E$. By the parametric equations (1.130), this function becomes a function of the vector $\mathbf{u}$:

$$g(\mathbf{u}) = \|B\mathbf{u} + \mathbf{a} - \mathbf{x}_0\|^2 = \|B\mathbf{u} + \mathbf{c}\|^2,$$

where $\mathbf{c} = \mathbf{a} - \mathbf{x}_0$. If we write the norm squared as the inner product of a vector with itself and expand this expression, we find that $g(\mathbf{u})$ is a quadratic function as in (1.120), where $A = B^t B$, $\mathbf{b} = 2B^t\mathbf{c}$, and $c = \mathbf{c}^t\mathbf{c}$. Since $B$ has rank $k$, $A$ is a positive definite symmetric matrix (Problem 1.82). Hence as in Section 1.12 the function $g$ has a unique minimizer, which can be found by equating its gradient to $\mathbf{0}$. The minimizer can be verified to be the foot of the perpendicular from $\mathbf{x}_0$ to the linear variety $E$, as in Fig. 1.27. Details are left to Problem 1.82.

We remark that the problem considered is a minimum problem with *side conditions* and can be treated by the method of Lagrange multipliers. This method is especially appropriate if the linear variety $E$ is described by a set of linear equations: $A\mathbf{x} = \mathbf{b}$ instead of by parametric equations (1.130); here $A$ is a $k \times n$ matrix of rank $k < n$. This case is considered in Section 2.2 and it is shown that the point $\mathbf{x}^*$ of $E$ closest to $\mathbf{x}_0$ is given by the equation:

$$\mathbf{x}^* = \mathbf{x}_0 + A^t(AA^t)^{-1}(\mathbf{b} - A\mathbf{x}_0). \tag{1.131}$$

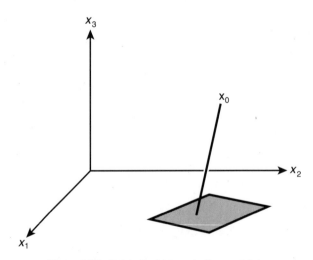

**Figure 1.27.** Point closest to $\mathbf{x}_0$ in linear variety.

## Application to Least Squares

The method of least squares is commonly used to fit a set of data by a function of prescribed type. Generally, one cannot fit the data exactly by such a function and hence tries to make the "total square error" as small as possible. For example, given data $(x_i, y_i)$ for $i = 1, \ldots, n$ one tries to find a function

$$f(x) = u_1 f_1(x) + \cdots + u_m f_m(x),   \tag{1.132}$$

where the $u_i$ are constants and the $f_i(x)$ are known functions, in order to provide the least square error

$$(f(x_1) - y_1)^2 + \cdots + (f(x_n) - y_n)^2.$$

Let $v_i = f(x_i)$ for each $i$, and let $B$ be the matrix $(b_{ij})$, where $b_{ij} = f_j(x_i)$ for $i = 1, \ldots, n$, $j = 1, \ldots, m$. Thus we have the vector equation

$$B\mathbf{u} = \mathbf{v}   \tag{1.133}$$

and we wish to choose $\mathbf{u}$ to minimize the total square error

$$(v_1 - y_1)^2 + \cdots + (v_n - y_n)^2 = \|\mathbf{v} - \mathbf{y}\|^2,   \tag{1.134}$$

which is the square of the distance in $\mathcal{R}^n$ from $\mathbf{v}$ to $\mathbf{y}$. Here $\mathbf{y}$ is known and $\mathbf{v}$ is an arbitrary point of the range of the linear mapping (1.133). But that range is the column space of the $n \times m$ matrix $B$; that is, it is the set of all linear combinations of the column vectors of $B$. If these column vectors are linearly independent, then the column space is a linear variety as in (1.65) with $\mathbf{a} = \mathbf{0}$ and is thus a subspace $W$ of $\mathcal{R}^n$ of dimension $m$. As remarked in Section 1.6, even when the columns are linearly dependent, the column space is such a linear variety, here a subspace, with dimension equal to the maximum number of linearly independent columns.

Accordingly, minimizing the total square error is equivalent to finding $\mathbf{u}$ in $\mathcal{R}^m$ such that $B\mathbf{u}$ minimizes the distance in $\mathcal{R}^n$ from $\mathbf{y}$ to the linear variety formed of the column space of $B$. As above, there is a unique point $\mathbf{v}^*$ in the linear variety that minimizes the distance from $\mathbf{y}$, and that point can be found, as above, by orthogonal projection of $\mathbf{y}$ onto the subspace. However, there may be more than one $\mathbf{u}$ in $\mathcal{R}^m$ for which $B\mathbf{u} = \mathbf{v}^*$. By linear algebra, uniqueness holds precisely when the columns of $B$ are linearly independent and thus form a basis of $W$.

***Example 1.*** Let $x_i = i$ for $i = 1, \ldots, 10$ and let the corresponding $y_i$ be the numbers 2.37, 2.82, 3.09, 3.48, 5.01, 3.29, 2.08, 1.58, 1.94, and 2.41. We seek a cubic polynomial

$$f(x) = u_1 x^3 + u_2 x^2 + u_3 x + u_4$$

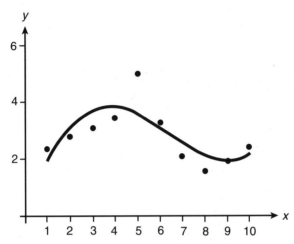

**Figure 1.28.** Fitting data by a cubic curve by least squares.

that fits the given data with least square error. By MATLAB we otain a
solution: $u_1 = 0.0254$, $u_2 = -0.4815$, $u_3 = 2.4993$, $u_4 = -0.0947$. This is
graphed in Fig. 1.28. (See Problem 1.83.)

## PROBLEMS

**1.74.** Find the global minimum of each function:
  (a) $y = x, 0 \leq x < \infty$.
  (b) $y = x^3 - 2x^2 + 5x - 1, 0 \leq x < \infty$.
  (c) $z = 3x^2 + y^2, x^2 + y^2 \geq 1$.
  (d) $z = e^{2x} + e^{-y}, x \geq 1, -1 \leq y \leq 1$.

**1.75.** (a)$\Rightarrow$(d). In each case verify that the function in the corresponding part
of Problem 1.74 is defined on an unbounded closed set and that the
function has limit $\infty$ as $\|\mathbf{x}\| \to \infty$.

*__1.76.__ (a) Let $E$ be an unbounded closed set in $\mathcal{R}^n$. Let $E_0$ the set of all points
$\mathbf{x}$ of $E$ such that $\|\mathbf{x}\| \leq r_1$, where $r_1$ is a positive number and let $E_0$ be
nonempty. Show that $E_0$ is a bounded closed set.

  (b) Show generally that the set $E_0$ of all points common to two intersecting closed sets is closed.

*__1.77.__ Show that $f(\mathbf{x}) = \|\mathbf{x} - \mathbf{x}_0\|$, the distance from $\mathbf{x}$ to a fixed point $\mathbf{x}_0$, is con-
tinuous in $\mathcal{R}^n$. [Hint: Use the principle that the composition of two
continuous functions is continuous. Show that here $f$ is the composition
of a polynomial in the $x_i$ and the function $\sqrt{u}$, for $u \geq 0$.]

**1.78.** Show that the function $f(\mathbf{x})$ of Problem 1.77 has limit $\infty$ as $\|\mathbf{x}\| \to \infty$. [Hint: By properties of the norm,

$$\|\mathbf{x}\| = \|\mathbf{x} - \mathbf{x}_0 + \mathbf{x}_0\| \leq \|\mathbf{x} - \mathbf{x}_0\| + \|\mathbf{x}_0\|,$$

and thus $f(\mathbf{x}) \geq \|\mathbf{x}\| - \text{const.}$]

**1.79.** Show that the square of the function $f(\mathbf{x})$ of Problem 1.77 is a strictly convex function in $\mathcal{R}^n$.

**1.80.** Find the distance from the given point to the given set:

(a) From $(3, 4)$ to the circle $x^2 + y^2 = 1$ in $\mathcal{R}^2$.

(b) From $(5, 1)$ to the parabola $y = 4x^2$ in $\mathcal{R}^2$.

(c) From $(1, 1, 1)$ to the circle $x^2 + y^2 = 1, z = 0$ in $\mathcal{R}^3$.

(d) From $(3, 3, 5)$ to the paraboloid $z = x^2 + y^2$ in $\mathcal{R}^3$. [Hint: Use cylindrical coordinates.]

**\*1.81.** Show that a linear variety of positive dimension in $\mathcal{R}^n$ is an unbounded closed set.

**1.82.** Let $B$ be an $n \times k$ matrix of rank $k$. Let $W$ be the subspace spanned by the columns of $B$.

(a) Show that the matrix $B^t B$ is a positive definite $k \times k$ matrix and is hence nonsingular (see Problem 1.49).

(b) Show that the closest point $\mathbf{x}^*$ to $\mathbf{x}_0$ in the linear variety (1.130) is given by

$$\mathbf{x}^* = \mathbf{a} - P\mathbf{c}, \quad P = B(B^t B)^{-1} B^t, \quad \mathbf{c} = \mathbf{a} - \mathbf{x}_0.$$

(c) By linear algebra (see Appendix A.18) $P$ is the orthogonal projection on the subspace $W$. Use this information to show that $\mathbf{x}^*$ is the foot of the perpendicular to $E$ through $\mathbf{x}_0$.

(d) Apply the result of part (b) to show that the solution of the least squares problem of Section 1.13 is given by

$$\mathbf{u} = (B^t B)^{-1} B^t \mathbf{y},$$

provided that $B$ has linearly independent columns.

**1.83.** Consider Example 1 in Section 1.13.

(a) Find the matrix $B = (b_{ij})$ corresponding.

(b) Show that the columns of $B$ are linearly independent and hence $B$ has rank 4. Conclude that the solution given is the only one.

(c) Fit the same data, using least squares, by a second-degree polynomial.

(d) Fit the same data, using least squares, by a function $f(x) = u_1 e^x + u_2 e^{-x}$.

**1.84.** Let $F(x, y)$ be as in Example 2 in Section 1.11. Show that $F$ is strictly convex in $\mathcal{R}^2$ and that $F$ has limit $+\infty$ as $||\mathbf{x}|| = (x^2 + y^2)^{1/2} \to \infty$. Thus $F$ has a unique global minimizer. [Hint: Use polar coordinates $r$, $\theta$, and show that $F(r\cos\theta, \sin\theta) \geq -5r + r^2 + (\frac{1}{2})r^4$.]

## 1.14   OTHER INNER PRODUCTS AND NORMS IN $\mathcal{R}^n$

In geometry the concepts of distance and angle are basic. We have thus far used only one formula for inner product: namely, the expression

$$\mathbf{x}^t\mathbf{y} = x_1 y_1 + \cdots + x_n y_n.$$

We will see that other formulas lead to satisfactory definitions of distance and angle.

We illustrate the idea by a 2-dimensional example. We use two coordinate planes, the $uv$-plane and the $xy$-plane, as in Fig. 1.29. We assume that distance and angle are measured as usual in the $xy$-plane and that the two planes are related by linear equations:

$$x = u + v, \qquad y = u + 2v. \tag{1.140}$$

Then the lines parallel to the axes in the $xy$-plane correspond to two sets of parallel lines in the $uv$-plane, as in Fig. 1.29. In general, there is a one-to-one correspondence between the points (or vectors) of the two planes. Equations (1.140) give $(x, y)$ in terms of $(u, v)$. From these equations we find that $u = 2x - y$, $v = -x + y$, giving $(u, v)$ in terms of $(x, y)$. We now use the

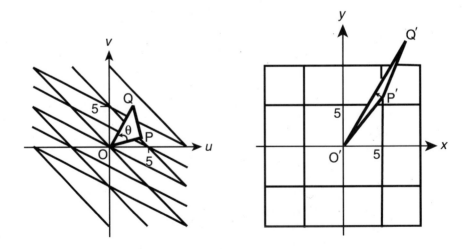

**Figure 1.29.** New geometry in a plane.

following way of measuring distance and angle in the $uv$-plane: For each pair of points in the $uv$-plane, we assign as distance the number $d$ equal to the distance between the corresponding pair of points in the $xy$-plane; similarly, for each pair of vectors in the $uv$-plane, we assign as angle $\theta$ the angle between the corresponding vectors in the $xy$ plane. Thus the distance between the points $P:(4, 1)$ and $Q:(3, 5)$ in the $uv$-plane is taken as the distance between the corresponding points $(5, 6), (8, 13)$ in the $xy$-plane, or $\sqrt{58}$. The angle $\theta$ formed by the vectors $OP$ and $OQ$ equals that between the vectors $(5, 6)$ and $(8, 13)$ in the $xy$-plane; with the help of a pocket calculator, we find that $\theta = 0.14$ radians. (See Problem 1.85.)

We remark that since in the $xy$-plane all the rules of Euclidean geometry are valid, they remain valid in the $uv$-plane as long as one measures distances and angles as above, by reference to the $xy$-plane.

The process described is in common use, with reference to mapmaking by aerial photography. Two streets of a town may be perpendicular, but usually they will not appear so in the photograph, and of course distances in the photograph differ from the true distances, and not just by a scale factor. To good approximation, the photograph produces a $uv$-plane that is related to the $xy$-plane being photographed by linear equations such as (1.140). Once one knows these equations, one can find the true distances and angles exactly as in the example.

Our example can now be generalized to $\mathcal{R}^n$. We let $(y_1, \ldots, y_n)$ be Euclidean coordinates for which the usual inner product and distance formulas are valid, and we introduce new coordinates $(x_1, \ldots, x_n)$ by an equation

$$\mathbf{y} = B\mathbf{x}, \tag{1.141}$$

where $B$ is an invertible $n \times n$ matrix. Since inner products determine norms and distances, we only need to specify how inner products are to be found in the coordinates $x_i$. As above, the inner product between vectors $\mathbf{a}$ and $\mathbf{b}$ in the $x$-coordinates must equal that between the corresponding $y$-vectors: $B\mathbf{a}$ and $B\mathbf{b}$. We write the inner product as $(\mathbf{a}, \mathbf{b})_A$. Thus our rule is

$$(\mathbf{a}, \mathbf{b})_A = (B\mathbf{a})^t B\mathbf{b} = \mathbf{a}^t B^t B\mathbf{b}. \tag{1.142}$$

We simplify the rule somewhat by writing $A = B^t B$:

$$(\mathbf{a}, \mathbf{b})_A = \mathbf{a}^t A\mathbf{b}. \tag{1.143}$$

The matrix $A = B^t B$ is then symmetric and it is also positive definite (Problem 1.82(a)). In fact, for every positive definite symmetric matrix $A$ of order $n$, we can write $A = B^t B$ in terms of an invertible matrix $B$; this follows from the Choleski decomposition of linear algebra (see Strang, 1976, p. 241, and Problem 1.86).

The rule (1.143) is all we need to retain of the process. However, since there are always a matrix $B$ and a change of coordinates (1.141) in the background, it is good to keep them in mind to explain what has been done, as in the aerial photograph. Furthermore the process allows us to reason that with the new inner product, all the rules of Euclidean geometry remain valid. In particular, we can state that

$$(\mathbf{a}, \mathbf{b})_A = (\mathbf{b}, \mathbf{a})_A,$$

$$(c_1\mathbf{a}_1 + c_2\mathbf{a}_2, \mathbf{b})_A = c_1(\mathbf{a}_1, \mathbf{b})_A + c_2(\mathbf{a}_2, \mathbf{b})_A, \tag{1.145}$$

$$(\mathbf{a}, \mathbf{a})_A > 0 \quad \text{except that} \quad (\mathbf{0}, \mathbf{0})_A = 0.$$

(See Problem 1.87.) The norm of a vector $\mathbf{a}$ can now be found as the (positive or zero) square root of the inner product of $\mathbf{a}$ with itself. We denote this norm by $\|\mathbf{a}\|_A$ and call it the $A$-norm of $\mathbf{a}$. Thus we have the rule

$$\|\mathbf{a}\|_A^2 = (\mathbf{a}, \ \mathbf{a})_A = \mathbf{a}^t A \mathbf{a}. \tag{1.146}$$

Furthermore we can now find angles by the usual rule:

$$(\mathbf{a}, \mathbf{b})_A = \|\mathbf{a}\|_A \|\mathbf{b}\|_A \cos\theta, \tag{1.147}$$

and two vectors are orthogonal when this angle is $\pi/2$ or when their inner product is 0. The Pythagorean theorem continues to hold in $A$-norms, as do the other rules for norms and inner products:

$$|(\mathbf{a}, \mathbf{b})_A| \le \|\mathbf{a}\|_A \|\mathbf{b}\|_A,$$

$$\|\mathbf{a}\|_A > 0 \quad \text{except} \quad \text{that} \ \|\mathbf{0}\|_A = 0, \tag{1.148}$$

$$\|c\mathbf{a}\|_A = |c| \|\mathbf{a}\|_A, \quad \|\mathbf{a} + \mathbf{b}\|_A \le \|\mathbf{a}\|_A + \|\mathbf{b}\|_A.$$

REMARK 1. A vector space with an inner product obeying the rules (1.145) is called an *inner product space*. We have shown that from each Euclidean vector space, with usual inner product, we can obtain an inner product space with the aid of a linear transformation (1.141), which can be regarded as a change of coordinates in the Euclidean vector space. Conversely, if a finite-dimensional vector space is an inner product space, with inner product $(\mathbf{a}, \mathbf{b})$ obeying the rules (1.145), then it can always be interpreted as $(\mathbf{a}, \mathbf{b})_A$ for an appropriate positive definite symmetric matrix $A$. The proof is left as an exercise (Problem 1.89). We note that a finite-dimensional inner product space is also called a finite-dimensional *Hilbert space*.

REMARK 2. A vector space with a norm $\| \ \|$ obeying the rules on the second and third lines of (1.148):

$$\|\mathbf{a}\| > 0 \quad \text{except} \quad \text{that} \ \|\mathbf{0}\| = 0,$$

$$\|c\mathbf{a}\| = |c| \|\mathbf{a}\|, \quad \|\mathbf{a} + \mathbf{b}\| \le \|\mathbf{a}\| + \|\mathbf{b}\|, \tag{1.149}$$

is called a *normed* vector space. Thus every inner product space is a normed vector space, with a norm defined as in (1.146) in terms of the inner product. However, there are norms in $\mathcal{R}^n$ for which there is no associated inner product. For example, in $\mathcal{R}^2$ one can assign to the vector $(x, y)^t$ the norm $|x| + |y|$, and this norm has no associated inner product; verification of these statements and generalizations are considered in Problem 1.92. Throughout, unless otherwise indicated, the notation $\|..\|$ will signify the Euclidean norm as in (1.61).

## 1.15   MORE ON MINIMUM PROBLEMS FOR QUADRATIC FUNCTIONS

Let

$$f(\mathbf{x}) = \mathbf{x}^t A \mathbf{x} - 2\mathbf{b}^t \mathbf{x} \tag{1.150}$$

in $\mathcal{R}^n$, where $A$ is a positive definite symmetric matrix of order $n$. Thus $f$ is a quadratic function, a quadratic form plus a linear function. Since $A$ is symmetric and nonsingular, we can write $\mathbf{b}$ as $A\mathbf{a}$ for some $\mathbf{a}$ in $\mathcal{R}^n$ and hence write

$$f(\mathbf{x}) = \mathbf{x}^t A \mathbf{x} - 2\mathbf{a}^t A \mathbf{x} = \|\mathbf{x}\|_A^2 - 2(\mathbf{a}, \mathbf{x})_A. \tag{1.151}$$

The function $f$ is strictly convex and, as in Section 1.12, it has a minimum in $\mathcal{R}^n$; the unique minimizer is obtained by solving the equation $A\mathbf{x} = \mathbf{b}$. It also has a minimum on each closed set $E$, since, as in Section 1.12, $f$ has limit $\infty$ as $\|\mathbf{x}\| \to \infty$; if $E$ is convex, the minimizer is unique.

In the convex case we can characterize the minimizer $\mathbf{x}^*$ in an interesting way: *It is the unique member of E such that for all $\mathbf{y}$ in E,*

$$(\mathbf{x}^* - \mathbf{a}, \mathbf{y} - \mathbf{x}^*)_A \geq 0 \qquad \text{for all } \mathbf{y} \text{ in } E. \tag{1.152}$$

Indeed, let $\mathbf{x}^*$ be the minimizer. Then for each $\mathbf{y}$ in $E$ the points $(1 - t)\mathbf{x}^* + t\mathbf{y} = \mathbf{x}^* + t(\mathbf{y} - \mathbf{x}^*)$ are in $E$ for $0 \leq t \leq 1$, and we find (see Problem 1.90) that

$$f(\mathbf{x}^* + t(\mathbf{y} - \mathbf{x}^*)) = f(\mathbf{x}^*) + 2t(\mathbf{x}^* - \mathbf{a}, \mathbf{y} - \mathbf{x}^*)_A + t^2 \|\mathbf{y} - \mathbf{x}^*\|_A^2. \tag{1.153}$$

For $t$ sufficiently small and positive, the right-hand side is greater than or equal to $f(\mathbf{x}^*)$ if and only if the inequality (1.152) holds. Therefore, since $\mathbf{x}^*$ is the minimizer, (1.152) must hold. Conversely, let (1.152) be valid for all $\mathbf{y}$ in $E$. Then we take $t = 1$ in (1.153) to conclude that for all $\mathbf{y}$ in $E, f(\mathbf{y}) \geq f(\mathbf{x}^*)$ and hence $\mathbf{x}^*$ is the minimizer.

REMARK 1. The inequality (1.152) is equivalent to the condition that $\cos \phi \geq 0$, where $\phi$ is shown in Fig. 1.30; the angle is measured in the geometry

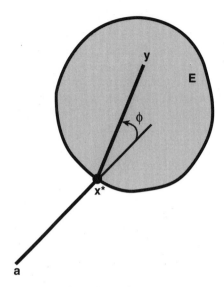

**Figure 1.30.** Significance of inequality (1.152).

of the $A$-norm, as in (1.147). The condition (1.152) for a minimizer is an example of a *variational inequality*. For further discussion of variational inequalities, see Ciarlet (1980).

If the set $E$ is a subspace of $\mathcal{R}^n$ (and is hence convex), then we can characterize the minimizer in another way: $\mathbf{x}^*$ is the minimizer of $f$ if and only if $\mathbf{x}^*$ is in $E$ and

$$(\mathbf{x}^* - \mathbf{a}, \mathbf{y})_A = 0 \qquad \text{for all } \mathbf{y} \text{ in } E. \tag{1.154}$$

In particular, one has

$$(\mathbf{x}^* - \mathbf{a}, \mathbf{x}^*)_A = 0. \tag{1.155}$$

(See Problem 1.90.)

REMARK 2. The results obtained here become more obvious if we observe that

$$f(\mathbf{x}) = \|\mathbf{x} - \mathbf{a}\|_A^2 - \|\mathbf{a}\|_A^2. \tag{1.156}$$

(See Problem 1.91.) Thus minimizing $f$ on $E$ is equivalent to finding the distance from $\mathbf{a}$ to $E$ in the $A$-geometry. Condition (1.154) says that the vector $\mathbf{x}^* - \mathbf{a}$ is perpendicular to each vector in the subspace or that $\mathbf{x}^*$ is the foot of the perpendicular from $\mathbf{a}$ to the subspace. It would in fact suffice to prove these results first for the case when $A$ is the identity matrix—that is, in usual

Cartesian coordinates—and then make a linear transformation (1.141) as in the previous section.

REMARK 3. From (1.156) we see that the level sets of $f$ are the same as the level sets of the function $g(\mathbf{x}) = \|\mathbf{x} - \mathbf{a}\|_A^2$. We now choose new Cartesian coordinates $x_i'$ in $\mathcal{R}^n$ as in (1.68), where the orthogonal matrix $B$ is chosen as in Section 1.8 so that $B^t A B$ is a diagonal matrix $\mathrm{diag}(\lambda_1, \ldots, \lambda_n)$, where the $\lambda_i$ are the eigenvalues of the matrix $A$. In these coordinates the function $g$ has value

$$\lambda_1 x_1'^2 + \cdots + \lambda_n x_n'^2,$$

where all the $\lambda_i$ are positive, since $A$ is positive definite. For $n \geq 2$ the level surfaces of this function are concentric ellipses, ellipsoids, or hyperellipsoids. Accordingly the fact that $f$ has limit $+\infty$ as $\mathbf{x} \to \infty$ is evident (see the discussion at the end of Section 1.12).

## Positive Semidefinite Case

We now allow the matrix $A$ to be positive semidefinite, so that its eigenvalues are required to be nonnegative, but some may be 0. If in fact one or more eigenvalues are 0, then $A$ is not positive definite and the discussion given above does not apply. We again ask whether the function $f$ has a minimizer. We assume here that $n$ is at least 2.

By an orthogonal substitution, as in Section 1.8, we can assume that

$$A = \mathrm{diag}(\lambda_1, \ldots, \lambda_n)$$

so that $f$ can be written as follows:

$$f(\mathbf{x}) = \lambda_1 x_1^2 + \cdots + \lambda_n x_n^2 - 2b_1 x_1 - \cdots - 2b_n x_n, \qquad (1.157)$$

where all $\lambda_j$ are nonnegative. If, say $\lambda_1 = 0$ and $b_1 \neq 0$, then for $x_2 = 0, \ldots, x_n = 0$ $f$ reduces to the linear function $-2b_1 x_1$, and therefore $f$ has no minimizer or maximizer; $f$ is not bounded above or below. This reasoning shows that $f$ can have a minimizer only if $b_j = 0$ whenever $\lambda_j = 0$. We assume that $f$ satisfies this condition. If $\lambda_j > 0$, then by completing the square, we can write the terms in $x_j$ as $\lambda_j(x_j - k_j)^2 + h_j$, where $k_j$ and $h_j$ are constants. If we do this for all such terms and make a translation in $\mathcal{R}^n$, then after renumbering the coordinates we can write $f$ thus:

$$f(\mathbf{x}) = \lambda_1 x_1^2 + \cdots + \lambda_k x_k^2 + w, \qquad (1.158)$$

where $w$ is a constant, $k \leq n$, and $\lambda_1 > 0, \ldots, \lambda_k > 0$. The equation (1.158) includes as special cases that in which $f$ reduces to the constant $w$ and that in which $k = n$, so that $A$ is positive definite. In the first of these cases, $f$ has infinitely many minimizers; in the second, it has a unique minimizer $\mathbf{x} = \mathbf{0}$. Otherwise, $1 < k < n$ and $f$ has infinitely many minimizers $(0, \ldots, 0, x_{k+1}, \ldots, x_n)$, where the last $n - k$ coordinates are arbitrary. In all these cases $f$ has the minimum $w$.

We can describe our conclusion as follows: When the matrix $A$ is positive semidefinite, either the function $f$ defined by (1.150) is unbounded below or else $f$ has a minimizer; the minimizer is unique only when $A$ is positive definite.

We can state further that if the quadratic function $f$ is bounded below, then $A$ is positive semidefinite and $f$ has a minimizer. Indeed we can assume that $f$ is expressed as in (1.157). If $A$ has a negative eigenvalue, say $\lambda_1$, then for $x_2 = 0, \ldots, x_n = 0$, $f$ is clearly not bounded below. Thus no negative eigenvalue can occur and $A$ must be positive semidefinite.

## PROBLEMS

**1.85.** Follow the procedure given at the beginning of Section 1.14 to determine distances and angles in the $uv$-plane by referring to the corresponding quantities in the $xy$-plane, using the equations (1.140) relating the two planes. Let $P:(4, 1)$, $Q:(3, 5)$, $R:(1, 7)$ be given in the $uv$-plane.

(a) Find the distance from $P$ to $Q$.

(b) Find the angle $POQ$.

(c) Find the angle $PQR$.

**1.86.** The Choleski decomposition expresses a positive definite symmetric matrix $A$ as $B^t B$, where $B$ is upper triangular.

(a) Find $B$ for $A = \begin{bmatrix} 1 & 1 \\ 1 & 2 \end{bmatrix}$.

(b) Find $B$ for $A$ the symmetric matrix of order 11 such that $a_{ii} = i$ for $i = 1, \ldots, 11$, $a_{i,i+1} = a_{i+1,i} = 1$ for $i = 1, \ldots, 10$, and $a_{ij} = 0$ otherwise. Use software such as MATLAB.

**1.87.** The rules (1.145) for $(\mathbf{a}, \mathbf{b})_A$ are a consequence of the way this inner product is defined in terms of the linear transformation (1.141). However, the rules can be deduced from the equation (1.143), regarded as a definition, where $A$ is a positive definite symmetric matrix. Deduce each of the following rules (1.145):

(a) First rule.

(b) Second rule.

(c) Third rule.

**1.88.** Use (1.146) as the definition of the $A$-norm in terms of the positive definite symmetric matrix $A$, and prove the following rules, with the aid of (1.145):

(a) Pythagorean theorem: If $(\mathbf{a}, \mathbf{b})_A = 0$, then $\|\mathbf{a} + \mathbf{b}\|_A^2 = \|\mathbf{a}\|_A^2 + \|\mathbf{b}\|_A^2$.

(b) The second rule in (1.148).

(c) The third rule in (1.148).

(d) The fourth rule in (1.148), with the aid of the first rule in (1.148).

REMARK. The first rule in (1.148) is called the *Schwarz inequality*. It is a consequence of the other rules. (See Kaplan, 1991, p. 49.)

*__1.89.__ Let an inner product $(\mathbf{a}, \mathbf{b})$ be defined in $\mathcal{R}^n$ such that the following rules hold: $(\mathbf{a}, \mathbf{b}) = (\mathbf{b}, \mathbf{a})$,

$$(c_1 \mathbf{a}_1 + c_2 \mathbf{a}_2, \mathbf{b}) = c_1 (\mathbf{a}_1, \mathbf{b}) + c_2 (\mathbf{a}_2, \mathbf{b}),$$

and $(\mathbf{a}, \mathbf{a}) > 0$ except for $(\mathbf{0}, \mathbf{0}) = 0$. Prove that there exists a positive definite symmetric matrix $A$ such that $(\mathbf{a}, \mathbf{b}) = (\mathbf{a}, \mathbf{b})_A$. [Hint: Let the vectors $\mathbf{e}_i$ be the standard basis of $\mathcal{R}^n$ (see Section 1.6), and let $a_{ij} = (\mathbf{e}_i, \mathbf{e}_j)$ for $i, j = 1, \ldots, n$. This defines the matrix $A$. Show that, if $\mathbf{a} = a_1 \mathbf{e}_1 + \cdots + a_n \mathbf{e}_n$ and $\mathbf{b} = b_1 \mathbf{e}_1 + \cdots + b_n \mathbf{e}_n$, then $(\mathbf{a}, \mathbf{b}) = \sum a_i b_j a_{ij} = \mathbf{a}^t A \mathbf{b}$.]

*__1.90.__ Let the function $f$ be defined by (1.151) in $\mathcal{R}^n$ under the conditions stated.

(a) Verify that $f$ satisfies (1.153).

(b) Use the characterization (1.152) of the minimizer $\mathbf{x}^*$ of $f$ on a convex set $E$ to show that if $E$ is also a subspace of $\mathcal{R}^n$, then the minimizer is also characterized by (1.154); show further that (1.154) then implies (1.155).

**1.91.** Show that (1.151) implies (1.156).

**1.92.** Define a norm in $\mathcal{R}^2$ by the equation $\|(x, y)^t\| = |x| + |y|$.

(a) Show that this norm obeys the rules (1.149).

(b) Show that this norm is not related to an inner product as in (1.146). [Hint: Equation (1.146) would make the norm squared a quadratic form. Thus one must show that there are no constants $a$, $b$, $c$ such that $\|(x, y)^t\|^2 = ax^2 + 2bxy + cy^2$.]

(c) Show that $\|(x, y)^t\| = \max(|x|, |y|)$ is also a norm in $\mathcal{R}^2$ and that this norm is also not associated with an inner product.

(d) Generalize the results of parts (a), (b), (c) to $\mathcal{R}^n$.

**1.93.** Show that each norm in $\mathcal{R}^n$ is a convex function, not strictly convex.

## 1.16   PHYSICAL APPLICATIONS

Many minimum problems arise as generalizations of the simple problem of mechanics considered in Example 6 at the end of Section 1.1. One seeks the equilibrium states of a physical system having a potential energy function $V$. Most commonly $V$ has one critical point, a minimizer of $V$, and that point describes the equilibrium state.

***Example 1.***   Let two particles of masses $m_1$ and $m_2$ move on a line subject to spring forces, as suggested in Fig. 1.31. Under appropriate assumptions, one is led to differential equations

$$m_1 \frac{d^2 x_1}{dt^2} - k^2(x_2 - 2x_1) = 0,$$

$$m_2 \frac{d^2 x_2}{dt^2} - k^2(a - 2x_2 + x_1) = 0.$$

(1.160)

There is a potential

$$V(x_1, x_2) = k^2(x_1^2 - x_1 x_2 + x_2^2 - ax_2)$$

and its critical point is the minimizer $x_1 = a/3$, $x_2 = 2a/3$, which describes the equilibrium state. (See Problem 1.94.) A similar conclusion is obtained for a system of many particles, connected to each other and to supports by springs. In fact, the Lagrangian theory of mechanics shows that for all such problems there is a potential function $V$ which has a unique minimizer, corresponding to the unique equilibrium state of the system.

If one generalizes Example 1 to the case of $n$ masses moving on a line, connected by springs, one can then pass to the limit: $n \to \infty$. The result is a partial differential equation, the one-dimensional *wave equation*. The limit process is carried out in Kaplan (1991, ch. 10), where it is pointed out that a reversal of the limit process leads to the idea of solving the wave equation approximately by solving the problem for $n$ particles with $n$ sufficiently large. The approximation involved here uses *difference expressions* as approximations

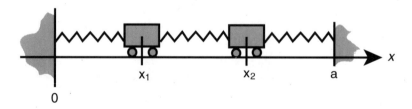

**Figure 1.31.** Example 1.

for derivatives. Similar results are obtained if one uses *finite element* approximations. (See Kaplan, 1981, ch. 14.)

The ideas just described apply to many physical problems. Physical laws lead to a description of the problem in terms of differential equations, most commonly, partial differential equations. In many cases the problem has an alternative description in terms of an *integral* that is minimized by the solution of the partial differential equation(s) satisfying appropriate boundary conditions. By means of finite differences or finite elements, the problem of minimizing the integral is reduced to that of minimizing a function of many variables, which is often required to satisfy side conditions (discussed in Chapter 2). In this way physical problems lead to a great variety of extremum problems for functions of several variables.

We give one further example and refer to Kaplan (1981, ch. 14) for other examples. We consider a membrane represented by a function $v = F(x, y)$ so that $v$ is the vertical displacement of the membrane from the $xy$-plane; we assume that the membrane is subject to an upward vertical pressure $f(x, y)$ and that the membrane is clamped along its boundary $C$ and thus $v = 0$ along the curve $C$ in the $xy$-plane. We find that the equilibrium state of the membrane minimizes the integral

$$J = \iint_R (\|\nabla v\|^2 + 2fv)dA, \tag{1.161}$$

where $R$ is the region in the $xy$-plane bounded by $C$. Various ways of approximating the integral (1.161) lead to minimum problems for a function of several variables. In Kaplan (1981, pp. 818–819) a *Rayleigh-Ritz* procedure is used to obtain a minimization problem for a quadratic function of $n$ variables. Similar results are found if one uses finite elements or finite differences.

We observe that the problems of minimization of integrals occurring here are typical problems of the *calculus of variations* (see Sections 1.1 and 1.5). There are many other physical problems that lead to such calculus of variations problems. In each case numerical solution of the problem is based on an approximation procedure such as those mentioned above and hence leads to a search for a minimizer of a function of several variables.

***Example 2.*** *A problem considered by Newton.* The objective is to find the shape of a solid bounded by a surface of revolution that will encounter minimum resistance when moved in the direction of its axis through a fluid. Under appropriate assumptions the problem leads to that of minimizing the integral

$$\int_a^b \frac{yy'^3}{1 + y'^2} dx \tag{1.162}$$

among functions $y(x)$ with given values at $a$ and $b$. (See Bliss, 1925, pp. 8–9.) If we subdivide the interval $[a, \, b]$ by equally spaced points

$$x_0 = a < x_1 < \cdots < x_n = b,$$

then we can approximate each function $y(x)$ by the piecewise linear function with value $y_i = y(x_i)$ at the $i$th subdivision point. We can then approximate the integral (1.162) by a sum

$$\sum_{i=1}^{n} \frac{y_i y_i'^3}{1 + y_i'^2} h, \tag{1.163}$$

where $y_i' = (y_i - y_{i-1})/h$ and $h = (b - a)/n$. Here $y_0$ and $y_n$ are known, and hence we are led to minimizing a function of $n - 1$ variables, $y_1, \ldots, y_{n-1}$. Other approximation procedures can be used.

In general, the formulas of physics define functions of several variables, and applications of these formulas can lead to maximum and minimum problems. For example, the intensity of heat radiated from a point source varies inversely as the square of the distance from the source. If there are several sources located in a plane, the $xy$-plane, one is led to a function

$$f(x, \, y) = \sum_{i=1}^{n} \frac{c_i}{r_i^2}$$

for the total intensity of heat at $(x, \, y)$, where the $c_i$ are positive constants and $r_i$ is the distance from $(x, \, y)$ to the $i$th source. By maximizing $f$ in a region $R$ not containing the sources, one is finding the warmest point in $R$.

## PROBLEMS

**1.94.** Verify all the details in Example 1 in Section 1.16.

**1.95.** Generalize the results of Example 1 of Section 1.16 to the case of three particles.

**1.96.** For Example 2 in Section 1.16, take $a = 0$, $b = 2$, and use the approximation (1.163) for the integral to be minimized. Consider the case $n = 2$, $y_0 = 1$, $y_2 = 0$, and use a computer to find the minimum.

## 1.17  BEST APPROXIMATION BY POLYNOMIALS

We consider continuous functions $f(x)$, $g(x)$, ..., of a real variable on an interval $[a, b]$. In this class of functions we can introduce an inner product by the definition

$$(f, g) = \int_a^b f(x)g(x)dx. \tag{1.170}$$

There is then a corresponding norm $\|f\| = \sqrt{(f,f)}$. One verifies that the usual properties of inner products and norms continue to be valid (Problem 1.98). Among our functions are the polynomials. We now formulate a minimum problem: For a given function $f(x)$ and a fixed positive integer $n$, find a polynomial $P(x)$ of degree at most $n$ that minimizes the number $z = \|f(x) - P(x)\|^2$, the "total square error." This problem can again be interpreted as one of finding the "least square error," but now the square error refers to all the values over an interval, rather than on a finite set as in Section 1.13.

This minimum problem again reduces to one in a finite-dimensional vector space, namely the space $\mathcal{R}^{n+1}$. We write a general polynomial of degree at most $n$ as

$$P(x) = u_0 + u_1 x + \cdots + u_n x^n. \tag{1.171}$$

Then by the definitions given we find (Problem 1.99) that our total square error is

$$z = \mathbf{u}^t A \mathbf{u} + \mathbf{b}^t \mathbf{u} + c, \tag{1.172}$$

where we use $\mathbf{u}$ and $\mathbf{b}$ to denote vectors of $\mathcal{R}^{n+1}$ and $c$ to denote a constant; $A$ is a symmetric matrix of order $n + 1$ with entries

$$a_{ij} = \int_a^b x^{i+j}dx = \frac{b^{i+j+1} - a^{i+j+1}}{i+j+1}$$

for $i, j = 0, 1, \ldots, n$. Also $\mathbf{b}$ is a constant vector with components

$$b_i = -2\int_a^b f(x)x^i dx = -2(f, x^i), \qquad i = 0, 1, \ldots, n,$$

and

$$c = \int_a^b f^2 dx = \|f\|^2.$$

Also the quadratic form $\mathbf{u}^t A \mathbf{u}$ is positive definite. It is equal to

$$\int_a^b P^2 dx = \|P\|^2 \geq 0$$

(Problem 1.100), and this integral can equal zero only when the polynomial $P(x)$ is identically zero, which in turn implies that all $u_i$ are 0.

From these calculations we see that our total square error $z$ is a quadratic function of $\mathbf{u}$ in $\mathcal{R}^{n+1}$, whose Hessian matrix is $2A$, a positive definite matrix. Therefore $z(\mathbf{u})$ is strictly convex in $\mathcal{R}^{n+1}$, and as at the end of Section 1.12, it has a unique minimizer $\mathbf{u}^*$ that can be found as the unique critical point of the function $z(\mathbf{u})$. (See Problem 1.101.)

REMARK. The polynomials of degree at most $n$ form an $(n+1)$-dimensional vector space, of which the polynomials $x^i$ for $i = 0, 1, \ldots, n$ form a basis and the $u_i$ are coordinates with respect to this basis. The inner product introduced by (1.170) is that associated with the matrix $A$, as in Section 1.14. We verify that for

$$P(x) = u_0 + u_1 x + \cdots + u_n x^n \quad \text{and} \quad Q(x) = v_0 + v_1 x + \cdots + v_n x^n, \quad (1.173)$$

one has

$$(P, Q) = \mathbf{u}^t A \mathbf{v} \qquad (1.174)$$

(Problem 1.102).

Instead of the basis $1, x, x^2, \ldots,$ one can use a basis obtained by orthogonalization from this one. The standard Gram-Schmidt orthogonalization process (see Isaacson and Keller, 1966, p. 199) leads to polynomials $Q_i(x)$, where each $Q_i$ has degree $i$. One can seek the minimizer of the total square error $z$ by expressing the general polynomial as a linear combination of the $Q_i$. Then the function $z$ has an expression like that in (1.172) except that the matrix $A$ is a diagonal matrix of order $n+1$. The critical point of the function $z(\mathbf{u})$ is then easily found.

The orthogonalization can be carried out for the infinite sequence $1, x, \ldots, x^n, \ldots,$ to yield an infinite sequence of polynomials $Q_i(x)$, with $Q_i$ of degree $i$, which are pairwise orthogonal: $(Q_i, Q_j) = 0$ for $i \neq j$. The orthonormal polynomials have been found and tabulated for the case $a = -1, b = 1$; they are constants times the *Legendre polynomials*. For other intervals a simple linear change of variable reduces the problem to that for the interval $[-1, 1]$.

The best approximation (with least total square error) to a function $f(x)$ over the interval $[-\pi, \pi]$ by a *trigonometric polynomial*

$$\frac{a_0}{2} + a_1 \cos x + b_1 \sin x + \cdots + a_n \cos nx + b_n \sin nx \qquad (1.175)$$

is found in exactly the same way, since the terms of this expression are already orthogonal over the interval.

Other systems of functions can be used. Also one can generalize the inner product (1.170) by introducing a weighting function $w(x)$ on the right:

$$(f, g) = \int_a^b f(x)g(x)w(x)dx. \tag{1.176}$$

Here the function $w(x)$ is nonnegative, with positive integral over the interval. The development is similar (see Problem 1.103).

### Case of Chebyshev Polynomials

When the interval is $[-1, 1]$ and the weighting function $w(x)$ is the function $(1 - x^2)^{-1/2}$, the process discussed leads to a system of orthogonal polynomials that are scalar multiples of the Chebyshev polynomials:

$$T_n(x) = \cos(n \arccos x), \quad n = 0, 1, 2, \ldots. \tag{1.177}$$

They have interesting extremum properties, which we describe without proof. (See Isaacson and Keller, 1966, ch. 5, for a full discussion.)

We fix $n \geq 1$ and, as above, consider the $(n + 1)$-dimensional vector space $\mathcal{R}^{n+1}$ of all polynomials of degree at most $n$. In this vector space we let $E$ be the set of all *monic* polynomials: that is, those of degree $n$ with leading coefficient 1: $P(x) = x^n + v_1 x^{n-1} + \cdots + v_n$. Then $E$ *is a convex set* (Problem 1.105(a)). On $E$ we consider the function $f$ which assigns to each $P$ the number

$$k = \max_{|x| \leq 1} |P(x)|. \tag{1.178}$$

We can consider $f$ as a function of $n$ variables, the coefficents $v_1, v_2, \ldots,$ of $P(x)$. Now we find that $f$ is a *convex function* on $E$; however, it is not strictly convex (Problem 1.105(c)). But it is known that $f$ has a unique minimizer, namely the polynomial $P(x) = 2^{1-n} T_n(x)$, and for this polynomial, $f$ has the value 1.

From (1.177) we find that $T_n(x)$ has $n$ roots, all in the interval $[-1, 1]$ (Problem 1.104). We now let $E_1$ be the subset of our convex set $E$ consisting of all polynomials of the form

$$(x - x_0)(x - x_1) \cdots (x - x_n), \tag{1.179}$$

where the $x_i$ are all in the interval $[-1, 1]$. Then $E_1$ is not convex (Problem 1.105(d)). However, the same function $f$ has a unique minimizer on $E_1$, and again the minimizer is the polynomial $P(x) = 2^{1-n} T_n(x)$.

The two results described have important applications in numerical analysis. The first result leads to the method of "economization of power series," which improves the accuracy of approximation of a function by its Maclaurin series; the second result shows that for interpolation of a smooth function by polynomials of degree $n$ over the interval $[-1, 1]$, the roots of $T_n(x)$ are generally the best choice of interpolation points. (See Isaacson and Keller, 1966, p. 267.)

## PROBLEMS

**1.97.** Find the approximation with least square error on the interval $[-1, 1]$:

(a) For $f(x) = e^x$ by a polynomial of degree at most 2.

(b) For $f(x) = x^5$ by a polynomial of degree at most 3.

**1.98.** Prove that the inner product (1.170) satisfies the rules:

(a) $(f, g) = (g, f)$.

(b) $(f, f) > 0$ except that $(f, f) = 0$ when $f(x) \equiv 0$.

(c) $(c_1 f_1 + c_2 f_2, g) = c_1(f_1, g) + c_2(f_2, g)$.

REMARK. The other properties of inner products and norms (see (1.145) and (1.148)) follow from these three properties. (See Kaplan, 1991, sec. 1.13.)

**1.99.** Justify the expression (1.172) for the total square error $z$ and the expressions for $a_{ij}$, $b_i$, and $c$ given.

**1.100.** Take $f$ to be identically 0 to show that in (1.172) $\mathbf{u}^t A\mathbf{u}$ equals $\|P\|^2$.

**1.101.** From (1.172) show that $z$ is minimized when $\mathbf{u} = \mathbf{u}^*$, the unique solution of the equation $2A\mathbf{u} + \mathbf{b} = \mathbf{0}$.

**1.102.** With $P$ and $Q$ as in (1.173), verify (1.174).

**1.103.** Prove the rules of Problem 1.98 for the inner product (1.176), under the hypotheses stated.

**1.104.** From (1.177) show that all the zeros of the polynomial $T_n(x)$ are in the interval $[-1, 1]$.

**1.105.** Let $E$ be the set of all monic polynomials of degree $n$.

(a) Show that $E$ is a convex set in $\mathcal{R}^{n+1}$, the vector space of all polynomials of degree at most $n$.

(b) Show that $E$ is a linear variety in $\mathcal{R}^{n+1}$ (and is hence a convex set, as in Section 1.7).

(c) Show that the function $f$ which assigns to each polynomial $P$ in $E$ the number $k$ in (1.178) is convex but is not strictly convex. [Hint: For the last assertion, consider $f$ on the line formed by the polynomials $x^n + v_n$.]

**(d)** Let $E_1$ be the subset of $E$ formed of the polynomials (1.179), with $|x_i| \leq 1$ for all $i$. Show that for $n \geq 2$ $E_1$ is not convex. [Hint: Show that the polynomials $P(x) = (x + 1)^n$ and $Q(x) = (x - 1)^n$ are in $E_1$ but the midpoint of the line segment joining them is not in $E_1$.]

## REFERENCES

Akhiezer, N. I. (1962). *The Calculus of Variations*, Blaisdell, New York.

Birkhoff, G., and MacLane, S. (1977). *A Survey of Modern Algebra*, 4th ed., Macmillan, New York.

Bliss, G. A. (1925). *Calculus of Variations*, Math. Assn. of America, Chicago.

Chern, S. S. (1967). "Curves and surfaces in Euclidean space," in *Studies in Global Geometry and Analysis*, 16–56, Math. Assn. of America, Englewood Cliffs, New Jersey.

Ciarlet. P. G. (1980). *The Finite Element Method for Elliptic Problems*, North-Holland, Amsterdam.

Coddington, E., and Levinson, N. (1955). *Theory of Ordinary Differential Equations*, McGraw-Hill, New York.

Hilbert, D., and Cohn-Vossen, S. (1952). *Geometry and the Imagination*, Chelsea, New York.

Isaacson, E., and Keller, H. B. (1966). *Analysis of Numerical Methods*, Wiley, New York.

Jensen, T. R., and Toft, B. (1995). *Graph Coloring Problems*, Wiley-Interscience, New York.

Kaplan, W. (1958). *Ordinary Differential Equations*, Addison-Wesley, Reading, MA.

Kaplan, W. (1981). *Advanced Mathematics for Engineers*, Addison-Wesley, Reading, MA.

Kaplan, W. (1991). *Advanced Calculus*, 4th ed., Addison-Wesley, Reading, MA.

Kaplan, W., and Lewis, D. J. (1970). *Calculus and Linear Algebra*, vol. 1, Wiley, New York.

Strang, G. (1976). *Linear Algebra and its Applications*, Academic Press, 1976. New York.

Struik, D. (1950). *Lectures on Classical Differential Geometry*, Addison-Wesley, Reading, MA.

Webster, R. (1994). *Convexity*, Oxford University Press, Oxford.

# 2

# Side Conditions

## 2.1 REVIEW OF VECTOR CALCULUS

In this section we review some standard material on calculus of functions of several variables. For a full treatment, one is referred to Kaplan (1991, ch. 2).

In Chapter 1 we emphasized functions $f(\mathbf{x})$, functions having real values, defined on sets in $\mathcal{R}^n$. We now consider sets of such functions:

$$y_1 = f_1(x_1, \ldots, x_n), \ldots, \quad y_m = f_m(x_1, \ldots, x_n), \tag{2.10}$$

all defined on the same set $E$ in $\mathcal{R}^n$. We can use a vector notation to describe the set of functions:

$$\mathbf{y} = \mathbf{f}(\mathbf{x}). \tag{2.10'}$$

Here $\mathbf{x}$ is in $\mathcal{R}^n$, $\mathbf{y}$ is in $\mathcal{R}^m$.

We say that (2.10) or (2.10') describes a *mapping* from the set $E$ in $\mathcal{R}^n$ to $\mathcal{R}^m$. The set $E$ is the *domain* of the mapping. The *range* of the mapping is the set $F$ of all values $\mathbf{y}$ taken on, as $\mathbf{x}$ varies over the set $E$. The concepts are illustrated by a linear mapping $\mathbf{y} = B\mathbf{x}$, where $B$ is an $m \times n$ matrix; here $E$ is all of $\mathcal{R}^n$, the range $F$ is a subspace of $\mathcal{R}^m$.

We assume that the concepts of continuity and differentiability of functions and the corresponding rules are known. We call the mapping (2.10) continuous if all the functions $f_i$ are continuous and call it differentiable if all these functions are differentiable; we call it continuously differentiable if all the functions have continuous partial derivatives.

For a differentiable mapping one has an $m \times n$ matrix $J$, which is formed of all the first partial derivatives of the functions:

$$A = (a_{ij}), \quad a_{ij} = \frac{\partial f_i}{\partial x_j}. \tag{2.11}$$

We call this the *Jacobian matrix* of the mapping (2.10). It can be abbreviated as $\mathbf{f_x}$ or as $\mathbf{y_x}$.

In the case $m = 1$, the Jacobian matrix is a row vector, which is the transpose of the *gradient* vector $\nabla f$ or grad $f$ (see the end of Section 1.6). The *directional derivative* of $f$ at the point $\mathbf{x}$ in the direction of the unit vector $\mathbf{u}$ is then $A\mathbf{u}$, where the entries in the Jacobian matrix $A$ are evaluated at $\mathbf{x}$.

If one takes differentials in (2.10), one obtains a set of linear equations relating the differentials $dy_1, \ldots, dy_m$ to $dx_1, \ldots, dx_n$. The matrix of this linear mapping is precisely the Jacobian matrix (2.11). We can abbreviate the linear equations as

$$dy = \mathbf{y_x} dx. \tag{2.12}$$

If the entries in the Jacobian matrix are evaluated at a chosen point of $E$, then this linear mapping approximates the given mapping near the chosen point; here the $dx_i$ are coordinates with origin at the chosen point, and the $dy_j$ are coordinates with origin at the corresponding point under the mapping (2.10). The mapping is illustrated for the case $n = 2$, $m = 3$ in Fig. 2.1.

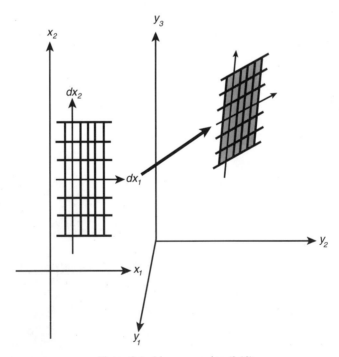

**Figure 2.1.** Linear mapping (2.12).

In the case $m = n$, (2.10) can be considered as a set of $n$ equations in the $n$ unknowns $x_1, \ldots, x_n$, and we could, in principle, solve them for these unknowns. That would provide an *inverse mapping*:

$$x_i = g_i(y_1, \ldots, y_n), \qquad i = 1, \ldots, n. \tag{2.13}$$

In the linear case, we are solving $\mathbf{y} = B\mathbf{x}$ for $\mathbf{x}$, and we know that that is possible precisely when $B$ is an invertible matrix. In the general nonlinear case, matters are not so simple. An *inverse mapping theorem* asserts that when the mapping (2.10) (with $m = n$) is continuously differentiable and the Jacobian matrix at a point $\mathbf{x}_0$ of $E$ is invertible, there is a continuously differentiable inverse mapping (2.13), providing a one-to-one correspondence between the points of a certain neighborhood of $\mathbf{x}_0$ and a neighborhood of $\mathbf{y}_0 = \mathbf{f}(\mathbf{x}_0)$. [Here, as in Section 1.6, a neighborhood of a point means an open set containing the point.] Furthermore, at corresponding points, the Jacobian matrix of the inverse mapping is the inverse of the Jacobian matrix of the initial mapping. The inverse mapping is locally unique; that is, two mappings such as (2.13), satisfying all the conditions described, must agree in some neighborhood of $\mathbf{y}_0$.

The *chain* rules for composite functions extend without change to functions of several variables. One verifies that these can be written in a concise vector form:

$$\mathbf{y_u} = \mathbf{y_x x_u} \tag{2.14}$$

(Problem 2.11). Here we have two differentiable mappings $\mathbf{y} = \mathbf{f}(\mathbf{x})$ and $\mathbf{x} = \mathbf{g}(\mathbf{u})$, and the Jacobian matrix on the left of (2.14) is that of the composite mapping $\mathbf{f}(\mathbf{g}(\mathbf{u}))$. We assume that the composite mapping is well defined in an open region of the $\mathbf{u}$-space. It is then a differentiable mapping with Jacobian matrix given by (2.14).

Closely related to the inverse mapping theorem is the *implicit function theorem*. One generalizes the equations (2.10) to a set of $m$ equations in $m + n$ unknowns $x_1, \ldots, x_n$ and $y_1, \ldots, y_m$:

$$\mathbf{F}(\mathbf{x}, \mathbf{y}) = \mathbf{0}. \tag{2.15}$$

Here the vector function $\mathbf{F}$ stands for $(F_1, \ldots, F_m)^t$, where each $F_i$ is a function of all $m + n$ unknowns. From the vector function $\mathbf{F}$ one can form *two* Jacobian matrices of interest here (there are others): $\mathbf{F_x}$, in which all the $y_j$ are held constant (and only the partial derivatives with respect to the $x_i$ appear), and $\mathbf{F_y}$, in which all the $x_i$ are held constant (and only the partial derivatives with respect to the $y_j$ appear). The vector function $\mathbf{F}$ is assumed to be defined and continuously differentiable in an open region $D$ of $\mathcal{R}^{m+n}$. The implicit function theorem then asserts that under proper conditions the implicit equation (2.15) can be solved for $\mathbf{y}$ as a function of $\mathbf{x}$. In detail, it asserts that such a solution $\mathbf{y} = \mathbf{f}(\mathbf{x})$ exists and is continuously differentiable in a neighborhood of a point

$\mathbf{x}_0$, with $\mathbf{f}(\mathbf{x}_0) = \mathbf{y}_0$, provided that $(\mathbf{x}_0, \mathbf{y}_0)$ is a point of $D$ at which (2.15) is satisfied and the Jacobian matrix $\mathbf{F_y}$ is invertible; as for the inverse mapping theorem, the solution is locally unique. As an example, one can solve the linear equation $A\mathbf{x} + B\mathbf{y} = \mathbf{0}$, where $A$ is $m \times n$ and $B$ is $m \times m$, provided that $B$ is invertible, to obtain $\mathbf{y} = -B^{-1}A\mathbf{x}$. This example is suggestive of the rule for the Jacobian matrix of the solution $\mathbf{y} = \mathbf{f}(\mathbf{x})$ for the general case:

$$\mathbf{F_x} + \mathbf{F_y}\mathbf{y_x} = \mathbf{0}, \quad \text{so that} \quad \mathbf{y_x} = -\mathbf{F_y}^{-1}\mathbf{F_x}.$$

**Example 1.**  For $m = 2, n = 3$, one has two equations:

$$F_1(x_1, x_2, x_3, y_1, y_2) = 0, \quad F_2(x_1, x_2, x_3, y_1, y_2) = 0.$$

If $(x_1^0, x_2^0, x_3^0, y_1^0, y_2^0)$ is a point of $\mathcal{R}^5$ that satisfies both equations and at which the Jacobian matrix

$$J = \begin{bmatrix} \dfrac{\partial F_1}{\partial y_1} & \dfrac{\partial F_1}{\partial y_2} \\[2ex] \dfrac{\partial F_2}{\partial y_1} & \dfrac{\partial F_2}{\partial y_2} \end{bmatrix}$$

is invertible, then there is a pair of functions

$$y_1 = f_1(x_1, x_2, x_3), \quad y_2 = f_2(x_1, x_2, x_3),$$

that satisfies both implicit equations identically in a neighborhood of $(x_1^0, x_2^0, x_3^0)$, at which the two functions $f_i$ have values $y_1^0, y_2^0$, respectively. Furthermore in this neighborhood the partial derivatives of the functions $f_i$ can be found from the rule that the derivative $\partial f_i / \partial x_j$ is the $ij$-entry of the matrix $-J^{-1}K$, where $K$ is the $2 \times 3$ matrix $(\partial F_i / \partial x_j)$.

REMARK. It should be noted that the implicit equations (2.15) are written in terms of two sets of variables, $x_i$ and $y_j$, simply in order to suggest that one can solve the $m$ equations for $m$ unknowns in terms of the remaining unknowns; which $m$ unknowns are chosen is arbitrary, provided only that the corresponding Jacobian matrix $J$ is invertible at the chosen point.

In particular, one can consider a set of implicit equations

$$F_i(x_1, \ldots, x_n) = 0, \quad i = 1, \ldots, m, \tag{2.16}$$

where now $1 \le m < n$. Under assumptions as above, one can solve these equations for $m$ of the unknowns in terms of the remaining ones. If $m = 1$ and $n = 3$, (2.16) is the familiar expression for a surface (e.g., a sphere) in

3-dimensional space and the implicit function theorem provides a representation of the surface in a form such as $z = f(x, y)$ near a point. For $m = 2$, $n = 3$, one has two surfaces intersecting in a curve, and the implicit function theorem provides a representation such as $y = f(x)$, $z = g(x)$ near a point; a more symmetric way of writing these equations is obtained by regarding $x$ as a parameter $t$; more generally, we have the representation of a curve as $x = f(t)$, $y = g(t)$, $z = h(t)$.

For the general case of (2.16), we consider the corresponding set in $\mathcal{R}^n$ as an $(n - m)$-*dimensional variety*. A 1-dimensional variety is a curve; a 2-dimensional variety is a surface, and so on. If the equations (2.16) can be written in matrix form as $A\mathbf{x} = \mathbf{b}$, where $A$ is $m \times n$, then we have a linear variety as in Section 1.6, and the result of solving the equations is the parametric representation (1.65).

In fact we have such a linear variety associated with our variety (2.16): namely the *tangent variety*

$$\mathbf{F_x} d\mathbf{x} = \mathbf{0}, \tag{2.17}$$

obtained by taking differentials in (2.16). Here the Jacobian matrix $J = \mathbf{F_x}$ is evaluated at a point $\mathbf{x}_0$ of the variety (2.16) at which the implicit function theorem is applicable and

$$d\mathbf{x} = (dx_1, \ldots, dx_n)^t = \mathbf{x} - \mathbf{x}_0$$

is the displacement of an arbitrary point $\mathbf{x}$ of the tangent variety from $\mathbf{x}_0$. The tangent variety is the $(n - m)$-dimensional linear variety containing all the lines tangent at $\mathbf{x}_0$ to curves in the surface passing through $\mathbf{x}_0$. Equation (2.17) asserts that the gradient vector of each $F_i$, at $\mathbf{x}_0$, is orthogonal to each such tangent line and hence is orthogonal to the tangent variety.

We are here assuming that at $\mathbf{x}_0$ the Jacobian matrix of the mapping $\mathbf{F}$ with respect to $m$ of the $x_i$ is invertible. This is equivalent to assuming that the Jacobian matrix $\mathbf{F_x}$ has rank $m$. This in turn implies that the row vectors of this Jacobian matrix, which are the transposed gradients of the $F_i$, are linearly independent. As we saw, these gradients are all orthogonal to the tangent linear variety at $\mathbf{x}_0$. By linear algebra, they form a basis for the subspace of $\mathcal{R}^n$ consisting of all vectors which are orthogonal to the tangent variety.

*Example 2.* In 3-dimensional space we consider the curve defined by two intersecting surfaces:

$$2xy - z = 0, \quad x^2 + y^2 - z^2 = 0.$$

Both equations are satisfied at the point $\mathbf{x}_0 : (1/\sqrt{2},\ 1/\sqrt{2},\ 1)$, so this is a point of the curve. The Jacobian matrix of the two given functions is the $2 \times 3$ matrix

$$\begin{bmatrix} 2y & 2x & -1 \\ 2x & 2y & -2z \end{bmatrix}$$

and at $\mathbf{x}_0$ this has rank 2. At $\mathbf{x}_0$ the tangent line is given by the linear equations

$$2y\,dx + 2x\,dy - dz = 0, \quad 2x + 2y - 2z\,dz = 0,$$

where $dx = x - x_0$, and so on, and the coefficients of $dx$, $dy$, $dz$ are evaluated at $\mathbf{x}_0$. (See Problem 2.4.)

### Parametric Equations of a Variety

We recall the standard parametric equations of a surface in space:

$$x = f(u,\ v), \quad y = g(u,\ v), \quad z = h(u,\ v).$$

Here the functions $f$, $g$, $h$ are continuously differentiable in an open region in the $uv$-plane. For fixed $v$, as $u$ varies, the point $(x,\ y,\ z)$ varies on a curve on the surface, with $u$ as "time" parameter, and the vector $(x_u,\ y_u,\ z_u)^t$ is a "velocity vector" tangent to the curve. There is a similar curve for $u$ held constant and a corresponding "velocity vector" $(x_v,\ y_v,\ z_v)^t$. For a useful parametric representation, we require that at each point of the surface, these two vectors are linearly independent. This is equivalent to requiring that the Jacobian matrix

$$\frac{\partial(f,\ g,\ h)}{\partial(u,\ v)}$$

has rank 2. In fact, from this condition, the implicit function theorem shows that the parametric equations provide a one-to-one correspondence between each neighborhood of a point $(u_0,\ v_0)$ in the parameter region and the points of the surface in a neighborhood of the corresponding point $(x_0,\ y_0,\ z_0)$. (See Fig. 2.2.)

The parametric representation of a given surface is not unique. We obtain an *equivalent* parametric representation, in the neighborhood of a point, as above, by a change of variables:

$$u = u(u',\ v'), \quad v = v(u',\ v')$$

providing a differentiable mapping, with differentiable inverse, providing a one-to-one correspondence between neighborhoods in the $uv$-plane and the $u'v'$-plane. In general, two parametric representations of the same portion of a

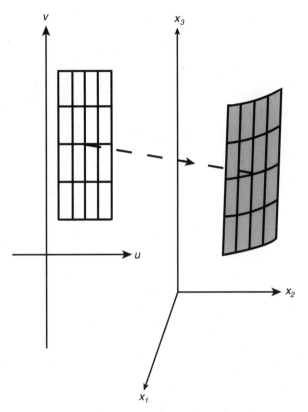

**Figure 2.2.** Parametric equations of a variety.

given surface (satisfying the rank condition given above) must be equivalent in appropriate neighborhoods.

For a surface $z = h(x, y)$, one can obtain such a parametric representation by taking $x = u$, $y = v$ and $z = h(u, v)$. More generally, if a surface is defined by an equation $F(x, y, z) = 0$, to which the implicit function theorem applies, then the equation can be solved for $z$ as $h(x, y)$, in a neighborhood of $(x_0, y_0)$ and again $x$ and $y$ can be taken as parameters in the neighborhood.

The discussion of parametrization for a surface extends in the obvious way to parametrizations of an $N$-dimensional variety, for $N > 2$. Let the variety be given by equations (2.16), with $n - m = N$. We are here interested only in the parametric representation of the variety (2.16) near a point $\mathbf{x}_0$ of the variety. There are $N$ parameters $u_1, \ldots, u_N$ and the parametric equations have the form

$$x_i = h_i(u_1, \ldots, u_N), \quad i = 1, \ldots, n. \tag{2.18}$$

Here $\mathbf{h}(\mathbf{u})$ is to be defined and continuously differentiable in a neighborhood of a point $\mathbf{u}_0$ of $\mathcal{R}^N$, with $\mathbf{h}(\mathbf{u}_0) = \mathbf{x}_0$, and (2.18) establishes a one-to-one correspondence between the points of this neighborhood and the points of the variety in a neighborhood of $x_0$. As for surfaces, we rule out degenerate cases by requiring that the Jacobian matrix $h_u$ have rank $N$. This implies that the *inverse mapping theorem* applies to some set of $N$ equations of the set (2.18). If, for example, it applies to the first $N$ equations, then they can be solved for $u_1, \ldots, u_N$ in terms of $x_1, \ldots, x_N$ (in an appropriate neighborhood of $u_0$), and then the remaining equations give $x_{N+1}, \ldots, x_n$ in terms of $x_1, \ldots, x_N$. This amounts to saying: There is a one-to-one differentiable mapping, with invertible Jacobian, of a neighborhood of $u_0$ onto a neighborhood of the point $(x_{0,1}, \ldots, x_{0,N})$ of $\mathcal{R}^N$, whose coordinates are the same as the first $N$ coordinates of $x_0$, and this converts the parametric representation (2.18) into an *equivalent* one with $x_1, \ldots, x_N$ as parameters. Here, as for surfaces, two parametrizations are equivalent if the parameters are related by a one-to-one diferentiable mapping with differentiable inverse. (See Problem 2.13.)

Whenever the implicit function theorem applies to (2.16) at $x_0$, such a parametric representation with $N$ of the $x_i$ as parameters is always available. The theorem permits solving (2.16) for, say, $x_{N+1}, \ldots, x_n$ as functions $h_{N+1}, \ldots, h_n$ of $x_1, \ldots, x_N$. By writing $x_i = u_i$ for $i = 1, \ldots, N$ one recovers a representation (2.18). For this particular parametrization, one sees at once that the Jacobian matrix $x_u$ has rank $N$.

***Example 3.*** For the sphere $x^2 + y^2 + z^2 = 1$ in $\mathcal{R}^3$, one has the parametrization

$$x = \cos u \cos v, \quad y = \cos u \sin v, \quad z = \sin u$$

for

$$\frac{-\pi}{2} < u < \frac{\pi}{2}, \quad -\pi < v < \pi.$$

The Jacobian matrix $h_u$ is the transpose of the following matrix:

$$\begin{bmatrix} -\sin u \cos v & -\sin u \sin v & \cos u \\ -\cos u \sin v & \cos u \cos v & 0 \end{bmatrix}.$$

One sees easily that it has rank 2 for each $(u, v)$. (See Problem 2.12.)

## PROBLEMS

**2.1. (a)** Find $dz$ for $z = xe^{x-y}$, and use it to evaluate $z$ at $x = 1.2$, $y = 1.1$ with aid of the known value $z = 1$ at $x = 1$, $y = 1$. Compare the result with the exact value of $z$.

**(b)** Proceed as in (a) for $z = x^2/y^3$.

**2.2.** Find the tangent plane to the surface at the point $(x, y, z)$ requested, verifying that the point is on the surface:

(a) $z = x^2 + 3xy - y^2$ at $(2, 1, 9)$.

(b) $z = x^2/(y^2 + 1)$ at $(4, 3, 8/5)$.

(c) $x^2 + y^2 + z^2 = 9$ at $(2, 2, 1)$.

(d) $xyz^2 - x + y - 3z = 0$ at $(1, 2, 1)$.

**2.3.** For each of the following functions $z = f(x, y)$ graph the level curve requested, find $\nabla z$ and verify that it is normal to the level curve at the point $(x, y)$ given:

(a) $z = x^2 + 2y^2$, curve $z = 3$ at $(1, 1)$.

(b) $z = xy$, curve $z = 1$ at $(1, 1)$.

**2.4.** Consider the curve of Example 2 in Section 2.1.

(a) Verify that the point $x_0$ is on the curve.

(b) Verify that the Jacobian matrix of the two functions is as stated and that it has rank 2 at $x_0$.

(c) Follow the procedure indicated to obtain the tangent line to the curve as the intersection of two planes through $x_0$.

**2.5.** Find the Jacobian matrix for the given vector function $y = f(x)$:

(a) $y_1 = 2x_1 - x_2 + x_3$, $y_2 = x_1 + 2x_2 - x_3$, $y_3 = x_1 - x_2 + 3x_3$.

(b) $y_1 = x_1^2 - x_2^2$, $y_2 = 2x_1x_2$.

**2.6.** Let two mappings from $\mathcal{R}^2$ to $\mathcal{R}^2$ be given: $z = x^2 - y^2$, $w = xy$; $x = u\cos v$, $y = u\sin v$. Represent the Jacobian of the composite mapping (taking $(u, v)$ to $(z, w)$) as the product of two Jacobian matrices as in (2.14) (but do not multiply out). Evaluate the Jacobian of the composite mapping for $u = 1$, $v = \pi/4$ by multiplying two numerical matrices. Check by carrying out the substitutions to express $z$ and $w$ in terms of $u$ and $v$ and then differentiating.

**2.7.** As in Problem 2.6, let two mappings be given: $z = f(x, y)$, $w = g(x, y)$; $x = p(u, v) = u^2 - uv$, $y = q(u, v) = uv + v^2$. Form the composite mapping, taking $(u, v)$ to $(z, w)$. Find its Jacobian matrix for $u = 1$, $v = 1$, if it is known that for $x = 0$, $y = 2$ one has $f_x = 1$, $f_y = 3$, $g_x = 5$, $g_y = -1$.

**2.8.** In each of the following cases let $z = f(u, v)$, $w = g(u, v)$ be defined implicitly by the given equations. Obtain expressions for

$$z_u, \ z_v, \ w_u, \ w_v, \ z_{uu}, \ z_{uv}, \ z_{vv}, \ w_{uu}, \ w_{uv}, \ w_{vv}.$$

(a) $uvw - wz + 1 = 0$, $uvz + zw - 1 = 0$.

(b) $u\cos w - v\sin z + 1 = 0$, $u\sin w + v\cos z - 1 = 0$.

**2.9.** Let the equations $x = u^2 - v^2$, $y = 2uv$ define a mapping from the $uv$-plane to the $xy$-plane.

(a) Show that the Jacobian matrix for $u = 2$, $v = 1$ is invertible, and hence, by the inverse mapping theorem, there is an inverse mapping taking a neighborhood of the corresponding point $x = 3$, $y = 4$ to a neighborhood of $u = 2$, $v = 1$.

(b) Find a linear approximation to the inverse mapping referred to in part (a).

**2.10.** Evaluate the directional derivative of the given function $f$ at the point stated and in the direction described:

(a) $f(x, y) = x^2 - xy$ at $(2, 1)$ in the direction of the vector $(3, -5)^t$.

(b) $f(x, y) = x^2 + y^2$ at a general point on the circle $x = \cos s$, $y = \sin s$ along the tangent vector in the direction of increasing $s$.

(c) $f(x, y, z) = e^x \sin(yz)$ at $(2, 2, 1)$ in the direction of the exterior normal to the sphere $x^2 + y^2 + z^2 = 9$.

**2.11.** Prove that the Jacobian matrix of the composition of two mappings is the product of the corresponding Jacobian matrices; that is, prove the rule (2.14) under the hypotheses stated.

**2.12.** Consider the surface of a sphere as in Example 3 of Section 2.1.

(a) Verify the expression for the Jacobian matrix of the parametrization given and that it has rank 2 for each $(u, v)$. Are there points of the sphere not included in the parametrization? Can they be included, without affecting the rank, by changing the region in the parameter plane?

(b) The parametrization given is obtained from spherical coordinates in space. Give a similar one based on cylindrical coordinates and carry out the discussion for it as in the text and as in part (a).

\*2.13. Let a parametrization (2.18) be given, where the Jacobian matrix $\mathbf{h}_u$ has rank $N$. Let $\mathbf{u} = \mathbf{u}(\mathbf{v})$ define an equivalent parametrization $\mathbf{x} = \mathbf{H}(\mathbf{v})$, with

$$\mathbf{H}(\mathbf{v}) = \mathbf{h}(\mathbf{u}(\mathbf{v})).$$

Show that the Jacobian matrix $\mathbf{H_v}$ also has rank $N$.

## 2.2  LOCAL MAXIMA AND MINIMA, SIDE CONDITIONS

Let a real function $f$ be defined and have continuous first and second partial derivatives in an open region $D$ in $\mathcal{R}^n$. Then from Chapter 1 we know that each local extremum of $f$ occurs at a critical point of $f$, at which all first partial derivatives of $f$ are 0, and thus $\nabla f$, the gradient vector of $f$, is $\mathbf{0}$ at the point (see

Section 1.6). Let $x_0$ be such a critical point, so $\nabla f = 0$ at $x_0$. Then as in Section 1.7 we can test for a maximum or minimum by examining the *Hessian matrix* of $f$, whose $ij$-entry is the second partial derivative $\partial^2 f / \partial x_i \partial x_j$. If at $x_0$ this matrix is positive definite, then $f$ has a strong local minimum at $x_0$. Similarly, if at $x_0$ the Hessian matrix is negative definite, then $f$ has a strong local maximum at the point. (If the Hessian matrix is nonsingular and *indefinite*, then there is neither a local minimum nor a local maximum at $x_0$, but a *saddle point*; see Section 1.9.)

Now let $f$ be given in $D$, as described, with continuous first partial derivatives in $D$, and let it be required to find the extrema of $f$ only on the set of points in $D$ which satisfy $m$ equations (with $m < n$):

$$g_j(x_1, \ldots, x_n) = 0, \qquad j = 1, \ldots, m. \tag{2.20}$$

These equations are termed *side conditions* and the problem is described as finding the extrema of $f$ subject to side conditions.

This problem is a special case of the problems considered in Chapter 1 in which one sought the extrema of $f$ on a subset $E$ of $\mathcal{R}^n$, for the points of $D$ satisfying (2.20) determine such a set $E$. In Chapter 1 little attention was paid to equations describing the set $E$. Here they are the main concern.

In Section 1.4 it is pointed out that in some cases the set $E$ can be represented by parametric equations and hence the problem can be reduced to one in fewer unknowns, with no side conditions (see Examples 2 and 3 in Section 1.4). In principle, this idea can be applied to (2.20). As in Section 2.1, under proper conditions, the implicit function theorem allows us to solve (2.20) for $m$ of the $x_i$ in terms of the remaining $x_i$, which can then be treated as parameters. The original function $f$ can then be expressed in terms of these parameters and we have a problem without side conditions, as in Section 1.4.

However, in many cases the program described is very difficult to carry out. For example, side conditions such as

$$x^3 + 3xy^2 + y^3 + xz^2 - z^3 - xz + 1 = 0,$$

$$y^3 + 3xyz + 2x^2z + 3xy + x - y + 2 = 0,$$

are very difficult to solve for two of the unknowns in terms of a third one. Despite the practical difficulties, we will find the parametrization to be valuable as a conceptual tool.

For the following discussion we will assume that the functions $g_j$ are all defined and have continuous first partial derivatives in the open region $D$ where $f$ is given, as above, and that at each point of $E$ (where all equations (2.20) are satisfied) the Jacobian matrix $\mathbf{g_x}$ has rank $m$. The rows of this matrix are the transposed gradients of the $g_j$; hence we are assuming that the $m$ gradient vectors $\nabla g_j$ are linearly independent at each point of $E$. The implicit function theorem applies at each point of $E$ and hence, as in Section 2.1, in a neighborhood of each point, $E$ is representable by parametric equations:

$$x_i = h_i(u_1, \ldots, u_{n-m}), \qquad i = 1, \ldots, n, \tag{2.21}$$

where the Jacobian matrix $\mathbf{h_u}$ has rank $N = n - m$; hence $E$ is an $(n-m)$-dimensional variety. The representation (2.21) could be obtained by solving (2.20) for $m$ of the $x_i$ in terms of the remaining ones, but could also be obtained in other ways (just as a circle can be represented by parametric equations such as $x = \cos u$, $y = \sin u$).

We now say that $f$ has a *critical point* at the point $\mathbf{x}_0$ of $E$ if the composite function $f(\mathbf{h}(\mathbf{u}))$ has a critical point at the point $\mathbf{u}_0$ for which $\mathbf{h}(\mathbf{u}_0) = \mathbf{x}_0$. Here we are using a parametric representation (2.21) in a neighborhood of $\mathbf{x}_0$. Thus at a critical point the gradient of the composite function $f(\mathbf{h}(\mathbf{u}))$ is $\mathbf{0}$. By the chain rule, this condition is equivalent to the condition that

$$f_{\mathbf{x}}\mathbf{h_u} = \mathbf{0} \qquad \text{for } \mathbf{u} = \mathbf{u}_0, \mathbf{x} = \mathbf{x}_0. \tag{2.22}$$

*The condition for a critical point is independent of the parametrization chosen.* Indeed, replacement of one parametrization by an equivalent one is achieved by replacing $\mathbf{h}(\mathbf{u})$ by $\mathbf{h}(\mathbf{u}(\mathbf{v})) = \mathbf{H}(\mathbf{v})$, where $\mathbf{u} = \mathbf{u}(\mathbf{v})$ is a one-to-one differentiable mapping with differentiable inverse; let $\mathbf{v}_0$ correspond to $\mathbf{u}_0$ under this mapping. Then by the chain rule the gradient of $f(\mathbf{H}(\mathbf{v}))$ is the transpose of

$$f_{\mathbf{x}}\mathbf{h_u}\mathbf{u_v}.$$

If we here set $\mathbf{v} = \mathbf{v}_0$, then the expression reduces to $\mathbf{0}$ by (2.22). Thus, if the condition for a critical point holds for one parameter, then it holds for an equivalent parameter.

One also verifies that a critical point of $f$ occurs at $\mathbf{x}_0$ if and only if, along each smooth curve $\mathbf{x} = \mathbf{x}(t)$ in $E$ passing through $\mathbf{x}_0$ for $t = t_0$, the function $f$ has a critical point as a function of a parameter on the curve; the curve is assumed to have a nonzero tangent vector at each point. (See Problem 2.21(b).) Such a curve can always be expressed as $\mathbf{x} = \mathbf{h}(\mathbf{u}(t))$, for $t$ sufficiently close to $t_0$, in terms of a parametrization (2.21). (See Problem 2.21(a).)

**Example 1.** The function $f$ has a critical point at the point $x = 1$, $y = 0$, $z = 0$ of the sphere of Example 3 in Section 2.1, precisely when, in terms of the parametrization of that example,

$$\phi(t) = f(\cos u(t) \cos v(t), \cos u(t) \sin v(t), \sin u(t))$$

has derivative 0 for $t = t_0$ for each choice of the continuously differentiable functions $u(t), v(t)$ with $[u'(t)]^2 + [v'(t)]^2 \neq 0$ and $u(t_0) = 0$, $v(t_0) = 0$.

We can now assert: *If the function f has a local maximum or minimum at a point $\mathbf{x}_0$ of E, then f has a critical point at $\mathbf{x}_0$.* Indeed, $f$ must also have a local extremum at $\mathbf{x}_0$ along each smooth curve in $E$ passing through $\mathbf{x}_0$ and has therefore a critical point as a function of a parameter on such a curve; as above, this means that $f$ has a critical point at $\mathbf{x}_0$. Thus as in Chapter 1, in order to find the local maxima and minima of $f$, we must first find the critical points of $f$ on $E$.

### Method of Lagrange Multipliers

It would appear from our discussion that one can seek the critical points only by relying on parametric equations for $E$. However, a very old procedure allows us to avoid this. This is the method of *Lagrange multipliers*. We describe the procedure and illustrate it and then explain why it is valid.

To find the critical points of $f(\mathbf{x})$ subject to the side conditions (2.20), one forms the $n$ equations:

$$\frac{\partial f}{\partial x_i} + \lambda_1 \frac{\partial g_1}{\partial x_i} + \cdots + \lambda_m \frac{\partial g_m}{\partial x_i} = 0, \qquad (2.23)$$

where $i = 1, \ldots, n$. These $n$ equations along with the $m$ equations (2.20) then serve as $n + m$ equations for the unknowns $x_i$ and $\lambda_j$. The solutions of these equations provide the critical points sought.

***Example 2.*** Find the extrema of the function $w = 2x + 2y + z$ on the sphere $E: x^2 + y^2 + z^2 = 1$. Here $n = 3$ and $m = 1$. The method of Lagrange multipliers leads us to the four equations:

$$2 + 2\lambda_1 x = 0, \quad 2 + 2\lambda_1 y = 0, \quad 1 + 2\lambda_1 z = 0, \quad x^2 + y^2 + z^2 = 1.$$

From these equations we find easily that $\lambda_1 = \pm 3/2$ and obtain the two critical points $(2/3, 2/3, 1/3)$ and $(-2/3, -2/3, -1/3)$ as possible minimizers or maximizers. Since $E$ is a bounded closed set, the function given has a global maximum and a global minimum on $E$. We find that $w = 3$ at the first point and $-3$ at the second and thus the first point is the unique maximizer, the second is the unique minimizer. (See Problem 2.16.) The problem has a simple geometric interpretation: The level surfaces of $w$ are parallel planes and the two points found are the points at which the level surface is tangent to the sphere. At such a point the gradient vector $\nabla w$ is normal to the sphere, and hence it is a constant times the gradient vector of the function $x^2 + y^2 + z^2 - 1$. That is exactly what the equations (2.23) say in this case; the constant is $-\lambda_1$ here.

The example explains the method. In general, at a critical point $\mathbf{x}_0$ of $f$ on $E$, $f$ must have a critical point $\mathbf{x}_0$ along each curve in the variety (2.20) passing through $\mathbf{x}_0$. This implies that the directional derivative of $f$ along the curve at $\mathbf{x}_0$ is 0, so the gradient of $f$ at $\mathbf{x}_0$ is orthogonal to the tangent to the curve at the point. Hence this gradient is orthogonal to the tangent variety of $E$ at $\mathbf{x}_0$ and, as at the end of Section 2.1, the gradient of $f$ must be a linear combination of the vectors $\nabla g_j$, all evaluated at $\mathbf{x}_0$. Conversely, at a point $\mathbf{x}_0$ of $E$ at which the equations (2.23) hold, the gradient of $f$ is orthogonal to the tangent linear variety of $E$ at $\mathbf{x}_0$, so $f$ has directional derivative 0 at $\mathbf{x}_0$ along each curve in $E$ passing through $\mathbf{x}_0$, and $f$ has a critical point at $\mathbf{x}_0$.

**Example 3.** *Point of a linear variety $E$ closest to a given point.* This is the problem treated in Section 1.13. The point $\mathbf{x}_0$ and the linear variety $E$: $A\mathbf{x} = \mathbf{b}$ are given, where the matrix $A$ has maximum rank. The function $f$ to be minimized is the squared distance function $\|\mathbf{x} - \mathbf{x}_0\|^2$, subject to the side condition $A\mathbf{x} = \mathbf{b}$. Application of the method of Lagrange multipliers leads to the two equations:

$$2(\mathbf{x} - \mathbf{x}_0) + A^t\lambda = \mathbf{0}, \quad A\mathbf{x} = \mathbf{b}.$$

Multiplication of the first equation by $A$ on the left and application of the second equation lead to an equation for $\lambda$. Substitution of this value in the first equation finally provides an expression for the minimizer $\mathbf{x}^*$:

$$\mathbf{x}^* = \mathbf{x}_0 + A^t(AA^t)^{-1}(\mathbf{b} - A\mathbf{x}_0).$$

As in Section 1.13 this is of course the foot of the perpendicular to $E$ through $\mathbf{x}_0$. When $E$ has dimension $n - 1$ (and is then termed a *hyperplane*—see Chapter 4), its equation can be written as $\mathbf{y}^t\mathbf{x} = w$, where $\mathbf{y}$ is a normal vector for $E$, and the nearest point is given by

$$\mathbf{x}^* = \mathbf{x}_0 + (\mathbf{y}^t\mathbf{y})^{-1}(w - \mathbf{y}^t\mathbf{x}_0)\mathbf{y}.$$

Verification of the details is left to Problem 2.17.

**Example 4.** *Extrema of a quadratic form on an ellipse.* We seek the extrema of $f(x, y) = \alpha x^2 + 2\beta xy + \gamma y^2$ (not identically 0) subject to the side condition $Q(x, y) \equiv ax^2 + 2bxy + cy^2 = 1$, where $Q$ is positive definite, so the side condition is the equation of an ellipse, as in Section 1.4. The method of Lagrange multipliers leads to the equations

$$2\alpha x + 2\beta y + \lambda(2ax + 2by) = 0, \quad 2\beta x + 2\gamma y + \lambda(2bx + 2cy) = 0, \quad (2.24)$$

along with the side condition. The two displayed equations are homogeneous linear equations in $x$ and $y$. They have only the trivial solution if the matrix of coefficients is invertible; but the trivial solution must be rejected, since it does not satisfy the side condition. Thus there are solutions only if $\det P = 0$, where

$$P = \begin{bmatrix} 2\alpha + 2\lambda a & 2\beta + 2\lambda b \\ 2\beta + 2\lambda b & 2\gamma + 2\lambda c \end{bmatrix}.$$

This leads to a quadratic equation for $\lambda$ with real roots $\lambda_1$, $\lambda_2$. For each root the lines (2.24) coincide and determine two diametrically opposite points on the ellipse. When the roots are distinct, one obtains two local (and global) maxima and two local (and global) minima. When the roots coincide, the function $f$ is constant on the ellipse. The details are left to Problem 2.18.

REMARK. This example has a generalization to $\mathcal{R}^n$, with $f$ a quadratic form in $n$ variables and $g$ a positive definite quadratic form in these variables. The next example is a special case. The general case is considered in Problem 2.20.

***Example 5.*** *Extrema of quadratic forms on the unit sphere.* We provide another aspect of the results of Section 1.8. We let the function $f$ be the quadratic form $x^t H x$, where $H$ is a constant symmetric matrix of order $n$. We seek the extrema of $f$ on the unit sphere $E$: $\|x\|^2 = 1$. We verify that the gradient of $f$ is the linear expression $2Hx$. Since there is only one side condition, we write $-\lambda$ for $\lambda_1$ (with a factor $-1$ chosen for reasons which will be clear). Thus the Lagrange multiplier method leads to the equations:

$$Hx = \lambda x, \quad \|x\|^2 = 1.$$

Accordingly we are seeking the unit eigenvectors of the symmetric matrix $H$. As shown in Section 1.8, one can choose a set of $n$ eigenvectors $x_i$ of $H$, with corresponding eigenvalues $\lambda_i$, forming an orthonormal system. Now we observe that for each eigenvector $x_i$ the function $f$ has the value

$$x_i^t H x_i = \lambda_i x_i^t x_i = \lambda_i.$$

Thus, if $\lambda_1$ is the largest eigenvalue, then $x_1$ must provide a local and global maximum for $f$ on the sphere. Similarly, if $\lambda_n$ is the smallest eigenvalue, then $x_n$ must provide a local and global minimum for $f$ on the sphere. In each of these cases, the negative of the vector is also a maximizer or minimizer. Furthermore it may happen that an eigenvalue is repeated. If the largest eigenvalue has multiplicity $k$, then all unit vectors that are linear combinations of the corresponding eigenvectors also are maximizers of $f$, and there is a similar statement for minimizers. The eigenvalues between the maximum and minimum eigenvalue provide critical points of $f$ but do not lead to maximizers or minimizers; at each of these points $f$ has a saddle point.

We see what is happening by considering a special case:

$$f(x,\ y,\ z) = 3x^2 + 2y^2 + z^2.$$

On the sphere $x^2 + y^2 + z^2 = 1$, the critical points of $f$ are found to be the 6 points $(\pm1,\ 0,\ 0)$, $(0,\ \pm1, 0)$, $(0,\ 0,\ \pm1)$, which are the unit eigenvectors of the matrix $H = \mathrm{diag}(3,\ 2,\ 1)$. At the first pair of points $f = 3$, at the second pair $f = 2$, and at the third pair $f = 1$. Clearly $f = 3$ and $f = 1$ only at the points given, so they are the maximizers and minimizers of $f$. However, $f$ has neither local maximum nor local minimum at the points $(0,\ \pm1,\ 0)$. These points lie on the circle $x^2 + y^2 = 1$, $z = 0$ and on the circle $y^2 + z^2 = 1$, $x = 0$. On the first circle all values of $f$ are greater than 2 except at the two points; on the second circle, all values of $f$ are less than 2 except at the two points. Figure 2.3 shows the level curves of $f$ on the portion of the sphere $x^2 + y^2 + z^2 = 1$ for which $y \geq 0$; these show the typical saddle-point configuration at $(0, 1, 0)$.

The special example indicates what is happening in the general case. As in Section 1.8 one can change axes in $\mathcal{R}^n$ in such a way that $f$ has the form

$$\lambda_1 x_1^2 + \cdots + \lambda_n x_n^2.$$

Then the reasoning of the example extends to show that all eigenvalues between the minimum and maximum ones lead to critical points which do not give local extrema of $f$. (See Problem 2.19.)

**Finding the Critical Points**

In many simple cases the Lagrange equations (2.23) can be solved without difficulty, as in Examples 2 and 3 above. However, in general, they are nonlinear in the $x_i$, and solving them can be a real challenge, calling on sophisticated techniques of numerical analysis. In Section 2.4 we indicate how in certain cases ordinary differential equations can be applied, as in Section 1.11.

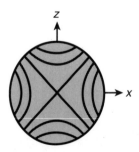

**Figure 2.3.** Level curves of $3x^2 + 2y^2 + z^2$ on the sphere $x^2 + y^2 + z^2 = 1$.

## PROBLEMS

**2.14.** Find the critical points of each of the following functions $z = f(x, y)$ with given side condition and determine whether each provides a local minimum or maximum. Also find the global maximum and minimum if either exists.

(a) $z = 2x + y$, $x^2 + y^2 = 1$.

(b) $z = 5x^2 - xy + y^2$, $x^2 + y^2 = 1$.

(c) $z = x^2 + y^2$, $xy^2 = 1$.

(d) $z = x^3 + y^3$, $xy^2 = 1$.

**2.15.** Find the critical points of the given function $f$ with given side conditions, and determine whether each provides a local minimum or maximum:

(a) $f(x, y, z) = x^4 + y^4 + z^4$, $x + y + z = 1$, $x^2 + y^2 + z^2 = 1$.

(b) $f(x, y, z, w) = xyzw$, $x^2 + y^2 = 1$, $z^2 + w^2 = 1$.

**2.16.** Carry out all the details of Example 2 in Section 2.2.

**2.17.** Carry out all the details of Example 3 in Section 2.2.

**2.18.** Carry out all the details of Example 4 in Section 2.2. [Hint: By continuity, the function $f$ has a global maximum and a global minimum on the ellipse. Hence the method of Lagrange multipliers gives the results.]

**2.19.** Carry out all the details of Example 5 in Section 2.2.

**2.20.** *Generalization of Example 4 to* $\mathcal{R}^n$. Let $A$ be a positive definite symmetric matrix of order $n$, and let $B$ be a symmetric matrix of order $n$. Seek the extrema of $f(\mathbf{x}) = \mathbf{x}^t B \mathbf{x}$ with the side condition $\mathbf{x}^t A \mathbf{x} = 1$. Show that the method of Lagrange multipliers leads to an eigenvalue problem that is equivalent to that of finding eigenvalues and eigenvectors of the matrix $A^{-1}B$. (In Kaplan, 1991, pp. 165–166, it is shown that all eigenvalues are real.)

**2.21. (a)** Under the assumptions of Section 2.2, show that each curve $\mathbf{x} = \mathbf{x}(t)$ on $E$ passing through $\mathbf{x}_0$ for $t = t_0$, with continuous nonzero derivative $\mathbf{x}'(t)$, can be represented, for $t$ sufficiently close to $t_0$, as $\mathbf{h}(\mathbf{u}(t))$, where $\mathbf{x} = \mathbf{h}(\mathbf{u})$ is a parametrization (2.21) and $\mathbf{u}'(t) \neq \mathbf{0}$. [Hint: Show this first for a parametrization using, say $x_1, \dots, x_N$, as parameters. It then follows for each equivalent parametrization.]

**(b)** Show, from the hypotheses stated in Section 2.2, that $f$ has a critical point at the point $\mathbf{x}_0$ of $E$ if and only if $f(\mathbf{x}(t))$ has a critical point at $t_0$ for each smooth curve as in part (a). [Hint: Represent the curve, for $t$ near $t_0$, as in part **(a)**.]

## 2.3   SECOND-DERIVATIVE TEST

For critical points of a function $f$ given in an open region of $\mathcal{R}^n$, we have seen in Chapter 1 that the Hessian matrix of $f$, formed of the second partial derivatives of $f$, allows one to test for local minima and maxima. We now extend this idea to a function $f$ with side conditions (2.20), as in the preceding section. We make the same hypotheses as in that section, except that we now assume that the functions involved have continuous second partial derivatives. From the results of Section 2.1, it follows that the functions defined implicitly will also have continuous second partial derivatives.

In detail, we let $\mathbf{x}_0$ be a critical point of $f$ and let the set $E$ be represented in a neighborhood of $\mathbf{x}_0$ by parametric equations (2.21), where the functions $h_i$ are defined and have continuous first and second partial derivatives in an open region $E_0$ of $\mathcal{R}^N$, with $N = n - m$. We let $F(\mathbf{u})$ be the composite function $f(\mathbf{h}(\mathbf{u}))$. As in Section 2.2, this function has a critical point at a point $\mathbf{u}_0$ in $E_0$, where $\mathbf{h}(\mathbf{u}_0) = \mathbf{x}_0$.

We now form the Hessian matrix of the function $F(\mathbf{u})$. As in Chapter 1, this matrix gives information about the nature of the critical point. However, the matrix depends on the particular parametrization (2.21) chosen! Nevertheless, we can proceed to get information from it.

To that end, we assume that in one particular parametrization the matrix $H$, evaluated at the critical point $\mathbf{u}_0$, is invertible, so $H(\mathbf{u}_0)$ does not have 0 as an eigenvalue. Since $H(\mathbf{u}_0)$ is symmetric, all its eigenvalues are real and a certain number of these, say $r$, are negative, while the remaining eigenvalues ($N - r$ in number) are positive. The number $r$ is called the *index* of the critical point $\mathbf{u}_0$ of the function $F$. Although the Hessian matrix itself changes when we change the parametrization, *the index of the critical point does not change*. A proof of this important rule is given at the end of this section. Because of this rule we can refer to the number $r$ as *the index of the critical point $\mathbf{x}_0$ from which we started*.

If now this index is 0, then there are no negative eigenvalues of $H$ at the critical point and this Hessian is positive definite. By continuity, this implies that $H$ is positive definite in a neighborhood of the critical point, so $F(\mathbf{u})$ is strictly convex in this neighborhood. As in Chapter 1, we conclude that $F$ has a strong local minimum at $\mathbf{u}_0$, so $f(\mathbf{x})$ has a strong local minimum at $\mathbf{x}_0$.

Similarly, if the index is $N$, then all eigenvalues of $H$ at the critical point are negative, so $H$ is negative definite, and $f$ has a strong local maximum at the critical point.

***Example 1.***   Let $f(x, y, z) = 3x^2 + 2y^2 + z^2$, where $x^2 + y^2 + z^2 = 1$. This example is treated at the end of Section 2.2, where the six critical points are found; they are the points of the sphere on the coordinate axes. We examine the critical point $(0, 0, 1)$ and use $x$, $y$ as parameters in a neighborhood of the point. We can write

$$x = u, \quad y = v, \quad z = \sqrt{1 - u^2 - v^2}$$

as the parametric representation. The critical point considered is at the point $u = 0$, $v = 0$ in the parameter plane. (Here $m = 1$, so $N = n - m = 3 - 1 = 2$.) In terms of $u$ and $v$, $f$ becomes

$$F(u, v) = 3u^2 + 2v^2 + 1 - u^2 - v^2 = 2u^2 + v^2 + 1.$$

Its Hessian matrix is the constant matrix $H = \text{diag}(4, 2)$, with index 0; $H$ is positive definite and the function $F$ has a strong local minimum at the critical point $(0, 0)$, and hence $f$ has a strong local minimum at $(0, 0, 1)$. At the critical point $(0, 1, 0)$ we use $x$ and $z$ as parameters. A similar procedure leads to the Hessian matrix $H = \text{diag}(2, -2)$ (Problem 2.22(b)). At this point $H$ has index 1 and there is a saddle point, as illustrated in Fig. 2.3.

**Example 2.** Let $f(x, y, z) = x^2 + y^2 + z^2$, where $x^3 + y^3 + z^3 + xyz = 4$. We can apply the method of Lagrange multipliers to seek the critical points. However, on symmetry grounds we expect that a point with $x = y = z$ would be a critical point. From the side condition we find that $(1, 1, 1)$ would be a candidate. At the point $f$ has gradient $(2, 2, 2)^t$, and if we write the side condition as $g(x, y, z) = 0$, then at the chosen point $g$ has gradient $(4, 4, 4)^t$. Thus the two gradient vectors at the point are parallel and the point $(1, 1, 1)$ is indeed a critical point.

The side condition is difficult to solve for one variable in terms of the others. Nevertheless, we can use the rules for implicit functions to treat two of the unknowns as parameters. We choose $x = u$ and $y = v$ as parameters and write $F(u, v) = u^2 + v^2 + z^2$, where the function $z(u, v)$ is defined by the side condition. By the chain rule and implicit differentiation (see Problem 2.23),

$$F_u = 2u + 2zz_u, \quad F_v = 2v + 2zz_v, \quad F_{uu} = 2 + 2z_u^2 + 2zz_{uu},$$

$$F_{uv} = 2z_v z_u + 2zz_{uv}, \quad F_{vv} = 2 + 2z_v^2 + 2zz_{vv},$$

$$3u^2 + 3z^2 z_u + vz + uvz_u = 0, \quad 3v^2 + 3z^2 z_v + uz + uvz_v = 0,$$

$$6u + 6zz_u^2 + 3z^2 z_{uu} + 2vz_u + uvz_{uu} = 0,$$

$$6z_u z_v + 3z^2 z_{uv} v + z + vz_v + uz_u + uvz_u v = 0,$$

$$6v + 6zz_v^2 + 3z^2 z_v v + 2uz_v + uvz_v v = 0.$$

If we now insert the values $u = 1$, $v = 1$, $z = 1$ in these equations, then we obtain linear equations for the first partial derivatives occurring. They can be solved for these derivatives, and the values can be substituted in the equations involving second partial derivatives. Again one has linear equations. They are easily solved to give $F_{uu} = F_{vv} = -1$, $F_{uv} = -0.5$. Thus the Hessian matrix of $F$ at the critical point has negative eigenvalues $-0.5$, $-1.5$, and the critical point has index 2. Therefore the function $f$ has a local maximum at the critical point $(1, 1, 1)$.

The procedure of the example clearly extends to the general case of a function $f$ in $\mathcal{R}^n$ with $m$ side conditions. If a critical point has been found, the corresponding Hessian matrix at the point can be computed easily; implicit differentiation leads to a set of *linear* equations for the first derivatives and a subsequent set of *linear* equations for the second derivatives. Thus all entries of $H$ are easily found. Computing the eigenvalues is then a standard computing problem, and except in degenerate cases, the nature of the critical point can be determined. The significant difficulty lies in solving the Lagrangian equations for the critical points.

## Invariance of the Index

We wish to show that the index of a critical point $\mathbf{x}_0$ does not depend on the particular parametrization chosen for a neighborhood of the point. One such parametrization is obtained from the implicit function theorem. Under the assumptions made, as in Section 2.2 we can solve the equations $g_j(x_1, \ldots, x_n) = 0, j = 1, \ldots, m$, for $m$ of the unknowns in terms of the remaining ones, which are treated as parameters, $N = n - m$ in number; we can denote these by $u_1, \ldots, u_N$ and then have the parametric equations (2.21). Here the $h_i$ are continuous and have continuous first and second partial derivatives on an open set $D$ of $\mathcal{R}^N$, containing the point $\mathbf{u}_0$, for which $\mathbf{h}(\mathbf{u}_0) = \mathbf{x}_0$.

All other such parametrizations are equivalent to the one chosen. Thus they can be obtained from this particular one by a one-to-one differentiable mapping:

$$v_k = v_k(u_1, \ldots, u_N), \qquad k = 1, \ldots, N,$$

with differentiable inverse, defined in a neighborhood of $\mathbf{u}_0$. Since we need continuous second partial derivatives, we assume that the functions $v_k(\mathbf{u})$ have such derivatives. The Jacobian matrix $J = \mathbf{u}_\mathbf{v}$ is assumed to be invertible at each point. We can think of this mapping as providing a change of coordinates. The function $F(\mathbf{u})$ becomes $G(\mathbf{v}) = F(\mathbf{u}(\mathbf{v}))$.

We now wish to compare the index of the critical point using $\mathbf{u}$ as parameter with that using $\mathbf{v}$; that is, the index of the critical point of $F(\mathbf{u})$ at $\mathbf{u}_0$ with that of $G(\mathbf{v})$ at $\mathbf{v}_0$. We have by the chain rule

$$\frac{\partial G}{\partial v_k} = \sum_{\alpha=1}^{N} \frac{\partial F}{\partial u_\alpha} \frac{\partial u_\alpha}{\partial v_k},$$

for $k = 1, \ldots, N$. For $\mathbf{u} = \mathbf{u}_0$ the derivatives of $F$ on the right are all 0, since $\mathbf{u}_0$ is a critical point of $F$. Hence the left sides are all zero for $\mathbf{v} = \mathbf{v}_0$. This confirms that $\mathbf{v}_0$ is a critical point of $G$, as is to be expected.

Next we differentiate again to obtain second partial derivatives. In differentiating the products on the right, we obtain terms having coefficients that are the *first* partial derivatives of $F$; as noted, these are all 0 at the critical point. Hence, for the second derivatives at the critical point, we can ignore these terms. The other terms involve the second derivatives of $F$, which are the entries in the Hessian matrix $H^u$ of $F$; similarly the second partial derivatives of the left hand side are entries in the Hessian matrix $H^v$ of $G$. Thus, with all quantities evaluated at the critical point, we find that

$$H^v_{kl} = \sum_{a=1}^{N} \sum_{\beta=1}^{N} H^u_{\alpha\beta} \frac{\partial u_\alpha}{\partial v_k} \frac{\partial u_\beta}{\partial v_l}$$

for $k, l = 1, \ldots, N$. These equations are equivalent to the statement that at the critical point

$$H^v = J^t H^u J.$$

This equation shows that if $H^u$ is invertible, then so is $H^v$; thus, if the index is defined for the $\mathbf{u}$-parametrization, then it is defined for the $\mathbf{v}$-parametrization. Furthermore, by linear algebra, two symmetric matrices $H^v$, $H^u$ related by an equation of this form, with $J$ invertible, have the same number of negative eigenvalues. (See Birkhoff and MacLane, 1977, p. 244; this is Sylvester's law of inertia.) Hence we conclude: *the index does not depend on the parametrization.*

## PROBLEMS

**2.22.** Consider Example 1 in Section 2.3.

    **(a)** Verify all the details leading to the conclusion that $f$ has a local minimum at $(0, 0, 1)$.

    **(b)** Follow the suggested procedure to show that there is a saddle point at $(0, 1, 0)$.

    **(c)** Show that there is a local maximum at $(1, 0, 0)$.

**2.23.** Verify all the details in the treatment of Example 2 in Section 2.3.

**2.24. (a)** Apply the method of Lagrange multipliers to Example 2 in Section 2.3, and show that there are seven critical points in all: namely the points $(1, 1, 1)$, $(\alpha, 0, 0)$, $(0, \alpha, 0)$, $(0, 0, \alpha)$, $(\beta, 2\beta, 2\beta)$, $(2\beta, \beta, 2\beta)$, $(2\beta, 2\beta, \beta)$, where $\alpha = 4^{1/3}$ and $\beta = (4/21)^{1/3}$.

    **(b)** Show that $f$ has a local minimum at $(0, 0, \alpha)$.

    **(c)** Show that $f$ has a saddle point at $(\beta, 2\beta, 2\beta)$.

**2.25.** One seeks the extrema of $f(x, y, z, w) = xye^{zw}$ with the side conditions $x^2 + y^2 = 2, z^2 + w^2 = 2$.

(a) Apply the method of Lagrange multipliers to show that there are 16 critical points $(\pm 1, \pm 1, \pm 1, \pm 1, )$.

(b) Take $x$ and $z$ as parameters and follow the procedure of Example 2 in Section 2.3 to determine whether $f$ has a local minimum or maximum at $(1, 1, 1, 1)$.

(c) Reduce the problem to one which has no side conditions by means of the parametrization $x = \sqrt{2} \cos \theta$, $y = \sqrt{2} \sin \theta$, $z = \sqrt{2} \cos \phi$, $w = \sqrt{2} \sin \phi$. Find all 16 critical points and determine whether $f$ has a local minimum or maximum at the point $(1, 1, 1, 1)$.

**2.26.** (*Invariance of the index*) With the symmetric matrix $H^u$, let $Q^u = \mathbf{u}^t H^u \mathbf{u}$ be the associated quadratic form. Let a *linear* change of parameter be made: $\mathbf{u} = J\mathbf{v}$, so the constant matrix $J$ is the Jacobian matrix $\mathbf{u}_v$. As in Section 2.3, $J$ is assumed to be invertible.

(a) Show that, after the substitution, $Q^u$ becomes $Q^v = \mathbf{v}^t H^v \mathbf{v}$, where $H^v = J^t H^u J$.

(b) The linear change of parameter is equivalent to introduction of new oblique axes in $\mathcal{R}^N$, with no change of origin. Use this idea and geometric reasoning to show for $N = 2$ that the index is invariant. [Hint: As in Section 1.4 the index is related to the type of conic section represented by the curve $Q(\mathbf{u}) = 1$.]

## 2.4 GRADIENT METHOD FOR FINDING CRITICAL POINTS

In Section 1.11 we showed how one could find the minimizer of a strictly convex function $F(\mathbf{x})$ by solving the system of ordinary differential equations:

$$\frac{d\mathbf{x}}{dt} = -\nabla F.$$

The same method can be used to find the maximizer for a strictly concave function, with the minus sign replaced by a plus sign. As remarked at the end of Section 1.11, one can also apply this procedure when $F$ is strictly convex (or concave) only in a neighborhood of a local maximizer or minimizer.

We can apply this reasoning to the case of a function $f$ with side conditions, as in the preceding sections. If $f$ has a critical point at $\mathbf{x}_0$ with index 0, so $\mathbf{x}_0$ is a local minimizer, then in terms of the parameter vector $\mathbf{u}$, $f$ becomes a function $F(\mathbf{u})$ which is strictly convex in a neighborhood of the parameter point $\mathbf{u}_0$ corresponding to $\mathbf{x}_0$. We can therefore form the vector differential equation

$$\frac{d\mathbf{u}}{dt} = -\nabla F \qquad\qquad (2.40)$$

and seek the minimizer by following a solution of this equation with increasing
$t$. If the initial point is close enough to the minimizer, then, since the level sets
of $F$ have a center at $\mathbf{u}_0$, the solution should approach the minimizer as $t$
increases.

This reasoning seems to require knowledge of the parametrization.
However, just as in computing the index, only the implicit equations are needed.
First, we can compute $\nabla F$ by the chain rule, as in Section 2.3. This depends on
first partial derivatives of the implicit functions. As we move along the solution
of (2.40) in the parameter space, the implicit functions become functions of $t$
whose derivatives can be found from the partial derivatives. We must therefore
adjoin to (2.40) a set of differential equations for the implicitly defined functions
(2.21). The combined set of differential equations provides the desired solution
of (2.40) in the parameter space. An example will make the process clear:

***Example 1.***   We again consider Example 2 of Section 2.3 and use the parametrization of that section in a neighborhood of the known local maximizer
$(1, 1, 1)$. Since we are seeking a maximizer, we form the equations (2.40) with a
plus sign:

$$\frac{du}{dt} = F_u = 2u + 2zz_u, \quad \frac{dv}{dt} = F_v = 2v + 2zz_v. \tag{2.41}$$

Here $z(u, v)$ is defined implicitly by the equation

$$u^3 + v^3 + z^3 + uvz = 4.$$

We find the derivatives $z_u$ and $z_v$ implicitly as in Section 2.3 and then have
the differential equation for $z$ as a function of $t$:

$$\frac{dz}{dt} = z_u \frac{du}{dt} + z_v \frac{dv}{dt}.$$

Here $du/dt$ and $dv/dt$ can be replaced by their values given above. Combining
the various expressions, we finally obtain the following three ordinary differential equations for $u, v, z$ as functions of $t$:

$$\frac{du}{dt} = 2u + 2zz_u, \tag{2.42}$$

$$\frac{dv}{dt} = 2v + 2zz_v, \tag{2.43}$$

$$\frac{dz}{dt} = 2uz_u + 2vz_v + 2z(z_u^2 + z_v^2), \tag{2.44}$$

where

$$z_u = -\frac{3u^2 + vz}{3z^2 + uv}, \quad z_v = -\frac{3v^2 + uz}{3z^2 + uv}. \tag{2.45}$$

We now choose an initial point, which must satisfy the implicit equations. The point selected is $(u, v, z) = (1.1, 0.9, 0.987348839)$, where the $z$-value is obtained by numerical solution of a cubic equation. The ordinary differential equations are then solved by the program ODE23 of MATLAB, using $t0 = 0$ and various choices of tolerance and final $t$. For example, with tolerance of $0.0001$ and final $t$ of 15, one finds that at $t = 15$ one has $u = 1.000055\ldots$, $v = 0.999950\ldots$, $z = 1.000000276\ldots$. Thus we have recovered the known local maximizer to good accuracy.

We can also experiment with other initial points. If we use the process for a local minimizer (minus sign in (2.40)) and the initial point $(1, -1, 1.79632190)$, with the $z$-value obtained as before, we obtain a solution which obviously converges to the point with $u = 0$, $v = 0$, $z = 1.584\ldots$. In fact there is a critical point $(0, 0, 4^{1/3})$, which we verify to be a local minimizer (Problem 2.23(b)). Here the MATLAB solution produces $u = 0.0000102\ldots$, $v = 0.0000102\ldots$, $z = 1.58415\ldots$. The $z$-value should be $1.5874\ldots$. In general, one should expect less accuracy in the values for the implicit functions; the values of the parameters are affected by the fact that one is approaching a minimizer, while there is no such stabilizing effect for the implicit functions. In general, there is no assurance that the process leads to a minimizer or maximizer; it cannot be known in advance whether the initial point is close enough to the critical point to be sure that we are inside the center configuration and that the parametrization assumed is valid. However, the process has value as a computational tool for attacking a difficult problem. Verification of the details for Example 1 and further exploration of this example are considered in Problem 2.28.

One difficulty that may arise is that of selecting the parameters. It can happen that one set of parameters is successful for a $t$-interval $[t_0, t_1]$ except for trouble at the end of the interval. One may be able to proceed further by selecting new parameters and finding the solution in these new parameters for an interval $[t_1', t_2]$, where $t_1' = t_1 - \delta$ and $\delta$ is a small positive number; the initial values at $t_1'$ should be obtained from the previous solution.

## PROBLEMS

**2.27.** Consider Example 1 of Section 2.3: $f(x, y, z) = 3x^2 + 2y^2 + z^2$, with side condition $x^2 + y^2 + z^2 = 1$.

(a) Apply the gradient method to find a local minimizer. [Hint: Take $u = x$, $v = y$ as parameters; let $F(u, v) = f(u, v, z(u, v))$. Obtain the differential equations $du/dt = -4u$, $dv/dt = -2v$, $dz/dt = (4u^2 + 2v^2)/z$. Start at $t = 0$, $(u, v, z) = (u_0, v_0, z_0) =$

(0.2, 0.2, 0.95916630),    which    satisfies    the    side    condition: $u_0^2 + v_0^2 + z_0^2 = 1$.]

**(b)** Try other initial points in **(a)**.

**(c)** Show that the differential equations of part (a) can be explicitly solved to yield

$$u = u_0 e^{-4t}, \quad v = v_0 e^{-2t}, \quad z = \pm[1 - u_0^2 e^{-8t} - v_0^2 e^{-4t}]^{1/2}$$

and hence conclude that for each initial point with $z_0 > 0$ the solution approaches (0, 0, 1) as $t \to +\infty$, and for each initial point with $z_0 < 0$ the solution approaches (0, 0, $-1$) as $t \to +\infty$.

REMARK. The treatment of the sign of $z$ requires some care. Why does one avoid $z_0 = 0$?

**2.28.** Consider Example 1 in Section 2.4.

**(a)** Verify all the steps leading up to the differential equations (2.42)–(2.44).

**(b)** Use MATLAB or other software to verify that the solution with initial point (1.1, 0.9, 0.987348839) leads to the local maximizer (1, 1, 1) as $t \to +\infty$.

**(c)** Obtain the ordinary differential equations for seeking a local minimizer and solve with initial point (1, $-1$, 1.79632190) to obtain the local minimizer (0, 0, $4^{1/3}$).

**(d)** Show that the global minimum of $f$ is attained at the point (0, 0, $\alpha$) of Problem 2.24**(b)**.

**(e)** To try to determine whether the only local maximizer found: (1, 1, 1) gives the global maximum of $f$, experiment with the differential equations (2.42)–(2.44) with different initial points to see whether the value $f(1, 1, 1) = 3$ is exceeded. (See part **(f)**.)

**(f)** Show that the set $E$ determined by the side condition is unbounded, and since $f$ is distance squared from the origin, conclude that $f$ has no global maximum on $E$, and (1, 1, 1) is only a local maximizer. [Hint: For each fixed $x$, $y$, the side condition is a cubic equation in $z$ that has at least one real root. Since $x^2 + y^2$ can be made arbitrarily large, there is a point of $E$ whose distance from the origin exceeds any chosen number $R$.]

## 2.5  APPLICATIONS

In many physical applications, side conditions arise naturally. For example, in problems of mechanics, particles or rigid bodies may be *constrained* to move in certain ways; such constraints lead to side conditions. An illustration is provided by the motion of a particle on a surface in space. If the force applied has a potential $V$ and thus the force is $-\nabla V$, then as in Section 1.1 each local minimizer of $V$ on the surface is a possible stable equilibrium state of the system. Thus one is led to a minimum problem for $V(x, y, z)$ subject to a side condition which is the equation of the surface.

In Section 1.16 it is shown how the solutions of certain physical problems minimize appropriate integrals and, through numerical methods, one is led to minimize functions of several variables. Often the solutions are required to satisfy certain *boundary conditions*; these in turn become side conditions for the functions of several variables.

We illustrate by considering a soap film problem. Assume that the surface of a soap film can be represented by a single-valued function $w(x, y)$ over a region $D$ bounded by a closed curve $\Gamma$ in the $xy$-plane. The value of $w(x, y)$ on the boundary $\Gamma$ is represented in terms of arc length $s$ as a continuous function $g(s)$. Pressure is then applied to one side of the soap film so that the surface is inflated to enclose a volume

$$V_0 = \int_D w(x, y)dA.$$

A constant surface tension model dictates that the surface energy

$$E(w) = \int_D \frac{T}{2}\left[\left(\frac{\partial w}{\partial x}\right)^2 + \left(\frac{\partial w}{\partial y}\right)^2\right]dA$$

attains its minimum value to achieve equilibrium with the pressure, where $T$ is the surface tension constant. We restate the problem as a constrained minimization:

$$\text{Minimize} \quad \int_D \frac{T}{2}\left[\left(\frac{\partial w}{\partial x}\right)^2 + \left(\frac{\partial w}{\partial y}\right)^2\right]dA$$

$$\text{subject to} \quad w(x, y) = g(s) \quad \text{on } \Gamma, \qquad\qquad (2.50)$$

$$\int_D w(x, y)dA = V_0.$$

Application of numerical methods, as suggested in Section 1.16, leads to a minimum problem for a quadratic function with linear side conditions.

There are many geometrical applications of maximum and minimum problems. We give some illustrations here, with emphasis on problems related to distance. We recall the discussion of distance from a point to a set and distance between two sets in Section 1.12. For that discussion it was important that the sets be *closed*. In the examples treated here the sets are determined by one or more equations of form: continuous function equals zero, and continuity alone implies that the sets are closed.

***Example 1.*** *Distance from a point to a curve.* We seek the point of the curve $x^4 + xy + y^4 = 1$ closest to the origin in the $xy$-plane. Thus we seek the minimum of the function $f(x, y) = x^2 + y^2$ subject to the given side condition. The Lagrange method leads to the equations

$$2x + \lambda(4x^3 + y) = 0, \quad 2y + \lambda(4y^3 + x) = 0,$$

along with the side condition. From these three equations we obtain the eight critical points: $(1, -1)$, $(-1, 1)$, $(0.7071, 0.7071)$, $(-0.7071, -0.7071)$, $(0.2691, 0.9290)$, $(0.9290, 0.2691)$, $(-0.2691, -0.9290)$, and $(-0.9290, -0.2691)$. By comparing the distances from the origin, we find that the last four points are the minimizers sought (whereas the first two are local and global maximizers). The details are left to Problem 2.29.

***Example 2.*** *Distance from a point to a surface.* We seek the point on the ellipsoid $x^2 + 2y^2 + 3z^2 = 1$ closest to the point $(1, 1, 1)$. Here the function being minimized is

$$f(x, y, z) = (x - 1)^2 + (y - 1)^2 + (z - 1)^2.$$

The Lagrange method leads to the equations:

$$2(x - 1) + 2\lambda x = 0, \quad 2(y - 1) + 4\lambda y = 0, \quad 2(z - 1) + 6\lambda z = 0$$

along with the side condition. The three displayed equations allow one to express $x$, $y$, $z$ as functions of $\lambda$. If one substitutes the expressions obtained into the side condition, one obtains an equation for $\lambda$, which can be written as an algebraic equation of degree 6. If this is solved (say by MATLAB), one obtains only two real roots: $0.703421805$ and $-2.175516386$. The first of these leads to the point $(0.58705366, 0.41548192, 0.32151597)$; the second leads to the point $(-0.85068997, -0.29841546, -0.18094474)$; as a check, one verifies that each point is on the ellipsoid. One sees at once that the first point is closer to $(1, 1, 1)$ than the second, so it is the minimizer sought; the second maximizes the distance. (See Problem 2.30.)

These examples were chosen to have equations which were straightforward to solve for the minimizer or maximizer. For most such problems that is not the case and one may have a difficult numerical search to conduct; in particular, the gradient method of the preceding section may be helpful.

***Example 3.*** *Shortest distance between two linear varieties.* Let two linear varieties in $\mathcal{R}^n$ be defined by equations $A_1 x = b_1$, $A_2 x = b_2$, where the matrices

$A_1$, $A_2$ have maximum rank. Finding the shortest distance between them is equivalent to finding the global minimum of the function $F(\mathbf{x}, \mathbf{y}) = \|\mathbf{x} - \mathbf{y}\|^2$ subject to the side conditions $A_1\mathbf{x} = \mathbf{b}_1$, $A_2\mathbf{y} = \mathbf{b}_2$. If the two linear varieties intersect, the global minimum of $F$ subject to the side conditions is 0. If they do not intersect, a minimizer exists but may fail to be unique. The existence of a minimizer is shown as follows: One can represent the two linear varieties by parametric equations (cf. (1.65)) in terms of two parameter vectors $\mathbf{u}_1$, $\mathbf{u}_2$. The function $F$ can then be expressed as a function of $\mathbf{u}_1$ and $\mathbf{u}_2$. One verifies that this function is a quadratic function. However, its second-degree term may be only positive semidefinite, not positive definite. We do know, from the definition of $F$, that $F > 0$ for all values of the parameter vectors. As in Section 1.15, this implies that $F$ has a minimizer but not necessarily a unique one. The conclusion is illustrated by the case of two straight lines in space: When they do not intersect, they may be skew and have a unique pair of closest points, or they may be parallel and have infinitely many pairs of closest points. When they intersect, they have just one point in common, or they may coincide, and so again there are infinitely many pairs of closest points.

We mention some additional classes of geometrical problems:

*Shortest distance between two curves.* For two curves in the plane, given by equations:

$$g(x, y) = 0, \quad h(x, y) = 0,$$

one is seeking the minimum of a function of *four* variables, say

$$f(x_1, y_1, x_2, y_2) = (x_2 - x_1)^2 + (y_2 - y_1)^2,$$

with two side conditions:

$$g(x_1, y_1) = 0, \quad h(x_2, y_2) = 0.$$

There is a similar formulation of the problem of the shortest distance between two surfaces in space, between two curves in space, between a curve and a surface in space. A curve in space can be described as the intersection of two surfaces in space, and so, with such a representation, the problem for two curves in space involves six coordinates and four side conditions. As in Example 3 there may fail to be a unique minimizer of the distance function; in fact, in the nonlinear case, there may be no minimizer, as is shown by the example of the two curves $e^x - y = 0$, $e^x + y = 0$ in the $xy$-plane. (See Problem 2.31.)

*Maximum area problems.* The standard calculus text has examples of finding the maximum area of certain figures (isosceles triangle, rectangle, etc.) inscribed in a circle. The problems become more difficult if one allows more general types of figures, such as arbitrary triangles or quadrilaterals. The area is a function of the coordinates of the vertices; for each vertex one has a side condition which states that the vertex is on the circle. One can generalize further by inscribing the polygons in other curves such as ellipses and by considering the analogous problems in space. (For the simpler problems one may be able to use parametrizations.) One can also consider perimeters or surface areas of the inscribed figures. (See Problem 2.32.)

### Application to differential geometry of surfaces

We consider a surface in space and curves in the surface, with attention restricted to a neighborhood of one point on the surface, with differentiability as needed. In calculus one develops the *Frenet formulas* for a curve in space. These assume that one uses arc length $s$ as parameter along the curve so that the "velocity vector" $d\mathbf{r}/ds$ is a unit tangent vector $\mathbf{T}$ to the path; here $\mathbf{r}$ is the position vector of a point $P$ on the path with respect to the origin. The acceleration vector is then $\mathbf{a} = d\mathbf{T}/ds$, and its length $\|\mathbf{a}\|$ is the curvature $\kappa$ at $P$. We assume that $\kappa \neq 0$ along the curve. Its reciprocal is $\rho$, the radius of curvature of the path. Then $\rho\mathbf{a} = \mathbf{N}$ is a unit normal vector to the path at $P$, called the *principal normal*. The vector $\mathbf{T} \times \mathbf{N} = \mathbf{B}$ is also a unit vector normal to the path at $P$; it is called the *binormal*. One shows that $d\mathbf{B}/ds = -\tau\mathbf{N}$; the scalar $\tau$ is the *torsion* of the path at $P$. The Frenet formulas are then as follows:

$$\frac{d\mathbf{T}}{ds} = \kappa\mathbf{N}, \quad \frac{d\mathbf{N}}{ds} = -\kappa\mathbf{T} + \tau\mathbf{B}, \quad \frac{d\mathbf{B}}{ds} = -\tau\mathbf{N}. \tag{2.51}$$

(See Kaplan, 1991, pp. 130–131, 221–222, for these formulas and (2.52)–(2.55) below. As in that reference we use the familiar dot product and cross product of vectors in $\mathcal{R}^3$.)

A surface in space is assumed represented by a parametric equation $\mathbf{r} = \mathbf{r}(u, v)$ in terms of parameters $u$ and $v$. In the calculus one shows that for a curve on the surface the element of arc length is given by the equation

$$ds^2 = E\,du^2 + 2F\,du\,dv + G\,dv^2, \tag{2.52}$$

where $E = \|\mathbf{r}_u^2\|$, $G = \|\mathbf{r}_v^2\|$, $F = \mathbf{r}_u \cdot \mathbf{r}_v$. Here $EG - F^2$ is assumed to be positive. Consequently the expression on the right of (2.52) is a *positive definite quadratic form*; it is called the *first fundamental form* of the surface. From this it follows that the vector $\mathbf{w} = \mathbf{r}_u \times \mathbf{r}_v$ has positive length and that $\mathbf{n} = (1/\|\mathbf{w}\|)\mathbf{w}$ is a unit normal vector for the surface.

A curve on the surface can now be given by a vector equation $\mathbf{r} = \mathbf{r}(u(s), v(s))$, where as above $s$ is arc length on the curve. We now find that for such a curve

$$\kappa\mathbf{N} \cdot \mathbf{n} = L\left(\frac{du}{ds}\right)^2 + 2M\frac{du}{ds}\frac{dv}{ds} + N\left(\frac{dv}{ds}\right)^2, \tag{2.53}$$

where $L = \mathbf{r}_{uu} \cdot \mathbf{n}$, $M = \mathbf{r}_{uv} \cdot \mathbf{n}$, $N = \mathbf{r}_{vv} \cdot \mathbf{n}$. The expression $L\,du^2 + 2M\,du\,dv + N\,dv^2$ is called the *second fundamental form* of the surface.

By (2.53) the curvature at $P$ of a curve on the surface through $P$ is the same for all such curves having the same principal normal $\mathbf{N}$ and unit tangent vector $\mathbf{T}$. Hence, to study the variation in the curvature among such curves, one can consider plane sections of the surface at $P$.

For each such section with fixed $\mathbf{T}$, the curve lies in a plane to which $\mathbf{T}$ and $\mathbf{N}$ are parallel (Fig. 2.4). (We exclude the case of the tangent plane at $P$.) Let $\theta$ be the angle between $\mathbf{N}$ and $\mathbf{n}$, with $0 \le \theta \le \pi$ and $\theta \ne \pi/2$, since the tangent plane is excluded. By (2.53), $\kappa \cos\theta$ has a constant value, since the right side of (2.53) does not depend on $\mathbf{N}$. This constant has absolute value denoted by $\kappa_n$, called the *normal curvature* of the surface for tangent direction $\mathbf{T}$. It is the curvature at $P$ of the section of the surface containing the tangent line, with direction $\mathbf{T}$, and the normal line to the surface at $P$; it occurs for $\theta = 0$ and for $\theta = \pi$.

We now study the variation of the normal curvature as $\mathbf{T}$ is varied. Thus we are studying the absolute value of the right side of (2.53) as the vector $(du/ds, dv/ds)^t$ is varied. If we write $\xi = du/ds$, $\eta = dv/ds$, then we seek the critical points of the quadratic function

$$f(\xi, \eta) = L\xi^2 + 2M\xi\eta + N\eta^2, \tag{2.54}$$

where

$$E\xi^2 + 2F\xi\eta + G\eta^2 = 1. \tag{2.55}$$

The last equation follows from (2.52); it describes an ellipse (or circle) in the $\xi\eta$-plane, since, as noted above, the left side is positive definite. Accordingly we

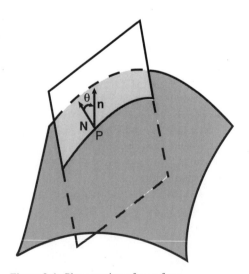

**Figure 2.4.** Plane section of a surface.

are studying a quadratic function on an ellipse, as in Example 4 of Section 2.2. As in that example, we find that $f$ has a global minimum $\lambda_1$ and a global maximum $\lambda_2$, with $\lambda_1 \leq \lambda_2$. If $\lambda_1 < \lambda_2$, then the extrema determine two different directions of the tangent line at $P$; these directions are called *directions of principal curvature*. By solving corresponding ordinary differential equations, one obtains two families of curves on the surface, called the *lines of curvature*. When $\lambda_1 = \lambda_2$, the function $f$ is constant on the ellipse; the surface is said to have an *umbilical point* at $P$. In general, $H = (\lambda_1 + \lambda_2)/2$ is called the *mean curvature* and $K = \lambda_1 \lambda_2$, the *Gaussian curvature* of the surface at $P$.

**Example 4.** Let the surface be the paraboloid $z = x^2 + 2y^2$. We use $x = u$ and $y = v$ as parameters, so $\mathbf{r}(u, v) = (u, v, u^2 + 2v^2)^t$. One finds that at the point corresponding to the parameter point $(u, v)$ the first fundamental form is

$$(1 + 4u^2)du^2 + 16uvdudv + (1 + 16v^2)dv^2,$$

and the second fundamental form is

$$(1 + 4u^2 + 16v^2)^{-1/2}(2du^2 + 4dv^2).$$

Thus for the parameter point $(0, 0)$, one seeks the extrema of $2\xi^2 + 4\eta^2$ for $\xi^2 + \eta^2 = 1$. One finds that $\lambda_1 = 2$, $\lambda_2 = 4$, and at the point the lines of curvature have directions given by the vectors $(1, 0, 0)^t$ and $(0, 1, 0)^t$. The mean curvature at the point is 3 and the Gaussian curvature is 8. (See Problem 2.33.)

## PROBLEMS

**2.29.** Consider Example 1 in Section 2.5.

(a) Verify the steps leading to the eight critical points given.

(b) Give reasoning to justify the assertion that the last four points are all closest to the origin.

(c) Show that the set $E$ defined by the side condition is a bounded closed set and hence the function $f$ has a global maximum on $E$, so the first four points are the corresponding maximizers. [Hint: As pointed out at the beginning of Section 2.5, the fact that $E$ is closed follows from continuity alone. To show that $E$ is bounded, observe that $E$ is a level set of a function having limit $\infty$ as $\|\mathbf{x}\| \to \infty$, as in the Remarks in Section 1.12.]

(d) Use the method of Section 2.3 to verify that each of the first four points is a local maximizer.

**2.30.** Verify the details for Example 2 in Section 2.5.

**2.31.** For each of the following minimum distance problems, formulate the problem as one for minimizing a function with given side conditions, but do not attempt to solve for the minimizer.

(a) Shortest distance between the two parabolas $y = -x^2 + 2x - 2$, $y = x^2 + 2x + 2$ in the $xy$-plane.

(b) Shortest distance between the line $x + 2y = 10$ and the ellipse $2x^2 + y^2 = 1$ in the $xy$-plane.

(c) Shortest distance between two skew lines in space. Assume that the two lines are given by two equations $A_1\mathbf{x} = \mathbf{b}_1$ and $A_2\mathbf{x} = \mathbf{b}_2$, where $A_1$ and $A_2$ are $2 \times 3$ matrices.

(d) Shortest distance between the line $x = t$, $y = 2t - 1$, $z = t + 5$ and the ellipsoid $3x^2 + 2y^2 + z^2 = 1$ in space.

(e) Shortest distance between the circle $x^2 + y^2 + z^2 = 1$, $x + y + z = 1$ and the circle $x^2 + y^2 + z^2 = 1$, $x + y + z = -1$ in space.

**2.32.** For the following maximum problems, formulate each as a problem for seeking the maximum of a function with side conditions, but do not attempt to solve it.

(a) Maximum area of a triangle inscribed in a circle of radius 1.

(b) Maximum area of a trapezoid inscribed in a circle of radius 1.

(c) Maximum area of a quadrilateral inscribed in a circle of radius 1.

(d) Maximum volume of a tetrahedron inscribed in a sphere of radius 1.

**2.33.** Consider the surface of Example 4 in Section 2.5.

(a) Verify the expressions given for the first and second fundamental forms.

(b) For the parameter point $(0, 0)$ verify the values given of $\lambda_1$, $\lambda_2$, the directions of curvature, the mean curvature, and the Gaussian curvature.

(c) Proceed as in part (b) for the parameter point $(1, 1)$.

## 2.6  KARUSH–KUHN–TUCKER CONDITIONS

Up to this point our side conditions have been in the form of equations. We now allow replacement of some equations by *inequalities*, of the form $h(x_1, \ldots, x_n) \leq 0$. Of course inequalities with $\geq$ can be turned into ones with $\leq$ by multiplication by $-1$. The problems with such side conditions occur in many applications, of which the most famous are those using *linear programming* to optimize industrial processes; here the functions occurring are all linear. In general, such problems are treated under the heading of *mathematical programming*, which is the main topic in Chapter 3. The side conditions are then referred to as *constraints*; the set $E$ of points satisfying all constraints is called the *set of feasible solutions*, sometimes denoted by $S_f$. At each point of $E$ a constraint is said to be *active* if it is valid at that point with the $=$ sign. The function to be maximized or minimized is called the *objective function*.

We now formulate an extremum problem for such constraints. We let the set $E$ (assumed nonempty) in $\mathcal{R}^n$ be defined by $N$ equations

$$g_i(\mathbf{x}) = 0, \qquad i = 1,, \ldots, N, \tag{2.60}$$

where $N < n$, and $M$ inequalities

$$h_j(\mathbf{x}) \le 0, \qquad j = 1, \ldots, M. \tag{2.61}$$

We seek the minimum of a function $f$ defined on $E$. We allow for the absence of the equations (2.60), and write $N = 0$ when this occurs. However, we assume that there is at least one inequality constraint.

## Comment on Notation

It is convenient to use vectors to represent a set of inequalities such as (2.61). In general, the inequality $\mathbf{a} \ge \mathbf{b}$ for vectors in $\mathcal{R}^n$ is equivalent to the $n$ inequalities $a_i \ge b_i$, $i = 1, \ldots, n$. Similarly $\mathbf{a} > \mathbf{b}$ is equivalent to the $n$ inequalities $a_i > b_i$.

REMARK. If the functions $h_j$ in (2.61) are all convex in $\mathcal{R}^n$, then each set

$$\{\mathbf{x} \mid h_j(\mathbf{x}) \le 0\}$$

is a sublevel set of a convex function and is therefore a convex set; the set of inequalities (2.61) then also describes a convex set. This statement follows from a general rule (Problem 2.37):

> If $E_1, \ldots, E_M$ are convex sets in $\mathcal{R}^n$, then the set $E$ of all points common to these $M$ sets (assumed to be nonempty) is convex.

The equations (2.60) do not generally describe a convex set; however, they do if each $g_i$ is a *linear* function or, more generally, a function of form $\mathbf{a}_i^t \mathbf{x} - b_i$, when (2.60) is equivalent to the one matrix equation $A\mathbf{x} = \mathbf{b}$ and (if nonempty) represent a linear variety in $\mathcal{R}^n$. By the general rule just stated, we conclude:

> The set $E$ of feasible solutions (assumed nonempty) is convex if $E$ can be described by the conditions
> $$A\mathbf{x} = \mathbf{b}, \quad h_j(\mathbf{x}) \le 0, \qquad j = 1, \ldots, M,$$
> where each $h_j$ is a convex function in $\mathcal{R}^n$.

Let $\mathbf{x}_0$ be a local minimizer of $f$ on $E$. If $N = 0$ and none of the inequality constraints (2.61) are active at $\mathbf{x}_0$, then the point $\mathbf{x}_0$ lies in the subset of $E$ defined by the $M$ inequalities $h_j(\mathbf{x}) < 0$. Typically this set is an open set in $\mathcal{R}^n$. Then, under suitable differentiability conditions, $\mathbf{x}_0$ is a *critical point* of $f$, at which the gradient of $f$ is $\mathbf{0}$. If there were no inequality constraints (2.61), and $N > 0$, so $E$ is defined by side conditions as in Section 2.2, then each minimizer of $f$ would again be a critical point of $f$, but would be determined by Lagrange equations, as in Section 2.2. The Karush–Kuhn–Tucker conditions apply to the general case when both inequalities and equations may occur.

To formulate these conditions, we let the functions $g_i$, $h_j$, and $f$ all be continuously differentiable in a neighborhood of a point $\mathbf{x}_0$ at which $f$ has a local minimum on $E$. At $\mathbf{x}_0$ let the gradient vectors of the $g_i$, together with those of the $h_j$ that are active at $\mathbf{x}_0$, form a linearly independent set of vectors. (We refer to this hypothesis below as *the linear independence assumption*.)

*Then scalars $\mu_i$ and nonnegative scalars $\lambda_j$ exist such that at $\mathbf{x}_0$,*

$$\nabla f + \sum_{i=1}^{N} \mu_i \nabla g_i + \sum_{j=1}^{M} \lambda_j \nabla h_j = \mathbf{0}, \tag{2.62}$$

*where each $\lambda_j$ for which $h_j < 0$ ( the constraint is not active at the point) is $0$.*

These are the Karush–Kuhn–Tucker conditions for the minimum problem stated. If "minimum" is replaced by "maximum," the conditions are modified by replacing each $\lambda_j$ by $-\lambda_j$.

In Section 2.8 we give a proof of the conditions. We remark that from the hypotheses, *an equation of form (2.62) must hold, with $\lambda_j = 0$ for the inactive constraints.* To show this, we let $m$ be the number of constraints active at $\mathbf{x}_0$. If $m < n$, then on the set defined by the active constraints, $f$ has a local minimum at $\mathbf{x}_0$; hence $f$ has a critical point at $\mathbf{x}_0$ and as in Section 2.2, multipliers exist so that (2.62) holds. If $m = n$, the linear independence assumption ensures that the gradient of $f$ is a linear combination of the gradients of the $g_i$ and of the active $h_j$, so that again (2.62) holds. The new assertion, to be proved in Section 2.8, is that, because of the local minimum at $\mathbf{x}_0$, all $\lambda_j$ are nonnegative.

***Example 1.*** For a function $f(x)$ on a closed interval $[a, b]$, the set $E$ is defined by two inequalities: $a - x \leq 0$ and $x - b \leq 0$ and $E$ is convex. The Karush–Kuhn–Tucker conditions assert that if the minimizer of $f$ occurs at a point $x_0$ interior to the interval, and hence neither constraint is active at the point, then $f'(x_0) = 0$. If $x_0 = a$, then the first constraint is active at $x_0$, and (2.62) asserts that $f'(x_0) - \lambda = 0$ for some nonnegative $\lambda$—that is, $f'(x_0) \geq 0$; this condition was pointed out in Section 1.1. There is a similar discussion of the case where $x_0 = b$.

***Example 2.*** We consider a function $f(x, y)$ with the constraints: $x = a$ and $y \leq b$. Thus the set $E$ is a ray parallel to the $y$-axis in the $xy$-plane, and $E$ is convex. With $g(x, y) = x - a$ and $h(x, y) = y - b$, we have $\nabla g = (1, 0)^t$ and $\nabla h = (0, 1)^t$ and these two vectors are linearly independent. If the minimizer occurs at $(a, y_0)$ with $y_0 < b$, then the second constraint is not active at the point and (2.62) asserts that at the minimizer $\nabla f$ is a multiple of $\nabla g$; that is, that $f_y = 0$ at the point. This is to be expected: Along $E$, $f$ is a function of $y$ alone, for $y < b$, and its minimum must occur at a critical point—that is, where $f_y = 0$. If the minimizer is the point $(a, b)$, then the second constraint is active at the point. Thus at the point

$$\nabla f + \mu(1, 0)^t + \lambda(0, 1)^t = \mathbf{0}$$

for some $\mu$ and some nonnegative $\lambda$; that is, at the point $f_x + \mu = 0$ and $f_y + \lambda = 0$. The first equation is no restriction; the second requires that at the point $f_y \leq 0$. This is to be expected from the discussion of Section 1.1 about minimizers at the end of an interval.

***Example 3.*** Now we take a function $f(x, y)$ with one constraint: $h(x, y) \leq 0$; if $h$ is convex in $\mathcal{R}^2$, then $E$ is convex. The set $E$ is a set in the $xy$-plane bounded by the level curve $h = 0$; since the gradient of $h$ points in the direction of *increasing* $h$, it points to the exterior of $E$, as in Fig. 2.5. Under the assumptions made, this gradient vector is not $\mathbf{0}$ at the minimizer if $h = 0$ at the point. Equation (2.62) asserts that if the minimizer is interior to $E$, so the constraint is not active, then $\nabla f$ is $\mathbf{0}$ at the point, as usual; if the minimizer is on the boundary of $E$, and hence the constraint is active, the condition asserts that $\nabla f$ is a multiple of $\nabla h$, where the multiplier is $-\lambda$, with $\lambda \geq 0$. Thus $\nabla f$ points to the *interior* of $E$; it can also be the zero vector. This

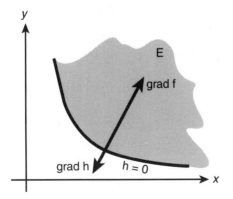

**Figure 2.5.** Set E defined by constraint $h(x, y) \leq 0$.

is to be expected, since $f$ should increase (or at least, not decrease) as one moves from the minimizer into $E$.

***Example 4.*** We seek the minimum of $f(x, y) = x^4 + x^3 y$ on the convex set $E$ defined by the two inequalities: $x^2 - 2x + 2 - y \leq 0$, $x^2 - 2x - 1 + y \leq 0$. The set $E$ is shown in Fig. 2.6. The Karush–Kuhn–Tucker conditions give the equations:

$$4x^3 + 3x^2 y + \lambda_1(2x - 2) + \lambda_2(2x - 2) = 0, \quad x^3 - \lambda_1 + \lambda_2 = 0. \qquad (2.65)$$

If at a minimizer neither constraint is active, then both $\lambda_j$ are 0. The second equation in (2.65) then gives $x = 0$, which is impossible in $E$. Thus this case cannot arise. If only the first constraint is active, then $\lambda_2 = 0$ in (2.65), and the first constraint is valid with $=$ instead of $\leq$. Elimination again leads to $x = 0$, so this case also is ruled out. If only the second constraint is active, then a similar reasoning leads to no candidates. Finally both constraints may be active. In that case one finds that the two points where the bounding parabolas of $E$ meet appear to be candidates; these are the points $(1 - \sqrt{1/2}, 3/2)$ and $(1 + \sqrt{1/2}, 3/2)$. One finds that the first of these leads to positive values for $\lambda_1$ and $\lambda_2$, whereas the second leads to negative values of both. Hence the first is the only possible minimizer, whereas the second can be a maximizer. We know that $f$ is continuous on the bounded closed set $E$ and hence has at least one minimizer and at least one maximizer. Therefore we have in fact found the unique minimizer and the unique maximizer. The details are left to Problem 2.35. Note that to completely justify the conclusions one must show that the linear independence assumption is verified at the minimizer and maximizer.

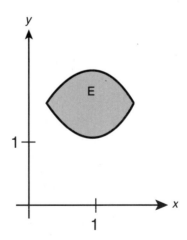

**Figure 2.6.** Convex set $E$ of Example 4.

## PROBLEMS

**2.34.** With the aid of the Karush–Kuhn–Tucker conditions solve the following extremum problems:

    **(a)** Find the maximum and minimum of $f(x) = ax$, where $a$ is a positive constant, for $x^2 - 1 \leq 0$.

    **(b)** Find the maximum and minimum of $f(x, y) = 3x - y$ for $x \geq 0$, $y \geq 0$, $x + y - 1 \leq 0$.

    **(c)** Find the maximum and minimum of $f(x, y) = x + y$ for $1 \leq x^2 + y^2 \leq 2$.

    **(d)** Find the minimum of $f(x, y) = x^2 + 2y^2$ for $1 \leq x + y \leq 2$. Why is there no maximum?

    **(e)** Find the maximum of $f(x, y) = xe^{-y}$ for $y = x^2$, $x \geq 0$.

    **(f)** Find the minimum and maximum of $f(x, y, z) = xyz$ for $x^2 + y^2 = 1$, $y^2 + z^2 = 1$, $x \geq 0$, $y \geq 0$, $z \geq 0$.

    **(g)** Find the maximum of $f(x, y, z) = x + 2y + 3z$ for $x \geq 0$, $y \geq 0$, $z \geq 0$, $x + y + z \leq 1$.

**2.35.** Provide all the details for Example 4 in Section 2.6. In particular, show that the linear independence assumption is satisfied at the minimizer and maximizer.

**\*2.36.** Give a proof of (2.62) for the case where $N = 0$ and $M = 1$ so there is only one constraint active at $x_0$ and (2.62) reduces to $\nabla f + \lambda_1 \nabla h_1 = 0$ at $x_0$; show that $\lambda_1$ cannot be negative.

**2.37. (a)** Let $E_1, \ldots, E_k$ be convex sets in $\mathcal{R}^n$ and let $E$ be the intersection of these sets (the set of all $x$ such that $x \in E_i$ for $i = 1, \ldots, k$). Show that, if $E$ is nonempty, then $E$ is a convex set.

    **(b)** Show that, if each $h_j$ is convex in $\mathcal{R}^n$, then (2.61) describes a set $E$ which, if nonempty, is convex.

    **(c)** Show that, if each $h_j$ is convex in $\mathcal{R}^n$, then (2.61) and the equation $Ax = b$ together describe a set $E$ which, if nonempty, is convex.

## 2.7  SUFFICIENT CONDITIONS FOR THE MATHEMATICAL PROGRAMMING PROBLEM

We continue the discussion of the mathematical programming problem of the preceding section: to find the minimum of a function $f$ defined on the subset $E$ of $\mathcal{R}^n$ characterized by equality and inequality constraints (2.60) and (2.61). The Karush–Kuhn–Tucker conditions are *necessary conditions* that $f$ have a local minimum at a point $x_0$. It is too much to expect that these conditions should also be sufficient conditions. In particular, it could happen that at $\mathbf{x}_0$ all the $\lambda_j$ are 0. Then (2.62) has the same form as Karush–Kuhn–Tucker conditions allowed for a *maximum*!

To rule out that possibility, one could require that at least one $\lambda_j$ be positive. However, the following example shows that this additional requirement does not ensure that $f$ has a local minimum at $\mathbf{x}_0$:

***Example 1.*** We seek the minimum of $f(x, y) = x - y^2$ subject to the constraints $h_1(x, y) = -x \leq 0$, $h_2(x, y) = -y \leq 0$. We verify that at $(0, 0)$ one has the Karush–Kuhn–Tucker condition

$$\nabla f + \lambda_1 \nabla h_1 + \lambda_2 \nabla h_2 = 0,$$

with $\lambda_1 = 1$ and $\lambda_2 = 0$; also the linear independence condition is satisfied. However, $f$ does not have a local minimum at $(0, 0)$, since for $x = 0$ $f$ reduces to $-y^2$, which has a local maximum as a function of $y$ at $y = 0$. (See Problem 2.38.)

The example suggests that we require that *all* $\lambda_j$ be positive at $\mathbf{x}_0$. However, the following example shows that again the requirement is not enough:

***Example 2.*** The function $f$ is the same as in Example 1, but the set $E$ is defined by just one constraint: $h_1(x, y) = -x \leq 0$. At $(0, 0)$ one finds that $\nabla f + \lambda_1 \nabla h_1 = 0$, with $\lambda_1 = 1$. Again the linear independence condition is satisfied. But $f$ again fails to have a local minimum at $(0, 0)$. (See Problem 2.39.)

The source of the difficulty is easily seen. By requiring that all $\lambda_j$ be positive, we make sure that the directional derivative of $f$ at $\mathbf{x}_0$ in a direction $\mathbf{u}$ "entering" the set $E$ is positive. By "entering" we here mean entering in a manner not tangent to the boundary of $E$; in detail, we need $(\nabla g_i)^t \mathbf{u} = 0$ and $(\nabla h_j)^t \mathbf{u} = k_j$ at $\mathbf{x}_0$, where, for all $h_j$ active at $\mathbf{x}_0$, $k_j \leq 0$ and at least one $k_j < 0$. By the Karush–Kuhn–Tucker condition (2.62), we verify that at $\mathbf{x}_0$

$$(\nabla f)^t \mathbf{u} = -\lambda_1 k_1 - \cdots - \lambda_M k_M > 0.$$

Thus making all $\lambda_j$ positive does make $f$ increase as one moves into $E$ from $\mathbf{x}_0$ along such a nontangent direction. The condition in no way limits the behavior of $f$ in a direction tangent to the boundary of $E$. At the end of this section, we point out a special case in which positivity of all the $\lambda_j$ does ensure that one has a local minimum at $\mathbf{x}_0$.

A convexity hypothesis can be introduced to ensure that we have a local minimum. In particular, if the feasible solution set $E$ is convex and $f$ is convex on $E$, then a point $\mathbf{x}_0$ satisfying the Karush–Kuhn–Tucker conditions must be a minimizer of $f$. Indeed, either no inequality constraint is active at $\mathbf{x}_0$ or else at least one inequality constraint is active at $\mathbf{x}_0$. In the first case, $\mathbf{x}_0$ is a critical point of $f$ and hence is a minimizer. In the second case, we let $\mathbf{x}_1$ be a second point of $E$ and let $\mathbf{x} = t\mathbf{a} + \mathbf{b}$ be the parametric equation of a line segment $\gamma$ joining the two points, with $t = 0$ at $\mathbf{x}_0$ and $t = 1$ at $\mathbf{x}_1$. Then along $\gamma$, $f$ becomes a convex function $F(t)$ with derivative $(\nabla f)^t \mathbf{a}$. Since all $g_i$ are constant in $E$, $(\nabla g_i)^t \mathbf{a} = 0$ for all $i$. By the Karush–Kuhn–Tucker conditions at $\mathbf{x}_0$, all $\lambda_j$ are nonnegative; furthermore $(\nabla h_j)^t \mathbf{a} \leq 0$ for all $j$, since $\gamma$ enters $E$ as $t$ increases. We conclude that $F'(0) \geq 0$. As in Section 1.2, we conclude that $F(0) \leq F(1)$, so $f(\mathbf{x}_0) \leq f(\mathbf{x}_1)$. Accordingly $\mathbf{x}_0$ is a minimizer of $f$, as asserted. Therefore, *when $f$ and $E$ are convex, the Karush–Kuhn–Tucker conditions are necessary and sufficient for a local minimizer.* As in general, for convex functions, each local minimizer is also a global minimizer.

From this result we can also obtain sufficient conditions for a local minimizer. Let $\mathbf{x}_0$ satisfy the Karush–Kuhn–Tucker conditions, and let a neighborhood of $\mathbf{x}_0$ exist such that the part $E_0$ of $E$ in this neighborhood is convex and such that $f$ is convex on $E_0$. Then again the Karush–Kuhn–Tucker conditions are necessary and sufficient that $\mathbf{x}_0$ be a local minimizer of $f$ on $E_0$.

As in Section 2.3, we can also develop a second derivative test for a local minimum. To that end, we make a suitable change of coordinates in $\mathcal{R}^n$:

$$u_i = u_i(x_1, \ldots, x_n), \qquad i = 1, \ldots, n. \tag{2.70}$$

The functions occurring are assumed to be defined in a neighborhood of $\mathbf{x}_0$ and to have continuous first and second derivatives. Furthermore we assume that the Jacobian matrix $J = (\partial u_i / \partial x_j)$ is invertible in this neighborhood so that, for suitable choice of the neighborhood, the inverse function theorem (Section 2.1) applies and there is a unique inverse mapping $\mathbf{x} = \mathbf{x}(\mathbf{u})$ whose Jacobian matrix at each $\mathbf{u}$ is the inverse of $J$ evaluated at the corresponding $\mathbf{x}$. Through the mapping each function $\phi(\mathbf{x})$ defined and differentiable in the neighborhood becomes a function $\Phi(\mathbf{u}) = \phi(\mathbf{x}(\mathbf{u}))$ defined in a neighborhood of $\mathbf{u}_0 = \mathbf{u}(\mathbf{x}_0)$. By the chain rule (Section 2.1) we verify (Problem 2.42) that

$$\nabla \Phi = (J^{-1})^t \nabla \phi. \tag{2.71}$$

Here the gradient on the left is formed of the partial derivatives with respect to the $u_i$; that on the right is formed of the partial derivatives with respect to the $x_i$. From (2.71) we conclude that if the Karush–Kuhn–Tucker conditions hold at $\mathbf{x}_0$, then they hold at $\mathbf{u}_0$ for the gradients of the functions obtained from the functions $f(\mathbf{x})$, $g_i(\mathbf{x})$, $h_j(\mathbf{x})$ by expressing $\mathbf{x}$ in terms of $\mathbf{u}$, with the same scalars $\mu_i$ and $\lambda_j$. We can say that *the Karush–Kuhn–Tucker conditions are invariant under coordinate transformations.*

We now introduce a particular coordinate transformation. We assume that the functions $f$, $g_i$, and $h_j$ have continuous first and second partial derivatives in a neighborhood of $\mathbf{x}_0$. As in Section 2.6 we let $m$ constraints be active at $\mathbf{x}_0$, where $m \leq n$, and we can assume the constraints numbered so that the active ones are those in (2.60) and (2.61) for $i = 1, \ldots, N$ and $j = 1, \ldots, L$, with

$N + L = m$. Here as in Section 2.6 we allow for $N = 0$ (i.e., no equality constraints), and we also assume that at least one inequality constraint is active at $\mathbf{x}_0$. We assume that the Karush–Kuhn–Tucker conditions hold at $\mathbf{x}_0$. Finally, we assume that the linear independence assumption is valid at $\mathbf{x}_0 = (x_1^0, \ldots, x_n^0)^t$.

Our coordinate transformation is then as follows:

$$u_i = g_i(x_1, \ldots, x_n) \qquad \text{for } i = 1, \ldots, N, \tag{2.72}$$

$$u_i = -h_{i-N}(x_1, \ldots, x_n) \qquad \text{for } i = N+1, \ldots, m, \tag{2.73}$$

$$u_i = x_i - x_i^0 \qquad \text{for } i = m+1, \ldots, n. \tag{2.74}$$

Here we have assumed coordinates in $\mathcal{R}^n$ numbered so that the gradients of the $n$ functions on the right sides of these equations are linearly independent at $\mathbf{x}_0$. This is possible, since the first $m$ of these gradients are linearly independent by assumption. The gradients of the coordinate functions $x_i$ are linearly independent, and hence, by linear algebra, $n - m$ of these can be adjoined to the previous $m$ gradients to form a set of $n$ linearly independent gradients. By proper numbering of the coordinates, we are led to the equations as stated. The linear independence of the gradients at $\mathbf{x}_0$ is equivalent to the invertibility of the Jacobian matrix of the equations at this point. Hence, as above, in an appropriate neighborhood of $\mathbf{x}_0$, there is a well-defined differentiable inverse mapping.

The $m$ constraints given lead to constraints in the coordinates $u_i$: namely $u_i = 0$ for $i = 1, \ldots, N$, $u_i \geq 0$ for $i = N+1, \ldots, m$. At $\mathbf{x}_0$ all these constraints are active; from this fact and the form of (2.74), we see that $\mathbf{x}_0$ corresponds to $\mathbf{u} = \mathbf{0}$.

We now choose $\delta$ positive and so small that the set $D_\delta$ of all $\mathbf{u}$ such that $|u_i| < \delta$ for all $i$ lies within the neighborhood of $\mathbf{u} = \mathbf{0}$ in which the inverse mapping is defined. The subset $E_\delta$ of $D_\delta$ on which the constraints of the preceding paragraph hold is a convex set. The function $f(\mathbf{x})$ becomes a function $F(\mathbf{u})$ defined on $D_\delta$ and satisfying the Karush–Kuhn–Tucker conditions at $\mathbf{u}_0 = \mathbf{0}$. Furthermore the constraints satisfy the linear independence assumption at this point, since the gradient vectors concerned are the standard basis vectors $\mathbf{e}_i$ of $\mathcal{R}^n$, for $i = 1, \ldots, m$. If now $F(\mathbf{u})$ is a convex function on $E_\delta$, then as above we can conclude that $\mathbf{u} = \mathbf{0}$ is a minimizer of $F$, so $f$ has a local minimum at $\mathbf{x}_0$. To ensure that $F$ is convex, we can require that the quadratic form $\mathbf{v}^t H \mathbf{v}$ is positive definite on the subspace $W$ which coincides with $\mathcal{R}^n$ if $N = 0$ and otherwise coincides with

$$\{\mathbf{v} \in \mathcal{R}^n \mid v_i = 0 \text{ for } i = 1, \ldots, N\},$$

where $H$ is the Hessian matrix of $F$ at $\mathbf{u} = \mathbf{0}$. When this holds, F is strictly convex on $E_\delta$, for $\delta$ sufficiently small and positive, $\mathbf{0}$ provides a strong local minimum for $F$, and $\mathbf{x}_0$ provides a strong local minimum for $f$.

We thus conclude: *Under the hypotheses stated, the coordinate transformation (2.72)–(2.74) converts $f(\mathbf{x})$ into a function $F(\mathbf{u})$. If the Hessian matrix H of F at $\mathbf{0}$ is such that the quadratic form $\mathbf{v}^t H \mathbf{v}$ is positive definite on the subspace W, then f has a strong local minimum at $\mathbf{x}_0$.*

The entries in the Hessian matrix of $F$ are its second partial derivatives. As in Section 2.3, these derivatives can be obtained, through implicit differentiation, from derivatives of $f$. Therefore application of the procedure is a straightforward calculation, requiring only the solution of linear equations and (to test for positive definiteness) computing the eigenvalues of a matrix.

***Example 3.*** We consider Example 4 of the preceding section, choosing as $\mathbf{x}_0$ the point $(1 - \sqrt{1/2}, 3/2)$, at which both constraints are active. We make the coordinate transformation $u = -x^2 + 2x - 2 + y$, $v = -x^2 + 2x + 1 - y$. The point $\mathbf{x}_0$ corresponds to the origin of the $uv$-plane. Since the linear independence assumption is valid at $\mathbf{x}_0$, the Jacobian matrix of this mapping is invertible at the point, but we can verify this fact directly (Problem 2.41). As in Example 2 of Section 2.3, we calculate

$$1 = -2xx_u + 2x_u + y_u, \quad 0 = -2xx_v + 2x_v + y_v,$$

$$0 = -2xx_u + 2x_u - y_u, \quad 1 = -2xx_v + 2x_v - y_v,$$

$$0 = -2x_u^2 - 2xx_{uu} + 2x_{uu} + y_{uu}, \quad 0 = -2x_v x_u - 2xx_{uv} + 2x_{uv} + y_{uv},$$

$$0 = -2x_v^2 - 2xx_{vv} + 2x_{vv} + y_{vv},$$

$$0 = -2x_u^2 - 2xx_{uu} + 2x_{uu} - y_{uu}, \quad 0 = -2x_v x_u - 2xx_{uv} + 2x_{uv} - y_{uv},$$

$$0 = -2x_v^2 - 2xx_{vv} + 2x_{vv} - y_{vv},$$

$$F_u = f_x x_u + f_y y_u = 4x^3 x_u + 3x^2 y x_u + x^3 y_u, \quad F_v = 4x^3 x_v + 3x^2 y x_v + x^3 y_v,$$

$$F_{uu} = 12x^2 x_u^2 + 4x^3 x_{uu} + 6xy x_u^2 + 6x^2 x_u y_u + 3x^2 y x_{uu} + x^3 y_{uu},$$

$$F_{uv} = 12x^2 x_u x_v + 4x^3 x_{uv} + 6xy x_u x_v + 3x^2 x_u y_v + 3x^2 y x_{uv} + 3x^2 x_v y_u + x^3 y_{uv},$$

$$F_{vv} = 12x^2 x_v^2 + 4x^3 x_{vv} + 6xy x_v^2 + 6x^2 x_v y_v + 3x^2 y x_{vv} + x^3 y_{vv}.$$

Here the values of $x$ and $y$ are known: $x = 0.29289322$, $y = 1.5$. The first four equations can thus be solved for $x_u, x_v, y_u, y_v$. We substitute the values found in the next six equations and obtain six linear equations for the six second partial derivatives $x_{uu}, x_{uv}, \ldots$. If all the known values are substituted in the expressions for the second partial derivatives of $F$, we obtain numerical values for these derivatives at the origin of the $uv$-plane. We find that

$$F_{uu} = 0.635184, \quad F_{vv} = 0.453204, \quad F_{uv} = 0.544194.$$

The eigenvalues of the Hessian matrix of $F$ at the origin are then found to be $1.095943$ and $-0.007554$. Hence the matrix is *not* positive definite and the method fails. Verification of the details is left to Problem 2.40.

### Discussion

One might have predicted the result of Example 3, since the Hessian matrix of the original function $f$ is not positive definite at $x_0$. However, the theorem on invariance of the index proved in Section 2.3 applies only at *critical points*, and $x_0$ is not a critical point. As shown below, a special first derivative test can be applied to this example, to establish the minimum. See also Problem 2.40(b).

REMARK 1. Let us suppose, as above, that the constraints active at $x_0$ are those for $i = 1, \ldots, m = N + L$ and that Lagrange multipliers $\mu_i$ and $\lambda_j$ exist so that (2.62) holds at $x_0$. Let $\lambda_1 < 0$ so that the Karush–Kuhn–Tucker conditions are *not* satisfied at $x_0$. We then assert that *if the constraint $h_1(x) \leq 0$ is allowed to become inactive, then $f$ takes on values less than $f(x_0)$.*

To show this, we make the change of coordinates (2.72)–(2.74) and consider the function $F(u)$. From the form of the constraints, the Lagrange multipliers $\mu_i$ and $\lambda_j$ satisfy the equation

$$\nabla F + (\mu_1, \ldots, \mu_N, \lambda_1, \ldots, \lambda_L, \ldots)^t = 0.$$

Thus at $u = 0$, $\partial F / \partial u_{N+1} = -\lambda_1 > 0$. Therefore $F$ decreases as $u_{N+1}$ decreases from 0, and $F$ takes on values less than $F(0)$ at feasible points on the negative $u_{N+1}$-axis near the origin. Thus our assertion follows.

### A Special First Derivative Sufficiency Condition

Examples 1 and 2 at the beginning of this section show that in general positivity of some or all of the $\lambda_j$ does not ensure that there is a local minimum. However, in one special case such a hypothesis is sufficient:

In the extremum problem of Section 2.6 for the function $f$, with side conditions (2.60), (2.61), let $N = 0$ and $M = n$. Let all $h_j$ be active at $x_0$, and let the linear independence assumption hold at $x_0$. Let the Karush–Kuhn–Tucker conditions (2.62) hold at $x_0$ with all $\lambda_j > 0$. Then $f$ has a strong local minimum at $x_0$.

To justify this assertion, we can use the mapping (2.72)–(2.74) to reduce our extremum problem to the case where, for each $j$, $h_j$ is the function $-x_j$ and the point $\mathbf{x}_0$ is the origin. As above, we consider only a neighborhood $|x_i| < \delta$ of the origin and the feasible solution set contains all points of this neighborhood for which $x_j \geq 0$ for $j = 1, \ldots, n$. The Karush–Kuhn–Tucker conditions state that at the origin $f$ has gradient $(\lambda_1, \ldots, \lambda_n)^t$ and all $\lambda_j$ are positive. Accordingly each partial derivative of $f$ is positive at the origin and hence, by continuity, is positive in a neighborhood of the origin; for proper choice of $\delta$, all first partial derivatives of $f$ are positive for $|x_j| < \delta$, for $j = 1, \ldots, n$. The mean value theorem of differential calculus (see Kaplan, 1991, p. 469) asserts that

$$f(\mathbf{x}_1) - f(\mathbf{x}_0) = (\nabla f(\mathbf{x}^*))^t(\mathbf{x}_1 - \mathbf{x}_0),$$

where $\mathbf{x}^*$ is an appropriate point on the line segment joining $\mathbf{x}_0$ and $\mathbf{x}_1$. If we apply this with $\mathbf{x}_0$ taken as the origin and $\mathbf{x}_1$ as any other point (necessarily feasible) with coordinates satisfying $0 \leq x_j < \delta$, then we conclude that $f(\mathbf{x}_1) > f(\mathbf{x}_0)$. Thus $f$ has a strong local minimum at the origin.

This new sufficiency condition provides a much simpler test for Example 3 (Example 4 of Section 2.6), for which $n = 2$ and there are precisely two inequality side conditions. We saw in Section 2.6 that the Karush–Kuhn–Tucker conditions provided just two candidates, at one of which both $\lambda_j$ are positive while at the other both $\lambda_j$ are negative. Hence, as we already know, one is a minimizer and one is a maximizer.

REMARK 2. By similar reasoning, we find the same conclusion if the condition $N = 0$, $M = n$ is replaced by the condition $M + N = n$.

## 2.8  PROOF OF THE KARUSH–KUHN–TUCKER CONDITIONS

A simple proof of the necessity of the Karush–Kuhn–Tucker conditions, under the linear independence assumption, is indicated in Problem 2.46. Here we give another proof, one that permits replacement of the linear independence assumption by a weaker assumption. The proof is based on the following theorem:

**Lemma of Farkas.**   Let $T$ be an $m \times n$ matrix. Let $\mathbf{b}$ be a vector of $\mathcal{R}^m$. Then there exists a $\mathbf{z}$ in $\mathcal{R}^n$ such that $T\mathbf{z} = \mathbf{b}$ and $\mathbf{z} \geq \mathbf{0}$ if and only if $\mathbf{b}^t\mathbf{y} \geq 0$ for all $\mathbf{y}$ in $\mathcal{R}^m$ such that $T^t\mathbf{y} \geq \mathbf{0}$.

A proof of this lemma is given below. We recall the vector notation for inequalities given in Section 2.6. To apply the lemma, we return to the hypotheses of Section 2.6 and assume first that there are no equality constraints (2.60). Then (2.62) is again valid at the minimizer $\mathbf{x}_0$, with no sum over $i$ and all $\lambda_j = 0$ for the constraints not active at $\mathbf{x}_0$. We wish to show that all $\lambda_j$ are nonnegative. From the inequality constraints (2.61) it follows that along a smooth path $\mathbf{x} = \mathbf{x}(t)$ which passes through $\mathbf{x}_0$ at $t = t_0$ and enters $E$ as $t$ increases past $t_0$, the derivative $\mathbf{v}_0 = \mathbf{x}'(t_0)$ must satisfy the following inequalities at $\mathbf{x}_0$:

$$(\nabla h_j)^t \mathbf{v}_0 \leq 0 \qquad \text{for } j = 1, \ldots, M \qquad (2.80)$$

for the constraints active at $\mathbf{x}_0$. Furthermore, we can show with the aid of the implicit function theorem that for every vector $\mathbf{v}_0$ satisfying (2.80) there is a path as described whose derivative at $t_0$ is $\mathbf{v}_0$ (see Problem 2.43). Since $\mathbf{x}_0$ is assumed to be a local minimizer of $f$, it follows that along such a path $f$ is nondecreasing in $t$ at $t_0$; that is, at $\mathbf{x}_0$, $(\nabla f)^t \mathbf{v}_0 \geq 0$. If we now let $T$ be the matrix whose column vectors are the vectors $-\nabla h_j$ for the constraints active at $\mathbf{x}_0$, all evaluated at $\mathbf{x}_0$, and let $\mathbf{b}$ be the vector $\nabla f$, then the Farkas lemma can be applied, with $\mathbf{v}_0$ playing the role of $\mathbf{y}$. Hence there is a $\mathbf{z}$ whose components are nonnegative numbers $\lambda_j$ corresponding to the active constraints at $\mathbf{x}_0$, such that (2.62) holds, with no sum over $i$ and all $\lambda_j = 0$ for the constraints not active at $\mathbf{x}_0$.

When equality constraints are present, the proof given can be modified to establish the analogous result. First of all, we can replace each equality constraint in (2.60) by two inequality constraints: $g_i = 0$ is replaced by $g_i \leq 0$ and $-g_i \leq 0$. By the linear independence assumption, (2.62) is valid as above except that the sum over $i$ appears. However, we can write $\mu_i \nabla g_i$ as $\nu_i \nabla g_i + \sigma_i \nabla(-g_i)$. Now for each smooth path $\mathbf{x} = \mathbf{x}(t)$, as above, $(\nabla g_i)^t \mathbf{v}_0 = 0$ at $\mathbf{x}_0$, since $g_i = 0$ along the path. This equation is equivalent to the two equations $(\nabla g_i)^t \mathbf{v}_0 \leq 0$ and $(-(\nabla g_i))^t \mathbf{v}_0 \leq 0$. Furthermore, for every vector $\mathbf{v}_0$ satisfying these inequalities for each $i$ and those relative to the active $h_j$, as above, there is a path $\mathbf{x} = \mathbf{x}(t)$, as above, with the corresponding vector $\mathbf{v}_0$ satisfying all the inequalities described (Problem 2.43), and again, since $\mathbf{x}_0$ is a local minimizer of $f$, $(\nabla f)^t \mathbf{v}_0 \geq 0$ at $\mathbf{x}_0$. Hence again the Farkas lemma can be applied; this time the matrix $T$ includes as columns the gradients of each $-g_i$ and of each $g_i$. The vector $\mathbf{z}$ includes corresponding *nonnegative* scalars $\nu_i$ and $\sigma_i$. Thus we have now shown that (2.62) is valid with $\mu_i = \nu_i - \sigma_i$ and all $\lambda_j$ corresponding to the active $h_j$ nonnegative. This completes the proof.

REMARK. The proof shows that instead of the linear independence assumption, one can simply assume that for each vector $\mathbf{v}_0$ satisfying the inequalities of the preceding paragraph, there is a smooth path as described with $\mathbf{x}'(t_0) = \mathbf{v}_0$. This new assumption was introduced in Kuhn and Tucker (1951) as the *constraint qualification*.

*Proof of the Farkas Lemma.* First we assume that, for given $\mathbf{b}$, there is a $\mathbf{z}$ with $T\mathbf{z} = \mathbf{b}$ and $\mathbf{z} \geq \mathbf{0}$. Then for $T^t \mathbf{y} \geq \mathbf{0}$ we have

$$\mathbf{b}^t \mathbf{y} = (T\mathbf{z})^t \mathbf{y} = \mathbf{z}^t T^t \mathbf{y} \geq 0.$$

(The inner product of two nonnegative vectors is nonnegative.)

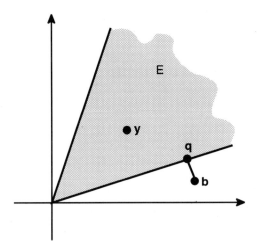

**Figure 2.7.** Proof of The Farkas lemma.

Next we assume that $\mathbf{b}^t\mathbf{y} \geq 0$ for all $\mathbf{y}$ in $\mathcal{R}^m$ such that $T^t\mathbf{y} \geq \mathbf{0}$, and we seek a nonnegative $\mathbf{z}$ in $\mathcal{R}^n$ such that $T\mathbf{z} = \mathbf{b}$. Let $E$ be the range of the mapping $T$ restricted to nonnegative vectors; that is, $E$ is the set of all $\mathbf{w} = T\mathbf{z}$ for $\mathbf{z}$ nonnegative. We must show that $\mathbf{b}$ is in $E$.

To this end, we verify that $E$ has the following properties: It is a closed, convex subset of $\mathcal{R}^m$, containing the origin; for each $\mathbf{w}$ in $E$, all vectors $t\mathbf{w}$ for $t \geq 0$ are in $E$. (See Problem 2.44.) Now as in Section 1.12 there is a point $\mathbf{q}$ of $E$ closest to $\mathbf{b}$ and as in (1.152) $(\mathbf{y} - \mathbf{q})^t(\mathbf{q} - \mathbf{b}) \geq 0$ for each $\mathbf{y}$ in $E$. (See Fig. 2.7.) Furthermore $(\mathbf{q} - \mathbf{b})^t\mathbf{q} = 0$. (See Problem 2.41(d).) From these results we conclude that for each nonnegative $\mathbf{z}$ in $\mathcal{R}^n$,

$$(T\mathbf{z})^t(\mathbf{q} - \mathbf{b}) = \mathbf{z}^t T^t(\mathbf{q} - \mathbf{b}) \geq 0.$$

But a vector whose inner product with all nonnegative vectors is nonnegative must itself be a nonnegative vector (Problem 2.45). Therefore $\mathbf{y} = \mathbf{q} - \mathbf{b}$ satisfies $T^t\mathbf{y} \geq 0$, and hence, by our assumption, $\mathbf{b}^t\mathbf{y} \geq 0$. But

$$\mathbf{b}^t\mathbf{y} = (\mathbf{b} - \mathbf{q})^t(\mathbf{q} - \mathbf{b}) = -\|\mathbf{b} - \mathbf{q}\|^2 \leq 0.$$

Therefore the equals sign must apply, and $\mathbf{b} = \mathbf{q}$, an element of $E$.                ∎

## PROBLEMS

**2.38.** For Example 1 in Section 2.7, verify that at $(0, 0)$ the linear independence assumption is satisfied and that the Karush–Kuhn–Tucker conditions give the equation given, with $\lambda_1 = 1$ and $\lambda_2 = 0$, but that $f$ does not have a local minimum at the point.

**2.39.** For Example 2 in Section 2.7, verify that at $(0, 0)$ the linear independence assumption is satisfied and that the Karush–Kuhn–Tucker conditions give the equation given, with $\lambda_1 = 1$, but that $f$ does not have a local minimum at the point.

**2.40.** Consider Example 3 of Section 2.7, which concerns the extremum problem of Example 4 of Section 2.6.

    **(a)** Verify the calculations given, leading to a Hessian matrix $H$ that is not positive definite.

    **(b)** Verify that the corresponding quadratic form $\mathbf{y}^t H \mathbf{y}$ has positive values for $\|\mathbf{y}\| = 1$ and $\mathbf{y} \geq 0$. Since $F_u \geq 0$ and $F_v \geq 0$ at $(0, 0)$ and $u \geq 0$ and $v \geq 0$ by the given constraints, one can conclude that $F$ has a local minimum at the origin. See Kaplan (1991, p. 164, prob. 10); for details, see Problem 2.48.

**2.41.** Seek the minimum of the function $f(x, y, z) = x^2 + y^2 + z^2$ subject to the constraints $x + 2y + 3z = 1$, $x \leq 0$, $y \geq 0$, $z \geq 0$ with the aid of the Karush–Kuhn–Tucker conditions, and use the method of Section 2.7, as in Example 3, to verify that the minimizer has been found. Interpret the result geometrically.

**2.42.** Verify (2.71), under the hypotheses stated.

**\*2.43.** In Section 2.7 it is asserted that under the linear independence assumption, for each vector $\mathbf{v}_0$ such that, at $\mathbf{x}_0$, $(\nabla g_i)^t \mathbf{v}_0 = 0$ for all $i$ and $(\nabla h_j)^t \mathbf{v}_0 \leq 0$ for all $j$, whenever $h_j$ is active at $\mathbf{x}_0$, there is a smooth path in $E$ such that $\mathbf{x}'(t_0) = \mathbf{v}_0$ and the path enters $E$ as $t$ increases past $t_0$. Prove this assertion. [Hint: The assertion will follow if we can find a path in $E$, with $\mathbf{x}'(t_0) = \mathbf{v}_0$, such that each $g_i$ is constant along the path and each $h_j$ is nonincreasing along the path as $t$ increases past $t_0$. To find such a path, use the mapping (2.72)–(2.74) of Section 2.7, and observe that under this mapping there is a one-to-one correspondence between the derivatives at $t_0$ of functions $\mathbf{x} = \mathbf{x}(t)$ with value $\mathbf{x}_0$ at $t_0$ and of the corresponding functions $\mathbf{u} = \mathbf{u}(t)$ with value $\mathbf{0}$ at $t_0$. Choose $\mathbf{x}_2(t) = \mathbf{x}_0 + (t - t_0)\mathbf{v}_0$ corresponding to $\mathbf{u} = \mathbf{u}_2(t)$; then choose $\mathbf{u} = \mathbf{u}_1(t) = (t - t_0)\mathbf{u}'_1(t_0)$, and let $\mathbf{x} = \mathbf{x}_1(t)$ be the corresponding path. For each function $\phi(\mathbf{x})$ that is differentiable at $\mathbf{x}_0$, corresponding to $\Phi(\mathbf{u})$, as in Section 2.7, use the relation $\Phi(\mathbf{u}_1(t)) = \phi(\mathbf{x}_1(t))$ to deduce that

$$(\nabla \Phi(\mathbf{0}))^t \mathbf{u}'_1(t_0) = (\nabla \phi(\mathbf{x}_0))^t \mathbf{x}'_1(t_0)$$

and conclude from the hypotheses concerning $\mathbf{v}_0$ that as $t$ increases past $t_0$ the path $\mathbf{u} = \mathbf{u}_1(t)$ enters the set $E_\delta$ of Section 2.7, so that the path $\mathbf{x} = \mathbf{x}(t)$ has the desired properties.

*2.44. Let $E$ be defined as in the proof of the Farkas Lemma in Section 2.8. Observe that the equation $\mathbf{w} = T\mathbf{z}$ defines a set in $\mathcal{R}^n$ which is either empty or else defines a linear variety which meets the first orthant (all coordinates nonnegative) of $\mathcal{R}^n$ precisely when $\mathbf{w}$ is in $E$.

(a) For each of the following choices of $T$, verify that $E$ is a closed set by showing that the set of $\mathbf{w}$ not in $E$ is open:

$$\begin{bmatrix} 1 & 1 \\ 2 & 2 \end{bmatrix}, \quad \begin{bmatrix} 1 & -1 \\ 2 & -2 \end{bmatrix}$$

(b) Proceed as in (a) to show that for $n = 2$ $E$ is a closed set.

(c) Proceed as in (a) to show that for $n = 3$ $E$ is a closed set.

REMARKS. The reasoning of parts (b) and (c) can be extended to all $n \geq 4$. The set $E$ has a simple structure as the "conical hull" of a finite set of vectors $\mathbf{w}_1, \ldots, \mathbf{w}_n$ (the column vectors of $T$) in $\mathcal{R}^m$. One can form $E$ by first forming the set $E_1$ of all $T\mathbf{z}$ with $\mathbf{z} \geq 0$ and $z_1 + \cdots + z_n = 1$; $E_1$ is the *convex hull* of the set of $n$ vectors, the smallest convex set containing all of the vectors. When the vectors are linearly independent, for $n = 2$ $E_1$ is a line segment and for $n = 3$ $E_1$ is a triangular region. See Problem 3.16. To obtain $E$ from $E_1$, one forms the set of all nonnegative multiples of the vectors in $E_1$. Thus $E$ is a ray, a sector, or a generalization of a sector to higher dimensions.

(d) $E$ is a convex set.

(e) If $\mathbf{y}$ is in $E$, then so is $t\mathbf{y}$ for each $t \geq 0$.

(f) If $\mathbf{q}$ is the point of $E$ closest to the point $\mathbf{b}$, then $(\mathbf{q} - \mathbf{b})^t \mathbf{q} = 0$.

2.45. Show that a vector of $\mathcal{R}^n$ whose inner product with all nonnegative vectors is nonnegative must itself be nonnegative.

2.46. *A proof of the Karush–Kuhn–Tucker conditions.* Use the change of coordinates (2.72)–(2.74) to prove that, under the linear independence assumption, a function $f$ having a local minimum at $\mathbf{x}_0$ satisfies the Karush–Kuhn–Tucker conditions at $\mathbf{x}_0$. [Hint: The function $f(\mathbf{x})$ becomes a function $F(\mathbf{u})$ having a local minimum at the origin on the set in $\mathcal{R}^n$ defined by the constraints $u_i = 0$ for $i = 1, \ldots, N$ and $u_i \geq 0$ for $i = N + 1, \ldots, m$, and these are the active constraints. As in Section 2.6, one has an equation

$$\nabla F(\mathbf{0}) + \sum_{i=1}^{N} \mu_i \mathbf{e}_i - \sum_{i=N+1}^{m} \lambda_i \mathbf{e}_i = \mathbf{0},$$

where as usual the $\mathbf{e}_i$ are the standard basis vectors of $\mathcal{R}^n$. Show that, since $F$ has a minimum at the origin, no $\lambda_i$ can be negative. Now use the invariance of the Karush–Kuhn–Tucker conditions under the change of coordinates to obtain the Karush–Kuhn–Tucker conditions for $f$ at $x_0$.]

**2.47. (a)** Equation (2.71) shows how the gradient of a function $\phi$ transforms under the change of coordinates (2.70), under the conditions described. Show that the second partial derivatives of $\phi$ transform in accordance with the rule

$$H^u = K^t(H^x - P)K, \quad P = (p_{ij}),$$

where $K = J^{-1}$, $H^u$ is the Hessian matrix of $\Phi$, $H^x$ is the Hessian matrix of $\phi$ and

$$p_{ij} = \sum_{k=1}^{n} \frac{\partial \Phi}{\partial u_k} \frac{\partial^2 u_k}{\partial x_i \partial x_j}.$$

[The results of Section 2.3 are a special case of these formulas and (2.71).]

**(b)** Apply the formula (2.71) and the formulas of part **(a)** to find the Hessian matrix of the function $F(u, v)$ in Section 2.7, Example 3.

**2.48. (a)** Let $f(\mathbf{x})$ be defined in a neighborhood $U$ of the origin in $\mathcal{R}^n$. Under appropriate assumptions of differentiability, along a line $\mathbf{x} = s\mathbf{v}$, where $\mathbf{v}$ is a unit vector, the function $f(\mathbf{x}(s))$ has a derivative

$$df/ds = (\nabla f)^t\mathbf{v} + s\mathbf{v}^t H\mathbf{v} + sg(\mathbf{x}(s)),$$

where $H$ is the Hessian matrix of $f$, $\nabla f$ and $H$ are evaluated at the origin, and $g(\mathbf{x})$ is a function defined in the neighborhood of the origin having limit 0 as $\mathbf{x}$ approaches the origin. [For $n = 2$, this assertion is equivalent to that made in Kaplan (1991, prob. 10, p. 164).] Show that if at the origin $\nabla f \geq \mathbf{0}$ and the quadratic form $\mathbf{v}^t H\mathbf{v}$ has a positive minimum for $\|\mathbf{v}\| = 1$ and $\mathbf{v} \geq \mathbf{0}$, then $f$ has a strong local minimum at the origin on the set $\{\mathbf{x}|\mathbf{x} \geq \mathbf{0}, \mathbf{x} \in U\}$.

**(b)** Generalize the result of part **(a)** by showing that if $f(\mathbf{x})$ satisfies the Karush–Kuhn–Tucker conditions at the origin for the constraints $x_i = 0$ for $i = 1, \ldots, N$ and $-x_i \leq 0$ for $i = N + 1, \ldots, m$, which are the active constraints, and at the origin the Hessian matrix $H$ of $f$ is such that the quadratic form $\mathbf{v}^t H\mathbf{v}$ has a positive minimum for $\|\mathbf{v}\| = 1$, $v_i = 0$ for $i = 1, \ldots, N$ and $v_i \geq 0$ for $i = N + 1, \ldots, m$, then $f$ has a strong local minimum at the origin.

**2.49.** Apply the results of Problems 2.47 and 2.48 to obtain the following *sufficient condition* based on the Karush–Kuhn–Tucker conditions: For the extremum problem of Section 2.6, with constraints (2.60) and (2.61), let the functions $f$, $g_i$ and $h_j$ have continuous first and second partial derivatives in a neighborhood of $x_0$ and let the linear independence assumption hold at $x_0$. Let the Karush–Kuhn–Tucker conditions hold at $x_0$, so that in (2.62) the $\lambda_j$ are nonnegative and for each inactive constraint at $x_0$ the corresponding $\lambda_j$ is 0. Let $H^u = K^t(H^x - P)K$, where $J$ is the Jacobian matrix of the transformation of coordinates (2.72)–(2.74), evaluated at $x_0$, $K = J^{-1}$, $H^x$ is the Hessian matrix of $f$ at $x_0$, $P = s_1 H_1 + \cdots + s_n H_n$, where for each $k$ $H_k$ is the Hessian matrix of $u_k(x)$ at $x_0$ and

$$(s_1, \ldots, s_n)^t = K \nabla f(x_0).$$

If the quadratic form $Q(v) = v^t H^u v$ has a positive minimum for $\|v\| = 1$, $v_i = 0$ for $i = 1, \ldots, N$ and $v_i \geq 0$ for $i = N+1, \ldots, m$, then $f$ has a strong local minimum at the origin.

REMARKS. For other sufficient conditions, see Polak (1997), Chapter 2, especially pp. 213–214. Finding the minimum of the quadratic form $Q(v)$ subject to the constraints specified can be a significant problem, in particular because in general the quadratic form may not be positive definite. In some cases showing that the minimum is positive may be easy; for example, in the case when the constraints are $v_i \geq 0$ for $i = 1, \ldots, n$ and all coefficients of the quadratic form are positive, as in Example 3 of Section 2.7. A simple practical way to approach the problem in an approximate way, if $n - N$ is not too large, is to evaluate $Q(v)$ for a large number of feasible vectors $v$ whose directions approximate, within some $\varepsilon$, all feasible directions. For example, for the constraints $v_i \geq 0$ for $i = 1, \ldots, n$, one could use the MATLAB matrix rand$(n, p)$; if $p$ is large, this matrix has a set of columns which should provide the desired approximating vectors $v$. If from this set one rejects the vectors $v$ which satisfy the inequalities $v_i \leq b$, for $0 < b < 1$ and $b$ close to 0, then the existence of a positive lower bound for the values $Q(v)$ for $v$ in the remaining set would (approximately) prove that $Q(v)$ has a positive lower bound on the set of feasible vectors for which $\|v\| = 1$. For $b$ small (and especially for $p$ large), the set of rejected vectors should be small. If for a particular $v$, the value of $Q(v)$ is *negative*, then the process should stop and the sufficiency test fails; this does not imply that the function $f$ does not have a local minimum at $x_0$, as the example $f(x) = x - x^2$ for $x \geq 0$ illustrates.

## REFERENCES

Birkhoff, G., and MacLane, S. (1977). *A Survey of Modern Algebra*, 4th ed., Macmillan, New York.

John, F. (1948). "Extremum problems with inequalities as subsidiary conditions," in *Studies and Essays Presented to R. Courant on His 60th Birthday*, Interscience, New York, pp. 187–204.

Kaplan, W. (1991). *Advanced Calculus*, 4th ed., Addison-Wesley, Reading, MA.

Kuhn, H. W., and Tucker, A. W. (1951). "Nonlinear programming," in *Proceedings of the Second Berkeley Symposium on Mathematical Statistics and Probability*, University of California Press, Berkeley, pp. 481–492.

Polak, E. (1997). *Algorithms and Consistent Approximations*, Springer-Verlag, New York, Berlin, Heidelberg.

# 3

# Optimization

Optimization is the goal of all practical design projects, although few achieve such an ultimate state of perfection. Commonly a design is finalized significantly short of its optimum as a result of a complex set of compromises made for economic, technical, or social reasons. The decision-making process is usually a mixture of analysis, experiments, and intuition. When the variables in a problem become numerous, experiments are expensive, intuition breaks down, and a systematic approach is necessary. Using the modern theory of optimization, one can often model the relationships between the variables mathematically and then, taking advantage of rapid computers, seek the optimal solution of the model problem numerically.

Many practical problems are naturally formulated as problems of minimization or maximization of (1) real functions of several real variables, or (2) of integrals depending on unknown functions. In the first case, one has a finite-dimensional problem, as in the previous chapters. In the second case one has a problem of the *calculus of variations*, an infinite-dimensional problem; however, methods of numerical analysis, especially those using finite elements, convert the problem into a finite-dimensional one. These finite-dimensional problems all fall in the area of *mathematical programming*. This area also includes maximum and minimum problems arising in operations research, notably those of linear programming.

In this chapter we consider some important problems of mathematical programming, treating them in order of increasing difficulty: quadratic, linear, convex, and general nonlinear problems.

## 3.1  CONVEXITY

We here review the ideas about convex sets and convex or concave functions presented in Chapters 1 and 2 and introduce some new properties.

A set $E$ in $\mathcal{R}^n$ is said to be convex if, for every pair of points in $E$, the line segment joining them is contained in $E$. A function $f$ defined on a convex set $E$ is convex if

$$f((1-t)\mathbf{x}_1 + t\mathbf{x}_2) \leq (1-t)f(\mathbf{x}_1) + tf(\mathbf{x}_2) \tag{3.10}$$

for $0 < t < 1$ and each pair of points $\mathbf{x}_1$, $\mathbf{x}_2$ in $E$. If in (3.10) the $\leq$ is replaced by $<$, then $f$ is said to be strictly convex in $E$. A function $f$ is said to be concave (strictly concave) if the function $-f$ is convex (strictly convex). The function $f$ is convex (strictly convex) in $E$ if and only if, on each line segment $\mathbf{x} = t\mathbf{a} + \mathbf{b}$ in $E$, $f$ is a convex (strictly convex) function of the parameter $t$. For a function $g(t)$ with continuous second derivative on an interval, $g$ is convex if and only if $g''(t) \geq 0$ on the interval, is strictly convex if $g''(t) > 0$ on the interval. This extends to $\mathcal{R}^n$ in terms of the Hessian matrix $H$ of a function, formed of its second partial derivatives: $f$ is convex on the convex set $E$ if and only if $H$ is positive semidefinite on $E$; if $H$ is positive definite on $E$, then $f$ is strictly convex on $E$.

If $f$ is a convex function on the convex set $E$ and $f$ takes on values less than the constant $c$, then the sublevel set $f(\mathbf{x}) \leq c$ is convex (as is the o-sublevel set on which $f < c$). The intersection of a finite number of convex sets is convex, and hence the subset of $\mathcal{R}^n$ defined by a finite set of linear equations and inequalities of the form $f_i(\mathbf{x}) \leq c$, where each $f_i$ is convex in $\mathcal{R}^n$, is convex. In particular, the following closed sets are convex: a point, a line segment in the plane, a triangular surface in space, a closed planar region bounded by a convex polygon, a solid tetrahedron, a solid cube, a solid octahedron, and their generalizations to $\mathcal{R}^n$, which we call *convex polyhedra*.

### Extrema of Functions

Each local minimum of a convex function is a global minimum. The minimizers of a convex function form a convex set on which the function is constant. A strictly convex function has at most one minimizer. Each local extremum of a differentiable function, defined on an open set $E$, occurs at a critical point of the function (where its gradient is zero). If the function is convex and $E$ is convex, then the critical point is a global minimizer; if the function is strictly convex on $E$, then it has a strong local minimum at the point. A function $f$ that is continuous on a closed, bounded set $E$ has a global minimum and maximum on $E$; if $E$ is closed and unbounded, and $f$ has limit $\infty$ as $\|\mathbf{x}\| \to \infty$, then $f$ has a global minimum on $E$. In particular, the distance from a fixed point to a variable point in a closed set $E$ has a global minimum, whether $E$ is bounded or not. This minimum is the distance from the point to set $E$. A function that is strictly convex in $\mathcal{R}^n$ has a global minimum if and only if it has limit $\infty$ as $\|\mathbf{x}\| \to \infty$. The norm and distance in these statements can be an $A$-norm and its associated distance, for a positive definite symmetric matrix $A$, as in Section 1.14.

## Extreme Points

An *extreme point* of the convex set $E$ is a point of $E$ that is the midpoint of no line segment whose endpoints are in $E$. Thus the extreme points of a closed triangular region are the vertices of the triangle; there is a similar statement for convex regions in the plane bounded by convex polygons, and the result generalizes to convex polyhedra in $\mathcal{R}^n$. One shows that each convex polyhedron has a finite number of extreme points, which are its vertices; the same statement applies to each unbounded convex set defined by a finite number of linear equations and linear inequalities (Problem 3.3). Every bounded closed convex set $E$ has an extreme point: for example, a point $\mathbf{x}_0$ of $E$ at maximum distance from the origin.

## Relative Interior, Relative Boundary

Let $E$ be a convex set in $\mathcal{R}^n$. Then $E$ may be contained in a straight line, a plane, or, in general, in a linear variety of dimension $d$. This is surely true for $d = n$. The smallest choice of $d$ is called the *dimension* of $E$. Thus a triangular surface in $\mathcal{R}^3$ has dimension 2. Let $V$ be a linear variety of smallest dimension containing $E$. There is only one such linear variety, for if two different linear varieties of dimension $d$ contained $E$, then their intersection would be a linear variety of dimension less than $d$ containing $E$. We define the *relative interior* and *relative boundary* of $E$ by considering only the portions of neighborhoods in the variety $V$. Thus a point $\mathbf{x}_0$ is a relative interior point of $E$ if for some $\delta > 0$ all $\mathbf{x}$ *in the variety $V$* within distance $\delta$ of $\mathbf{x}_0$ are contained in $E$; $\mathbf{x}_0$ is a relative boundary point of $E$ if for every $\delta > 0$ there is a point *in $V$* and in $E$ within distance $\delta$ of $\mathbf{x}_0$ and there is a point *in $V$* and not in $E$ within distance $\delta$ of $\mathbf{x}_0$. Every point of $E$ is either a relative interior point of $E$ or a relative boundary point of $E$, not both.

If, in particular, $E$ is a triangular surface in $\mathcal{R}^3$ and the linear variety $V$ is the $x_1 x_2$-plane, then we think of $E$ as a set in this plane and the relative boundary consists of the points on the edges of the triangle, the relative interior consists of the points inside the triangle in the usual sense. The same applies to each triangular surface $E$ in $\mathcal{R}^3$. However, for such a triangular surface in space, there are no interior points by the usual definition; every point is a boundary point. The example of the triangle is typical of convex polyhedra, in general: The relative boundary of a convex polyhedron consists of the points on the sides, faces, and so on, of the polyhedron; the remaining points of the polyhedron form its relative interior (a nonempty set, also a convex set).

We remark that the convex set $E$ can consist of a single point, which is itself a linear variety of dimension 0. We then consider this point to be the relative interior of $E$; there are no relative boundary points.

Let $E$ be a nonempty convex set in $\mathcal{R}^n$. If $E$ is $n$-dimensional, then the interior of $E$ is an open region and is itself a convex set (Problem 3.5(a)). If $E$ is of dimension less than $n$, then its interior is empty, but its relative interior is nonempty and again forms a convex set (Problem 3.5(b)).

## Further Properties

If $E_1$ is a nonempty convex set in $\mathcal{R}^m$ and $E_2$ is a nonempty convex set in $\mathcal{R}^n$, then the set of all points in $\mathcal{R}^{m+n}$ whose first $m$ coordinates determine a point in $E_1$ and whose last $n$ coordinates determine a point of $E_2$ is convex (Problem 3.7). There is a similar statement for sets in $\mathcal{R}^N$ with $N = n_1 + \cdots + n_k$.

If the function $f$ is convex (strictly convex) on the convex set $E$, then $f$ is also convex (strictly convex) on each convex subset of $E$ (Problem 3.6). Let the function $f$ be convex in $\mathcal{R}^n$, and take on values less than $c$ so that the set on which $f < c$ is nonempty. Let $E$ be the sublevel set $f \le c$. Then the relative boundary of $E$ is the level set $f = c$ and the relative interior of $E$ is the o-sublevel set $f < c$ (Problem 3.8). In this case $E$ has dimension $n$ and the word "relative" can be omitted. If one adjoins to the inequality $f \le c$ $m$ linear equations, given by a matrix equation $Ax = b$ (added constraints), then typically the subset $E_1$ of $E$ satisfying all these constraints is a convex set of dimension $n - m$; this is surely the case if $A$ has rank $m$ and at least one interior point of $E$ satisfies all $m$ equations. An example is given by the set $E_1$ defined by the constraints

$$x_1^2 + x_2^2 + x_3^2 \le 1, \quad x_1 + x_2 + x_3 = 0$$

in $\mathcal{R}^3$. Here $E$ is a solid sphere, $A$ has rank 1, and $E_1$ is a circular disk. The relative boundary of $E_1$ is a circle, the remaining points of $E_1$ form its relative interior. (See Problem 2(e).)

In general, if $f$ and $g$ are (strictly) convex on a convex set $E$, then so is $\beta f + \gamma g$ for every choice of the positive constants $\beta$ and $\gamma$; in fact, if $f$ is strictly convex and $g$ is convex on $E$, then $\beta f + \gamma g$ is strictly convex on $E$ (Problem 3.9). Furthermore, if $B$ is a nonsingular matrix of order $n$ and $f$ is (strictly) convex in $\mathcal{R}^n$, then so also is $g(\mathbf{y}) = f(B\mathbf{y})$ (Problem 3.10). More generally, if $f$ is convex in $\mathcal{R}^n$, then $f(\mathbf{a} + C\mathbf{y})$ is convex in $\mathcal{R}^m$ for each constant vector $\mathbf{a}$ and $n \times m$ matrix $C$; if further $m \le n$ and $C$ has rank $m$, then there is a similar statement for strict convexity (Problem 3.11).

## Convexity of Norms and Related Functions

In $\mathcal{R}^n$, the Euclidean norm $\|\mathbf{x}\|$ and the $A$-norm $\|\mathbf{x}\|_A$ and their squares are convex functions; in fact each allowed norm in $\mathcal{R}^n$ is convex (end of Section 1.14). Furthermore each quadratic function

$$f(\mathbf{x}) = \tfrac{1}{2}\mathbf{x}^t A\mathbf{x} + \mathbf{b}^t\mathbf{x} + c, \tag{3.11}$$

with positive definite $A$, is strictly convex (Section 1.12). Under appropriate conditions the function $g = \sqrt{f}$ is strictly convex or convex. It is shown in

Section 1.12 that the quadratic function $f$ has a global minimum in $\mathcal{R}^n$. If this minimum is positive, then $g$ is strictly convex in $\mathcal{R}^n$. Indeed, along a line $\mathbf{x} = \mathbf{x}_0 + s\mathbf{u}$, with distance $s$ as parameter and $\mathbf{u}$ a unit vector, $g$ becomes a function

$$z = \sqrt{ps^2 + qs + r}$$

with $p > 0$ (Problem 3.12(a)). The function under the square root sign has a positive minimum and $z$ is continuous for all $s$, with continuous first and second derivatives. One verifies that $z''(s) > 0$ (Problem 3.12(b)) and hence strict convexity follows on each line and hence in general. If the global minimum of $f$ is 0, one can still form $g(\mathbf{x})$; one verifies that it is convex in $\mathcal{R}^n$. In fact, $g$ is strictly convex on each line not passing through the unique minimizer of $f$ (Problem 3.12(c)). From the rules given above, it can be concluded that each linear combination $\beta g_1 + \gamma g_2$ of two such functions, with positive coefficients and different minimizers $\mathbf{x}_1$, $\mathbf{x}_2$, is strictly convex on every line except the one through the two minimizers, and in general, every linear combination $\beta_1 g_1 + \cdots + \beta_k g_k$ of $k$ such functions with positive coefficients, is strictly convex in $\mathcal{R}^n$ if the $k$ minimizers do not lie on a line.

### Extrema of Linear Functions

Let $E$ be a closed convex set in $\mathcal{R}^n$, and let $f$ be linear (not constant) on $E$. If $f$ has a global minimum on $E$, then this minimum is taken on only at relative boundary points of $E$ (Problem 3.13(a)). If $E$ is also bounded, then the global minimum of $f$ is taken on at some extreme point of $E$ (Problem 3.13(b)).

### PROBLEMS

**3.1.** Find the extreme points of the set defined by the given equations and inequalities:

(a) $x + y + z = 1, x \geq 0, y \geq 0, z \geq 0$ in $\mathcal{R}^3$.

(b) $0 \leq x \leq 1, -x - 1 \leq y \leq x + 1$ in $\mathcal{R}^2$.

(c) $x + y + z = 0, -1 \leq x \leq 1, -1 \leq y \leq 1, -1 \leq z \leq 1$ in $\mathcal{R}^3$.

**3.2.** For each set find the dimension, boundary, interior, relative boundary, and relative interior:

(a) As in Problem 3.1(a).

(b) As in Problem 3.1(b).

(c) As in Problem 3.1(c)

(d) $2x^2 + y^2 + 3z^2 \leq 1, z \geq 0$ in $\mathcal{R}^3$.

(e) $x_1^2 + x_2^2 + x_3^2 \leq 1, x_1 + x_2 + x_3 = 0$ in $\mathcal{R}^3$.

**3.3.** A convex polyhedron can be defined as a nonempty bounded convex set $E$ defined by a finite number of linear equations and a finite number of linear inequalities: $\mathbf{a}_i'\mathbf{x} \le b_i$, $i = 1, \ldots, N$, in $\mathcal{R}^n$.

    **(a)** Show that a convex polyhedron has a finite number of extreme points. [Hint: The finite set of linear equations determines a linear variety containing $E$. Without loss of generality, one can assume that this linear variety is given by equations: $x_p = 0, \ldots, x_n = 0$ and hence reduce the question to one in which the linear equations are absent. Let $\mathbf{x}_0$ be an extreme point of $E$. Then a certain number of the constraints must be active at $\mathbf{x}_0$. If these constraints are active along a line segment having $\mathbf{x}_0$ as a relative interior point, then $\mathbf{x}_0$ could not be an extreme point; why not? Hence $\mathbf{x}_0$ must be the unique solution of the corresponding set of linear equations. This conclusion shows that there can be at most a finite number of extreme points. Show that there must be at least one extreme point by showing that a point of $E$ at maximum distance from the origin is such a point.]

    **(b)** Verify that the proof of part **(a)** applies even when $E$ is allowed to be unbounded, except that there may be no extreme point.

**3.4.** Let $E$ be a convex set in the plane.

    **(a)** Show that if $E$ is 2-dimensional, then the interior of $E$ is an open region and is a convex set. [Hint: The interior of $E$ is an open set, possibly empty. Here one must show that it is nonempty and that each two points of the interior can be joined by a broken line lying in the interior. The set $E$ cannot lie on a line and hence contains 3 points not on a line. By convexity, $E$ contains the triangle determined by these three points (as a convex polyhedron) and hence contains a $\delta$-neighborhood of one point of $E$. Use this result and convexity to show that the interior of $E$ is a convex open region.]

    **(b)** Let $E$ be 1-dimensional. Show that the interior of $E$ is empty but the relative interior of $E$ is a nonempty convex set.

**3.5.** Let $E$ be a convex set in $\mathcal{R}^n$. Generalize the results of Problem 3.4 by showing the following (see Problem 4.15):

    **(a)** If $E$ is $n$-dimensional, then the interior of $E$ is a convex open region.

    **(b)** If $E$ is of dimension $m$, $1 \le m < n$, then the interior of $E$ is empty and the relative interior of $E$ is a nonempty convex set.

**3.6.** Let $F \subset E \subset \mathcal{R}^n$ where $E$ and $F$ are convex. Show that if function $f$ is (strictly) convex on $E$, then it is (strictly) convex on $F$.

**3.7.** Let $E_1$ be a nonempty convex set in $\mathcal{R}^m$, and $E_2$ a nonempty convex set in $\mathcal{R}^n$. Let $E$ be the set of all points in $\mathcal{R}^{m+n}$ whose first $m$ coordinates represent a point in $E_1$ and whose last $n$ coordinates represent a point of

$E_2$. Show that $E$ is a convex set in $\mathcal{R}^{m+n}$. [Hint: Represent the points of $\mathcal{R}^m$ by coordinates $(x_1, \ldots, x_m)$ and those of $\mathcal{R}^n$ by coordinates $(y_1, \ldots, y_n)$ so that the points of $E$ are represented by $(\mathbf{x}, \mathbf{y})$. Verify that the equations of a line segment joining points $\mathbf{x}_1$, $\mathbf{x}_2$ in $E_1$ and those of a line segment joining points $\mathbf{y}_1$, $\mathbf{y}_2$ in $E_2$ together are the equations of a line segment joining the points $(\mathbf{x}_1, \mathbf{y}_1)$, $(\mathbf{x}_2, \mathbf{y}_2)$ in $E$.]

**3.8.** Let the function $f$ be convex in $\mathcal{R}^n$, and let $f$ take on values less than $c$. Let $E$ be the sublevel set $f \leq c$ so $E$ is convex. Let $A$ be an $m \times n$ matrix of rank $m$.

    **(a)** Show that the interior of $E$ is the set on which $f < c$ and that the boundary of $E$ is the level set $f = c$.

    **(b)** Let the equation $A\mathbf{x} = \mathbf{b}$ be satisfied at an interior point of $E$. Show that the set $E_1$ defined by this equation and the inequality $f(\mathbf{x}) \leq c$ has dimension $n - m$.

**3.9.** Let $\beta$ and $\gamma$ be positive constants, $E$ a convex set in $\mathcal{R}^n$, and $f(\mathbf{x})$, $g(\mathbf{x})$ convex functions on $E$.

    **(a)** Show that $\beta f + \gamma g$ is convex on $E$.

    **(b)** Show that if $f$ is strictly convex, then $\beta f + \gamma g$ is strictly convex on $E$.

**3.10.** Let $B$ be an $n \times n$ nonsingular matrix, and let $f$ be convex (strictly convex) in $\mathcal{R}^n$. Show that $f(B\mathbf{y})$ is convex (strictly convex) in $\mathcal{R}^n$.

**3.11.** Let $C$ be an $n \times m$ matrix with rank $m$. Show that convexity (strict convexity) of a function is preserved under the (affine) transformation $\mathbf{x} = \mathbf{a} + C\mathbf{y}$; that is, if $f(\mathbf{x})$ is convex (strictly convex) in $\mathcal{R}^n$, then $f(\mathbf{a} + C\mathbf{y})$ is convex (strictly convex) in $\mathcal{R}^m$. Discuss what happens if $C$ has rank less than $m$.

**\*3.12.** Let $f(\mathbf{x})$ be defined as in (3.11), with positive definite matrix $A$, and let $g(\mathbf{x}) = \sqrt{f(\mathbf{x})}$.

    **(a)** Show that if the global minimum of $f$ is positive, then along each line $\mathbf{x} = \mathbf{x}_0 + s\mathbf{u}$, with distance $s$ as parameter and $\mathbf{u}$ a unit vector, $g$ becomes a function $z(s)$ of the form stated, with $p > 0$.

    **(b)** Show further, under the assumptions of **(a)**, that $z''(s) > 0$.

    **(c)** Show that if the global minimum of $f$ is 0, then one can only assert that $z''(s) > 0$ along each line not passing through the unique minimizer of $f$; along a line passing through the minimizer, show that $z(s)$ is continuous and piecewise linear, with one corner, and that $g$ is convex along such a line. Hence $g$ is convex.

**3.13.** Let $E$ be a closed convex set in $\mathcal{R}^n$, and let function $f$ be linear ($f = \mathbf{c}^t \mathbf{x}$ for some constant vector $\mathbf{c}$) and not constant on $E$, with global minimum at $\mathbf{x}_0$ on $E$.

**(a)** Show that $x_0$ is a relative boundary point of $E$.

**(b)** Show that, if $E$ is bounded, then $f$ takes its global minimum at some extreme point of $E$.

**3.14.** The rules for norms are given in (1.149). A semi-norm is required to satisfy all these rules except the rule: norm $= 0$ only for the zero vector. Show that every norm or semi-norm of vectors in $\mathcal{R}^n$ is a convex function.

**3.15.** Let $A$ be an $n \times n$ matrix. Show that if $A$ is positive definite, then $\sqrt{x^t A x}$ is a norm, but if $A$ is only positive semidefinite, then it is a semi-norm (see Problem 3.14; see also the end of Section 1.15). [Hint: As in Section 1.8 one can assume coordinates chosen in $\mathcal{R}^n$ so that $x^t A x = \lambda_1 x_1^2 + \cdots + \lambda_k x_k^2$, for some $k$, $0 \leq k < n$, where the $\lambda_i$ are positive.]

**3.16.** A *convex combination* of $k$ points $x_1, \ldots, x_k$ in $\mathcal{R}^n$ is a linear combination $c_1 x_1 + \cdots + c_k x_k$ for which each coefficient is nonnegative and the sum of the coefficients is 1. The *convex hull* of the set of $k$ points is the set of all convex combinations of the points. More generally, the convex hull of a set $F$ is the set of all convex combinations formed from all finite subsets of $F$. Find the convex hull for each of the following choices of $F$:

**(a)** Two points in space.

**(b)** Three points in space not on a line.

**(c)** Four points in space not on a plane.

**(d)** The union of a point and a line segment, not containing the point, in space.

**(e)** The union of the point $(0, 0, 1)$ and the set $z = 0$, $x^2 + y^2 \leq 1$ in $xyz$-space.

**\*3.17.** Prove that the convex hull (see Problem 3.16) of a set $F$ is a convex set.

## 3.2   MATHEMATICAL PROGRAMMING, DUALITY

We recall the concepts of mathematical programming (Sections 2.6–2.8). One studies extremum problems for a function $f$ defined on a set $E$ in $\mathcal{R}^n$ which is defined by equations and inequalities:

$$g_i(x) = 0, \quad i = 1, \ldots, N, \tag{3.20}$$

$$h_j(x) \leq 0, \quad j = 1, \ldots, M. \tag{3.21}$$

These side conditions are referred to as constraints; each one is said to be active at $\mathbf{x}$ if it is valid at $\mathbf{x}$ with the equals sign. The set $E$ is called the feasible solution set or $S_f$, the function $f$ is the objective function.

Two such extremum problems are said to be *dual* problems if they can be expressed in the following form:

(i)   Minimize $\psi(\mathbf{x})$ on $E \subset \mathcal{R}^n$,

(ii)  Maximize $\phi(\mathbf{y})$ on $F \subset \mathcal{R}^m$, where both extrema are attained, and

$$\min_{\mathbf{x} \in E} \psi(\mathbf{x}) = \max_{\mathbf{y} \in F} \phi(\mathbf{y}). \tag{3.22}$$

One often starts with one of the two problems and calls it the *primal problem* and then seeks a suitable second problem, so that together they form a dual pair of form (i), (ii); the second problem is then called the *dual problem*. The assertion that they satisfy the condition (3.22) is called the *duality theorem* for the particular case.

***Example 1.*** $\psi(x_1, x_2) = x_1^2 + x_2^2 - 2x_1 + 4x_2$, $E : x_1 + x_2 = 1$, $\phi(y) = -(y^2/2)$ $-2y - 5$, $F = \mathcal{R}^1$. We easily find (Problem 3.18) that $\psi$ attains its global minimum of $-3$ and $\phi$ attains its global maximum of $-3$. The example appears artificial, but it is obtained by a general method described in Section 3.4 below; see Example 1 in Section 3.4.

In this chapter we give procedures for finding duals for several classes of optimization problems. One can always create a dual of (i) trivially by taking $F = E$ and $\phi(\mathbf{y}) = 2\mu - \psi(\mathbf{y})$, where $\mu$ is the global minimum of $\psi$ (Problem 3.22). The challenge is to find a dual that is essentially different from the primal problem. In the last chapter the powerful methods of Fenchel and Rockafellar are presented; they provide duals for a great many problems of importance in mathematical programming.

When a dual pair is known, two kinds of benefits may follow. First, one has two different problems with the same optimum value and one may be easier to solve than the other. The minimizer $\mathbf{x}^*$ for (i) may not be related to the maximizer $\mathbf{y}^*$ for (ii) (if these are unique), but if they are related, knowledge of one may help in determining the other. Secondly, one can seek approximate solutions $\mathbf{x}'$, $\mathbf{y}'$ of the two problems. Because of (3.22), necessarily $\psi(\mathbf{x}') > \phi(\mathbf{y}')$ (for the case of equality, see below), and the true common value (3.22) must lie between these two numbers. Thus one has precise *upper and lower estimates* for the extreme values. Such estimates are rare in numerical analysis. Thus finding a dual is often of considerable value.

We remark that if a dual pair, as defined above, is known, and

$$\psi(\mathbf{x}') = \phi(\mathbf{y}') \tag{3.24}$$

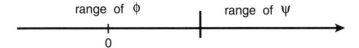

**Figure 3.1.** Ranges for dual programs.

for some $\mathbf{x}'$ in $E$ and some $\mathbf{y}'$ in $F$, then $\mathbf{x}'$ must be a minimizer $\mathbf{x}^*$ of $\psi$ and $\mathbf{y}'$ must be a maximizer $\mathbf{y}^*$ of $\phi$. Also, if we can show that

$$\psi(\mathbf{x}) \geq \phi(\mathbf{y}) \tag{3.25}$$

for all $\mathbf{x}$ in $E$ and all $\mathbf{y}$ in $F$ and that (3.24) holds for one $\mathbf{x}'$ in $E$ and one $\mathbf{y}'$ in $F$, then we have a dual pair (i), (ii). Typically the values of $\psi$ on $E$ fill an interval of the real axis, as do those of $\phi$ on $F$; (3.25) then asserts that the $\psi$-interval lies to the right of the $\phi$-interval, and (3.24) makes the intervals have a common endpoint (Fig. 3.1).

For many of the optimization problems considered in this chapter, duality will be established with the aid of the *Karush–Kuhn–Tucker conditions* of Sections 2.6–2.8. For convenience we restate these here, including the "constraint qualification" of Section 2.8.

Let the functions $f$, $g_i$, and $h_j$ all have continuous first partial derivatives in a neighborhood of a point $\mathbf{x}_0$ at which $f$ has a local minimum. Let the linear independence assumption hold at $\mathbf{x}_0$: The gradients of the $g_i$ and of the $h_j$ that are active at the point are linearly independent. Then scalars $\mu_i$ and nonnegative scalars $\lambda_j$ exist such that at $\mathbf{x}_0$

$$\nabla f + \sum_{i=1}^{N} \mu_i \nabla g_i + \sum_{j=1}^{M} \lambda_j \nabla h_j = \mathbf{0}, \tag{3.26}$$

where each $\lambda_j$ for which $h_j(\mathbf{x}_0) < 0$ (the constraint is not active at the point) is 0. A reversal of the inequality signs for the $h_j$ or a replacement of minimum by maximum changes the signs of the $\lambda_j$.

In the statement of the conditions the linear independence assumption can be replaced by the following: For each vector $\mathbf{v}_0$ of $\mathcal{R}^n$ such that at $\mathbf{x}_0$, $\mathbf{v}_0^t \nabla g_i = 0$ for all $i$ and $\mathbf{v}_0^t \nabla h_j \leq 0$ for those $h_j$ that are active at $\mathbf{x}_0$, there is a smooth curve $\mathbf{x} = \mathbf{x}(t)$ with $\mathbf{x}(t_0) = \mathbf{x}_0$, satisfying all constraints (3.20), (3.21), and the condition $\mathbf{x}'(t_0) = \mathbf{v}_0$.

## 3.3  UNCONSTRAINED QUADRATIC OPTIMIZATION

Linear algebraic and differential equations are the models of many physical systems or their first approximations. These linear models enjoy a quite complete theoretical foundation and an array of powerful methods for analyzing them. In general, they can be shown to arise as solutions of optimization problems for *quadratic* functions or functionals. In mechanics the principles of minimum potential energy, of least action, and of maximum dissipation provide the source for many such functions or functionals. If the problem is linear, the quantity to be optimized is quadratic.

The passage from quadratic optimization problems to linear equations is also inherent in mathematics. For example, one learns from the calculus that the process of minimizing or maximizing a real quadratic function $\psi(x) = ax^2 + bx + c \, (a \neq 0)$ leads to a linear equation. This one variable function always has a unique minimum (or maximum) and a unique minimizer (or maximizer). In a finite-dimensional vector space, the analogous problem is *unconstrained quadratic optimization*; it again leads to a linear equation for the minimizing (or maximizing) vector. The analysis is a little more involved, but it is still the simplest class of optimization problems and we will begin there. The topic is already considered in Sections 1.12–1.15. The results given here amplify these sections somewhat.

## Quadratic Functions in $\mathcal{R}^n$

We write the function as

$$\psi(\mathbf{x}) = \tfrac{1}{2}\mathbf{x}^t A\mathbf{x} - \mathbf{b}^t\mathbf{x} + c, \tag{3.30}$$

where $\mathbf{x} \in \mathcal{R}^n$ is the variable, $A$ is $n \times n$, $\mathbf{b} \in \mathcal{R}^n$ and $c \in \mathcal{R}$ are constant data, and the matrix $A$ is symmetric. The factor $\tfrac{1}{2}$ and the signs are chosen for convenience. We further assume that $A$ is *positive definite* and thus as in Section 1.8 all its eigenvalues are positive and $A$ is nonsingular. As in Section 1.12 the function $\psi$ is strictly convex in $\mathcal{R}^n$, which is a closed convex set. We verify below that $\psi$ has a local minimum at a point $\mathbf{x}^*$. Since $\psi$ is strictly convex, it has no other local minimizer and attains its global minimum at $\mathbf{x}^*$. To find $\mathbf{x}^*$, as in Section 1.12 we seek the critical points of $\psi$, which are the solutions of the equation

$$\nabla\psi \equiv A\mathbf{x} - \mathbf{b} = \mathbf{0}. \tag{3.31}$$

Since $A$ is nonsingular, (3.31) has a unique solution $\mathbf{x}^* = A^{-1}\mathbf{b}$. This solution is the minimizer sought, for $\psi$ has the positive definite Hessian matrix $A$ at $\mathbf{x}^*$.

As a check, we can also reason as follows: For each nonzero vector $\mathbf{y}$ in $\mathcal{R}^n$, a calculation (Problem 3.28) shows that

$$\psi(\mathbf{x}^* + \mathbf{y}) = \psi(\mathbf{x}^*) + \tfrac{1}{2}\mathbf{y}^t A\mathbf{y} > \psi(\mathbf{x}^*). \tag{3.32}$$

Thus $\psi$ attains its unique minimum at $\mathbf{x}^*$.

Similarly one shows (Problem 3.23(a)) that $\psi$ is concave and has a unique maximizer if $A$ is negative definite. If $A$ is indefinite and nonsingular (i.e., $A$ has both positive and negative eigenvalues), then $\psi$ still has a unique critical point, but it is a saddle point (Section 1.9) providing neither local minimum nor local maximum. If 0 is an eigenvalue and hence $A$ is singular, the critical point, if it exists, is not unique. If 0 is an eigenvalue and if all nonzero eigenvalues are positive, then one verifies that $A$ is *positive semidefinite*: That is, $\mathbf{x}^t A \mathbf{x} \geq 0$ for all $\mathbf{x}$, and thus $\psi$ is convex but not strictly convex. One concludes that every critical point (if there is one) gives the same global minimum of $\psi$. As in Sections 1.3 and 1.9 every critical point of $\psi$ is a minimizer; as in Section 1.15 a minimizer exists precisely when $\psi$ is bounded below. (See Problem 3.23(b).)

The various cases of the preceding paragraphs are illustrated for $n = 2$ by the functions $x^2 + y^2$, $x^2 - y^2$, $x^2 + y$, $x^2 + 2xy + y^2$. We give further illustrations:

***Example 1.*** $\psi = 5x^2 + 2xy + y^2 - 6x + 2y + 7$. Here $A = \begin{bmatrix} 10 & 2 \\ 2 & 2 \end{bmatrix}$ and $\mathbf{b} =$ $(6, -2)^t$. One verifies that $A$ has positive eigenvalues and is hence positive definite. One finds that the unique minimum is attained at $x = 1$, $y = -2$, where $\psi = 2$. (See Problem 3.24(i).)

***Example 2.*** $\psi = x^2 + 4xy + 4y^2 + x - 2y$. One verifies that $A$ is positive semidefinite, but $\psi$ has no critical point (Problem 3.24(g)).

REMARK. When $A$ is positive definite, the function (3.30) can be written as $(\frac{1}{2})\|\mathbf{x} - \mathbf{x}^*\|_A^2 +$ constant in terms of the $A$-norm, as in Section 1.15 (Problem 3.25). Thus minimizing $\psi$ is equivalent to finding the point closest to $\mathbf{x}^*$ in terms of the $A$-norm, and that point is of course $\mathbf{x}^*$ itself; the minimum value of $\psi$ is the constant.

## Dual Problem

The primal problem of minimizing the function (3.30) in $\mathcal{R}^n$ has various duals. For example, the problem

$$\text{maximize} \quad \phi(\mathbf{y}) = -\tfrac{1}{2}\mathbf{y}^t A^{-1}\mathbf{y} + \mathbf{b}^t A^{-1}(\mathbf{y} - \mathbf{b}) + c \qquad (3.33)$$

in $\mathcal{R}^n$ is such a dual (see Problem 3.26).

## PROBLEMS

**3.18.** (Example 1 in Section 3.2.) Given $\psi(x_1, x_2) = x_1^2 + x_2^2 - 2x_1 + 4x_2$, verify that $\psi_{\min} = -3$ is attained at $(x_1, x_2) = (2, -1)$ under the constraint $x_1 + x_2 = 1$. Find the maximum of $\phi(y) = -\frac{1}{2}y^2 - 2y - 5$ for $y$ unrestricted. Show the ranges of $\psi$ and $\phi$ as subsets of the real line and note the minimum of $\psi$ and maximum of $\phi$.

**3.19.** Given $\psi(x) = 2x^2 - 8x + 5$ and $E = \{x \mid 0 \le x \le 4\}$, find $\min_{x \in E} \psi$. Let $\phi = -6 - (2y^2 - 8y + 5)$, and find its maximum for $0 \le y \le 4$. Show that these are dual problems.

**3.20.** Show that the following are dual problems:

(a) Find the minimum of the function $\psi(x_1, x_2, x_3)$, equal to the distance squared from the origin in $\mathcal{R}^3$ to the point $\mathbf{x}$ on the plane $ax_1 + bx_2 + cx_3 + d = 0$; find the maximum of $\phi(y) = -(\frac{1}{4})(a^2 + b^2 + c^2)y^2 + dy$. [This type of problem is treated in Section 3.4.]

*(b) (A generalization of part (a)) Show that the following is a pair of dual problems: In $\mathcal{R}^n$ find the minimum of $\psi(\mathbf{x}) = \|\mathbf{x}\|_2^2$ on the linear variety $B\mathbf{x} - \mathbf{g} = 0$, where the $m \times n$ matrix $B$ has rank $m < n$; find the maximum in $\mathcal{R}^m$ of $\phi(\mathbf{y}) = -(\frac{1}{4})\|B^t\mathbf{y}\|_2^2 - \mathbf{g}^t\mathbf{y}$.

**3.21.** (Duality in linear programming—see Sections 3.6 and 3.7) Verify that the following are dual problems: Find the minimum of $\psi(x, y) = 3x - 2y$ for $x \ge 0$, $y \ge 0$, $x + 2y \le 4$, $5x + y \le 3$; find the maximum of $\phi(z, w) = -4z - 3w$ for $z \ge 0$, $w \ge 0$, $z + 5w \ge -3$, $2z + w \ge 2$. [Hint: Graph the sets $E$ and $F$ and consider the level curves (straight lines) of $\psi$ and $\phi$ on these sets.]

**3.22.** Let $F = E \subset \mathcal{R}^n$. Show that if $\psi$ has the global minimum $\mu$ on $E$, then the problems where one would minimize $\psi$ on $E$ and maximize $\phi(\mathbf{y}) = 2\mu - \psi(\mathbf{y})$ on $F$ are dual problems.

**3.23.** (a) Let the matrix $A$ in (3.30) be negative definite. Show that the function has a unique maximizer.

(b) Let $A$ be positive semidefinite. Show that every critical point of $\psi$ (if there is one) gives the same global minimum of $\psi$.

**3.24.** Locate the critical points, if there are any, of each of the following quadratic functions of $x$ and $y$, and determine whether each provides a global maximum or minimum or is a saddle point:

(a) $x^2 + y^2$.

(b) $x^2 - y^2$.

(c) $x^2 + y$.

(d) $x^2 + 2xy + y^2$.

(e) $x^2 + 2xy + 2y^2$.

(f) $x^2 + 2xy - y^2$.

(g) $x^2 + 4xy + 4y^2 + x - 2y$.

**(h)** $2x^2 + y^2 + 4x - 6y.$

**(i)** $5x^2 + 2xy + y^2 - 6x + 2y + 7.$

**(j)** $x^2 - 2xy + y^2 - 2x + 2y.$

**3.25.** With the aid of the $A$-norm of Section 1.14, show that the function $\psi$ of (3.30), with positive definite $A$, can be written as $(\frac{1}{2})\|\mathbf{x} - \mathbf{x}^*\|_A^2 + \text{const}.$

**3.26.** Show that (3.33) is a dual for the problem of minimizing (3.30) with positive definite $A$.

**3.27.** Show that, for each positive integer $m$, a dual to the problem of minimizing (3.30) with positive definite $A$ is given by the problem:

$$\text{maximize} \quad \phi(\mathbf{y}) = -\tfrac{1}{2}\mathbf{y}^t\mathbf{y} + c - \tfrac{1}{2}\mathbf{b}^t A^{-1}\mathbf{b} \quad \text{in} \quad \mathcal{R}^m.$$

**3.28.** From (3.30) with $A$ positive definite, deduce (3.32) for each nonzero $\mathbf{y}$.

## 3.4   CONSTRAINED QUADRATIC OPTIMIZATION IN $\mathcal{R}^n$

### Quadratic Programs

A quadratic program (QP) is a problem of seeking the minimum or maximum of a quadratic function of $n$ variables subject to a finite number of constraints (side conditions) in the form of linear equations and linear inequalities. Thus a typical QP (for a minimum) can be written as follows:

$$\begin{aligned} \textit{minimize} \quad & \psi(\mathbf{x}) = \tfrac{1}{2}\mathbf{x}^t A\mathbf{x} - \mathbf{r}^t\mathbf{x} + c \\ \textit{subject to} \quad & \mathbf{b}_i^t\mathbf{x} = g_i, \qquad i = 1, \ldots, N, \\ \textit{and to} \quad & \mathbf{b}_i^t\mathbf{x} \geq g_i, \qquad i = N+1, \ldots, m. \end{aligned} \qquad (3.40)$$

Here $\psi$ is a quadratic function in $\mathcal{R}^n$, $A$ is a symmetric matrix of order $n$, which we assume to be positive definite, $\mathbf{r}$ is a given vector of $\mathcal{R}^n$, and $c$ is a scalar. We allow $N$ to be 0 so that there are no equality constraints, but we usually assume that there is at least one inequality constraint. We further assume that $N < n$ and that the $\mathbf{b}_i$ in the equality constraints are linearly independent, that the set $S_f$ of feasible points is nonempty, and that at each point $\mathbf{x}$ of $S_f$ the vectors $\mathbf{b}_i$ of the constraints active at $\mathbf{x}$ (i.e., for which $\mathbf{b}_i^t\mathbf{x} = g_i$) are linearly independent.

Optimization problems of this type arise naturally in geometry. For example, finding the shortest distance from a point $(x_0, y_0, z_0)$ to a half-space defined by $b_1x + b_2y + b_3z - g \geq 0$ is such a problem with the objective function $\psi(x, y, z) = (x - x_0)^2 + (y - y_0)^2 + (z - z_0)^2$ (the square of the distance). Expressing the problem in the form of (3.40), we have $A = 2I$, $\mathbf{r}^t = (2x_0, 2y_0, 2z_0)$, $c = x_0^2 + y_0^2 + z_0^2$, $N = 0$, $m = 1$, $\mathbf{b}_1 = (b_1, b_2, b_3)^t$, $g_1 = g$, and $c = x_0^2 + y_0^2 + z_0^2$. If the given point is contained in the half-space, it is the minimizer for a zero distance. Otherwise, the minimizer lies on the boundary of the halfspace.

REMARK. We observe that every QP for a minimum is a minimization problem for a strictly convex function on a closed convex set $E$ (see Section 3.1). Hence a local minimum provides a global minimum, and there is at most one minimizer. As in Section 1.15 the problem is equivalent to that of finding the point of $E$ closest to a given point of $\mathcal{R}^n$ in terms of the distance associated with the $A$-norm. Therefore there is a unique minimizer.

Problems in linear theories of elasticity, fluid dynamics, heat transfer, and electromagnetism, with respective inequality conditions typically arising from frictional, thermodynamic, and saturation phenomena, can be approximated by (3.40) with various interpretations for the constant matrices and vectors. The dimensions of these problems are large and constraints numerous. An efficient algorithm is essential. (For example, see O'Leary and Yang, 1978.)

## Quadratic Program with Equality Constraints

If all constraints are equalities, we have the problem

$$\text{minimize} \quad \psi(\mathbf{x}) \quad \text{subject to} \quad B\mathbf{x} = \mathbf{g}, \qquad (3.40')$$

where $\psi$ is as in (3.40), $B$ is $N \times n$, with rows $\mathbf{b}_i^t$ for $i = 1, \ldots, N$, and $\mathbf{g} = (g_1, \ldots, g_N)^t$. In accordance with the linear independence assumption, we assume further that $B$ has maximum rank $N$ and that $N < n$. (We can even allow $N = n$; the formulas to follow are valid in that case also—see Section 2.6.)

This problem is treated in Section 1.15 in terms of the $A$-norm. Here we solve the problem formally by the Lagrange multiplier method and then give an interpretation to the process. We first construct the Lagrangian function

$$L(\mathbf{x}, \mathbf{y}) = \tfrac{1}{2}\mathbf{x}^t A\mathbf{x} - \mathbf{r}^t\mathbf{x} + c + \mathbf{y}^t(B\mathbf{x} - \mathbf{g}),$$

using a vector Lagrangian multiplier $\mathbf{y} \in \mathcal{R}^N$; the $y_i$ here are the same as the $\lambda_i$ of Section 2.2. We equate to zero the partial derivatives of $L(\mathbf{x}, \mathbf{y})$ with respect to the components of $\mathbf{x}$ and $\mathbf{y}$ and obtain the equations:

$$\begin{aligned} \nabla_{\mathbf{x}}L &= A\mathbf{x} - \mathbf{r} + B^t\mathbf{y} = \mathbf{0}, \\ \nabla_{\mathbf{y}}L &= B\mathbf{x} - \mathbf{g} = \mathbf{0}. \end{aligned} \qquad (3.41)$$

Since $A$ is symmetric and positive definite, so is $A^{-1}$ (Problem 3.38(a)). Since $B$ has maximum rank, $BA^{-1}B^t$ is symmetric and positive definite (see Problem

3.38(b), and also Problem 1.82). We hence obtain the unique solution of equations (3.41):

$$\mathbf{y}^* = (BA^{-1}B^t)^{-1}(BA^{-1}\mathbf{r} - \mathbf{g}),$$
$$\mathbf{x}^* = A^{-1}(\mathbf{r} - B^t\mathbf{y}^*).$$
(3.42)

Here $\mathbf{x}^*$ is the unique minimizer of the problem (3.40') and the meaning of $\mathbf{y}^*$ will now be explored.

Motivated by the second equation of (3.42), we define a new function,

$$\phi(\mathbf{y}) = L\big(A^{-1}(\mathbf{r} - B^t\mathbf{y}), \ \mathbf{y}\big)$$

obtained by replacing $\mathbf{x}$ by $A^{-1}(\mathbf{r} - B^t\mathbf{y})$ in $L(\mathbf{x}, \mathbf{y})$. After calculation, we find (Problem 3.32)

$$\phi(\mathbf{y}) = -\tfrac{1}{2}(\mathbf{r} - B^t\mathbf{y})^t A^{-1}(\mathbf{r} - B^t\mathbf{y}) - \mathbf{g}^t\mathbf{y} + c. \tag{3.43}$$

If $\mathbf{y} = \mathbf{y}^*$, then $\mathbf{x} = \mathbf{x}^*$, and hence

$$\phi(\mathbf{y}^*) = L(\mathbf{x}^*, \ \mathbf{y}^*) = \psi(\mathbf{x}^*). \tag{3.44}$$

We claim that $\phi(\mathbf{y}^*)$ is in fact the global maximum of $\phi(\mathbf{y})$ for $\mathbf{y} \in \mathcal{R}^N$. Thus

$$\min_{B\mathbf{x}=\mathbf{g}} \psi(\mathbf{x}) = \max_{\mathbf{y} \in \mathcal{R}^N} \phi(\mathbf{y}). \tag{3.45}$$

Thus we have obtained a *duality theorem*, as in Section 3.2, for the problem (3.40') with equality constraints. The Lagrangian multiplier vector $\mathbf{y}$ plays the role of the dual variable here. Like the primal variable $\mathbf{x}$, the dual variable $\mathbf{y}$ also carries physical or geometrical meaning in the various fields that (3.43) models. As in Section 3.1 the roles of primal and dual can be interchanged. In physical applications the problem formulated from basic principles is usually called primal.

To justify (3.45), we expand (3.43) into a standard form of quadratic function

$$\phi(\mathbf{y}) = -\tfrac{1}{2}\mathbf{y}^t(BA^{-1}B^t)\mathbf{y} + \mathbf{d}^t\mathbf{y} + k$$

with a constant vector $\mathbf{d} = BA^{-1}\mathbf{r} - \mathbf{g}$ and a constant scalar $k = c - \tfrac{1}{2}\mathbf{r}^t A^{-1}\mathbf{r}$. Since $-BA^{-1}B^t$ is negative definite, we conclude from Section 3.3 that $\phi(\mathbf{y})$ has a unique maximizer defined by the equation $-(BA^{-1}B^t)\mathbf{y} + \mathbf{d} = 0$ which yields the same solution

$$\mathbf{y}^* = (BA^{-1}B^t)^{-1}(BA^{-1}\mathbf{r} - \mathbf{g})$$

as that in (3.42); it is the maximizer of the dual problem.

From the second equation (3.42) we see that $\mathbf{x}^*$ can be found from $\mathbf{y}^*$. From the first equation (3.41) we can obtain $\mathbf{y}^*$ from $\mathbf{x}^*$.

**Example 1.** Minimize $\psi(x_1, x_2) = x_1^2 + x_2^2 - 2x_1 + 4x_2$, where $x_1 + x_2 = 1$. This is the same minimum problem as that in Example 1 in Section 3.2. It is left as an exercise (Problem 3.29) to apply the method of this section to find $\mathbf{x}^*$, the dual problem and $\mathbf{y}^*$.

We now reformulate our primal problem as a maximization problem:

$$maximize \quad \phi(\mathbf{y}) = -\tfrac{1}{2}\mathbf{y}^t A\mathbf{y} + \mathbf{r}^t\mathbf{y} + c$$
$$subject\ to \quad B\mathbf{y} = \mathbf{g}, \tag{3.46}$$

where $A$, $\mathbf{r}$, $c$, $B$, $\mathbf{g}$ are as above. Then its dual becomes

$$minimize \quad \psi(\mathbf{x}) = \tfrac{1}{2}\mathbf{x}^t A'\mathbf{x} - \mathbf{d}^t\mathbf{x} + h, \tag{3.47}$$

where $A' = BA^{-1}B^t$, $\mathbf{d} = \mathbf{g} - BA^{-1}\mathbf{r}$ and $h = -\tfrac{1}{2}\mathbf{r}^t A\mathbf{r} + c$. This pair has an application in mechanics:

**Example 2.** A linear spring with elastic constant $k$ is fixed at one end and is stretched by a force $P$ at the other. The loaded end displaces an amount $x$. For this simple problem, the equilibrium equation is $y = P$, where $y$ is the internal force in the spring. In mechanics language, quantities representing forces and moments are called static; quantities describing motions are called kinematic. The static and kinematic solutions to this simple problem are, respectively,

$$y = P \quad and \quad x = \frac{P}{k}.$$

These results can be formalized as the solutions to a pair of dual optimization problems in the form of (3.46) and (3.47). Since the static and kinematic variables are each a scalar variable in this problem, the corresponding matrices and vectors in (3.46) and (3.47) become $1 \times 1$. They are $A = [1/k]$, $B = [1]$, $\mathbf{r} = (0)$, $c = 0$, $\mathbf{g} = (P)$, and $\mathbf{d} = (P)$. We minimize the potential energy

$$\psi(x) = \tfrac{1}{2}x[k]x - Px$$

and maximize the "complementary energy" $\phi(y)$ subject to equilibrium constraint,

$$\phi(y) = -\frac{1}{2}y\begin{bmatrix}1\\k\end{bmatrix}y, \quad y - P = 0.$$

It is easy to verify the duality relation, $\psi_{min} = -P^2/(2k) = \phi_{max}$.

### Generalization to a Case of Nonpositive Definite Matrix $A$

Mathematicians often relax or weaken the premise of a theorem to produce a stronger assertion. If $A$ in (3.40') is positive definite only in the null space of $B$, an $(n - N)$-dimensional subspace of $\mathcal{R}^n$ (see Section 1.6), then the problem still has a unique solution. Indeed, $\psi$ is strictly convex on the closed convex set (linear variety) defined by the equation $Bx = g$ and has a local minimum on this set (Problem 3.34). We can find the solution by Lagrange multipliers. Again we are led to (3.41). This is equivalent to a system of $n + N$ linear equations in $n + N$ unknowns. We can verify that the coefficient matrix of this system:

$$\begin{bmatrix} A & B^t \\ B & O \end{bmatrix}$$

is nonsingular by showing that the corresponding homogeneous system:

$$Ax + B^t y = 0$$

$$Bx \qquad = 0$$

admits only the trivial solution. Solutions of the second equation define the null space of $B$. Multiplying the first equation by $x^t$, one obtains $x^t Ax + (Bx)^t y = 0$, or $x^t Ax = 0$. Since $A$ is positive definite for all $x$ in the null space of $B$, $x^t Ax = 0$ implies that $x = 0$. Substituting this trivial $x$ into the first equation, one obtains $B^t y = 0$ and concludes that $y = 0$, since $B$ has maximum rank. Thus (3.41) has a unique solution, which provides the global minimum for (3.40').

### Case of Euclidean Distance

If the objective function $\psi$ is the function $(\frac{1}{2})x^t x$, then our QP becomes the problem of minimizing the Euclidean distance from the origin of $\mathcal{R}^n$ to the convex set $E$ in $\mathcal{R}^n$ defined by the conditions (3.40). For the case of equality constraints, the formulas (3.42) derived above become

$$y^* = -(BB^t)^{-1}g, \quad x^* = -B^t y^*. \tag{3.42'}$$

A QP program (3.40) can always be reduced to one with this special $\psi$. As in Section 1.14 the Choleski process can be applied to write $A = C^t C$ for a nonsingular matrix $C$ and the substitution $u = Cx$ reduces the quadratic term to the form $(\frac{1}{2})u^t u$. Then an appropriate translation (in effect, "completing the square") eliminates the term of first degree. The new objective function differs from one of form $(\frac{1}{2})x^t x$ by an added constant, which has no effect on the minimizer. (See Problem 3.35.)

## 3.5 QP WITH INEQUALITY CONSTRAINTS, QP ALGORITHM

Now we return to the inequality constraint case of (3.40); some of the constraints may be equations. There is again a unique minimizer, as remarked above.

Locating the minimizer under inequality constraints requires more work. The problem is to identify which of the inequalities attain the status of an equation—that is, becomes active—at the optimal solution point. The original equality constraints, if present, are always active. It is possible that none of the inequalities becomes active at the solution point; in such a case the optimal solution of the unconstrained QP (or the QP with the original equality constraints if present) satisfies all constraints and is therefore the solution of the given QP. The following example illustrates the cases which can arise:

***Example 1.*** A simple QP in $\mathcal{R}^2$ has the form

$$minimize \quad \psi(\mathbf{x}) = x_1^2 + x_2^2 - 6x_1 - 4x_2$$

$$subject\ to \quad x_1 \geq 0,\ x_2 \geq 0,\ x_1 + x_2 \geq \alpha.$$

For $\alpha = 4$ one observes that the unconstrained optimal solution $\mathbf{x}^* = (3, 2)^t$ satisfies all constraints and gives $\psi_{min} = -13$. For $\alpha = 6$ the third constraint is violated by such a solution. Therefore the minimizer must be on the boundary of the region determined by the three linear inequalities (Fig. 3.2). One observes that it is on the boundary line $x_1 + x_2 = 6$ and that $\mathbf{x}^* = (3.5, 2.5)^t$, $\psi_{min} = -12.5$. Accordingly, the stated problem has the same solution as the one with the same quadratic function $\psi$ and the single active constraint, $x_1 + x_2 = 6$; such a problem can be solved by the method for equality constraints. The dual variable for the problem is a scalar and $\phi_{max} = -12.5$ at $y^* = -1$. (See Problem 3.36.)

The example can also be explained in geometric terms. The quadratic function can be rewritten as $\psi = (x_1 - 3)^2 + (x_2 - 2)^2 - 13$ which is the squared distance of $(x_1, x_2)$ from the point $(3, 2)$ minus 13. As in Section 3.2 we denote the region defined by the constraints by $S_f$, the *feasible solution set*, shown as the shaded region in Fig. 3.2. The problem of minimizing $\psi$ is equivalent to finding the point in $S_f$ closest to $(3, 2)$. Such an interpretation applies to the whole class of problems being considered, as pointed out in the Remarks in the third and last paragraphs of Section 3.4.

### Active Set Method

This is a procedure developed by Fletcher and others (see Fletcher, 1971, and 1987). We describe the ideas behind it, illustrate it by an example, and then summarize it as an algorithm.

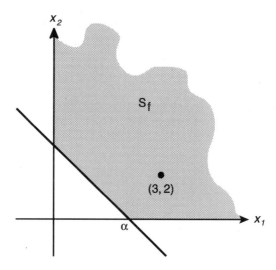

**Figure 3.2.** Example of quadratic program.

Let us assume that a feasible point **x** has been found. (We comment below on how it can be determined.) If in fact **x** is the minimizer **x\***, that is verified and the procedure stops. Otherwise, the procedure then leads to a next feasible point, which we here call **z**, such that $\psi(\mathbf{z}) < \psi(\mathbf{x})$. To obtain **z**, one first determines which constraints are active at **x**. One then seeks the minimizer of $\psi$ subject to the corresponding set of equality constraints; that is, subject to $\mathbf{b}_i'\mathbf{x} = g_i$ for $i = 1, \ldots, N$ and for those $i > N$ for which equality holds at **x**; we refer to these constraints as "the initial active set." The minimizer is found as in Section 3.4 (along with the corresponding set of Lagrange multipliers).

It could happen that **x** itself is the minimizer of $\psi$ subject to the initial active set of constraints. In that case one examines the Lagrange multipliers associated with the corresponding inequality constraints. If none of them is positive, then the Karush–Kuhn–Tucker conditions apply, and the results of Section 2.7 show that $\mathbf{x} = \mathbf{x}^*$. (We must use "positive" here rather than "negative," since the side conditions have $\geq$ rather than $\leq$, as in Section 2.7.) If one of them, say $y_1$, is positive, then, as in Remark 1 in Section 2.7, $\psi$ has a smaller minimum on the subset of $S_f$ defined by the initial active set modified by removing the constraint associated with $y_1$, that is, by allowing this constraint to become inactive. If there are several positive multipliers, the largest is chosen for this step. The minimizer of $\psi$ with the modified active set may well fall outside $S_f$. However, one chooses the vector from **x** to this minimizer as a *search direction* **s** and proceeds along that direction until one reaches the boundary of $S_f$. This produces the desired feasible point **z**, with $\psi(\mathbf{z}) < \psi(\mathbf{x})$.

If **x** is not the minimizer of $\psi$ subject to the initial active set of constraints, then one finds that minimizer as in Section 3.4. If that minimizer is a feasible point, then one chooses it as **z** and proceeds as in the preceding paragraph. If the minimizer falls outside $S_f$, one again uses it to determine a search direction

**s.** This leads to a point **z** on the boundary of $S_f$ at which at least one additional inequality constraint has become active. In either case $\psi(\mathbf{z}) < \psi(\mathbf{x})$ as desired.

Thus we have a procedure for successively decreasing the value of $\psi$. Starting with the feasible point $\mathbf{x}^{(0)}$, one obtains a sequence $\mathbf{x}^{(k)}$ of feasible points at which $\psi$ decreases steadily as $k$ increases. In Fletcher, (1971, 1987, ch. 10) it is shown that except in certain degenerate cases, the sequence terminates in a finite number of steps and produces the minimizer sought.

We note here that if the initial active set of constraints contains no constraints for $i > N$ and the initial feasible point **x** minimizes $\psi$ subject to the initial active set of constraints, then **x** is the minimizer sought for our QP. This follows at once from the Karush–Kuhn–Tucker conditions.

***Example 2.*** Minimize $\psi(x, y) = (\tfrac{1}{2})(x^2 + y^2)$ subject to the constraints

$$-3x + y \geq -11, \quad -x - 3y \geq -17, \quad x - y \geq -7, \quad 2x + 3y \geq 11.$$

By graphing, we find that the set $S_f$ is a region in the $xy$-plane bounded by a quadrilateral with vertices $(4, 1)^t$, $(5, 4)^t$, $(-1, 6)^t$, $(-2, 5)^t$. If we choose $(2, 5)^t$ as initial feasible point $\mathbf{x}^{(0)}$, then only the second constraint is active; the minimizer of $\psi$ subject to this constraint is the feasible point $\mathbf{x}^{(1)} = (1.7, 5.1)^t$. There is just one Lagrange multiplier, which we find to be 1.7, which is positive. Hence we drop the one equality constraint and minimize $\psi$ subject to no constraints. This leads to the origin, which is not feasible (fourth constraint violated). We search along the line segment from $(1.7, 5.1)^t$ to the origin, and find that we leave $S_f$ at the point $\mathbf{x}^{(2)} = (1, 3)^t$. At this point only the fourth constraint is active; the corresponding minimizer is $\mathbf{x}^{(3)} = (22/13, 33/13)^t$. The one Lagrangian multiplier is negative, so $\mathbf{x}^{(3)}$ is the minimizer sought.

If the initial feasible point chosen were the vertex $(5, 4)^t$, then the initial active set would contain two constraints, which together uniquely determine the initial point **x**, and hence it is the minimizer of $\psi$ subject to the two constraints. As pointed out in Section 2.6, the two corresponding Lagrange multipliers are obtained by simply expressing $\nabla\psi$ at **x** as a linear combination of the two (linearly independent) gradients of the functions appearing in the side conditions; as noted in Section 3.4, the formulas (3.42) remain valid and also provide the Lagrange multipliers.

For the point $(5, 4)^t$, both multipliers are found to be positive, and the larger one arises from the second constraint; hence that is the one to be dropped. The details for this choice are left to Problem 3.37(b).

## Finding a Feasible Point

The problem of finding a feasible point **x** to initiate the procedure can be reduced to a problem in linear programming (LP) and solved by the standard methods for LP; see the end of Section 3.7.

*A QP Algorithm*

1. By dividing the inequalities in (3.40) by appropriate positive scalars, achieve that for each inequality, $\mathbf{b}_i$ has Euclidean norm 1.
2. Set $k = 0$.
3. Let $\mathbf{x}^{(0)}$ be a given feasible point and $\mathcal{A}$ the set of indices $i$ for the initial set of active constraints.
4. Find the minimizer $\mathbf{v}^{(k)}$ of $\psi$ with constraints (3.40) for $i$ in $\mathcal{A}$ and the corresponding Lagrange multiplier vector $\mathbf{y}^{(k)}$. If $\mathbf{x}^{(k)} \neq \mathbf{v}^{(k)}$, set $\mathbf{s}^{(k)} = \mathbf{v}^{(k)} - \mathbf{x}^{(k)}$. Go to step 6.
5. Let $y_q^{(k)}$ be the maximum of $y_i^{(k)}$ for $i$ in $\mathcal{A}$ and $i > N$. If there are no such values of $i$ or if $y_q^{(k)} \leq 0$, then $\mathbf{x}^{(k)}$ is the minimizer sought and the procedure stops. Otherwise, remove $q$ from $\mathcal{A}$, find the minimizer $\mathbf{w}^{(k)}$ of $\psi$ subject to the new set of active constraints and set $\mathbf{s}^{(k)} = \mathbf{w}^{(k)} - \mathbf{x}^{(k)}$.
6. Set

$$\mathbf{x}^{(k+1)} = \mathbf{x}^{(k)} + \alpha \mathbf{s}^{(k)}, \quad \alpha = \min\left(1, \min_{i \in \mathcal{B}} \beta_i\right),$$

where $\beta_i = (g_i - \mathbf{b}_i^t \mathbf{x}^{(k)})/(\mathbf{b}_i^t \mathbf{s}^{(k)})$ and $\mathcal{B} = \{i \mid i > N, \mathbf{b}_i^t \mathbf{s}^{(k)} > 0, i \notin \mathcal{A}\}$. If $\alpha = \beta_q < 1$, add index $q$ to $\mathcal{A}$. Replace $k + 1$ by $k$, and go to step 4.

REMARKS. Step 1 of the algorithm has the effect of eliminating the ambiguity in the Lagrange multipliers resulting from multiplying the active constraints by arbitrary positive scalars. The algorithm requires repeated solution of a QP with only equality constraints. The formulas of Section 3.4 for such a QP can be replaced by others, taking advantage of matrix updating, which greatly improve the procedures. (See Fletcher, 1987, ch. 10; Gill and Murray, 1978; and Moré and Wright, 1993, ch. 6.)

**PROBLEMS**

**3.29.** Consider Example 1 in Section 3.4 as a case of the problem (3.40′), and use the formulas (3.42) to find the minimum; also use (3.43) to derive the dual problem.

**3.30.** Verify that the formulas (3.42) and (3.43) are applicable, and use them to find the minimizer, $\mathbf{y}^*$, and the dual problem:

(a) $A = \begin{bmatrix} 3 & 1 \\ 1 & 1 \end{bmatrix}$, $\mathbf{r} = \begin{bmatrix} 5 \\ 2 \end{bmatrix}$, $c = 2$, $B = \begin{bmatrix} 3 & 1 \end{bmatrix}$, $g = [1]$.

(b) $A = \begin{bmatrix} 3 & 0 & 0 \\ 0 & 5 & 2 \\ 0 & 2 & 3 \end{bmatrix}$, $\mathbf{r} = \begin{bmatrix} 1 \\ 1 \\ 1 \end{bmatrix}$, $c = 7$, $B = \begin{bmatrix} 2 & 1 & 1 \\ 3 & 2 & 4 \end{bmatrix}$, $g = \begin{bmatrix} 2 \\ 5 \end{bmatrix}$.

**3.31.** Carry through the derivation of (3.42) from (3.41).

**3.32.** Derive (3.43) from the definition of $\phi(\mathbf{y})$ in terms of the Lagrangian function $L$.

**3.33.** Consider the problem (3.40′) with

$$A = \begin{bmatrix} 3 & 2 \\ 2 & 1 \end{bmatrix}, \quad \mathbf{r} = \begin{bmatrix} 5 \\ 4 \end{bmatrix}, \quad c = 1, \quad B = [5 \ \ 1], \quad g = [2].$$

Show that $A$ is not positive definite in $\mathcal{R}^2$ but is positive definite on the null space of $B$. Use the method of Lagrangian multipliers to find the global minimum.

**3.34.** Consider the problem (3.40′) in the case when $A$ is known to be positive definite only on the null space of $B$.

(a) Show that $\psi$ is strictly convex on the linear variety $B\mathbf{x} = \mathbf{g}$. [Hint: A line in this linear variety has vector equation $\mathbf{x} = \mathbf{x}_0 + s\mathbf{u}$, where $\mathbf{u}$ is a unit vector in the null space of $B$.]

(b) Show that $\psi$ has a unique local minimum on the linear variety. [Hint: By the results of Section 1.6, the points of the linear variety can be represented by the equation $\mathbf{x} = \mathbf{x}_0 + W\mathbf{v}$, where $W$ is an $n \times k$ matrix of rank $k = n - N$ and $\mathbf{v}$ is an arbitrary vector in $\mathcal{R}^k$. Show that when $\psi$ is expressed in terms of $\mathbf{v}$, the problem (3.40′) becomes a problem of form (3.30), with positive definite matrix $A$.]

**3.35.** In (3.40) let $A$ be a positive definite matrix, and let $A = C^t C$ (Choleski decomposition). Let $\mathbf{u} = C\mathbf{x}$, $\mathbf{v} = \mathbf{u} - (C^{-1})^t\mathbf{r}$. Show that $\psi$ becomes $(\frac{1}{2})\mathbf{v}^t\mathbf{v} + \text{const}$ and find the constant.

**3.36.** Carry out all the details in the discussion of Example 1 in Section 3.5.

**3.37.** Apply the QP algorithm of Section 3.5 to find the minimizer for the QP of Example 2 in Section 3.5 for the following choices of initial feasible point; find the $\mathbf{x}^{(k)}$ and $\mathbf{y}^{(k)}$ by the formulas of Section 3.4.

(a) $\mathbf{x}^{(0)} = (2, 5)^t$ as in the text.

(b) $\mathbf{x}^{(0)} = (5, 4)^t$.

(c) $\mathbf{x}^{(0)} = (1, 4)^t$.

**3.38.** Let $A$ be a positive definite symmetric matrix of order $n$. Let $B$ be an $m \times n$ matrix of rank $m < n$.

(a) Show that $A^{-1}$ is also symmetric and positive definite.

(b) Show that $BA^{-1}B^t$ is also symmetric and positive definite.

(c) Show that the shortest Euclidean distance from the origin to the linear variety $B\mathbf{x} = \mathbf{g}$ in $\mathcal{R}^n$ is

$$d = [\mathbf{g}^t(BB^t)^{-1}\mathbf{g}]^{1/2}.$$

## 3.6 LINEAR PROGRAMMING

A nonconstant linear function does not have a finite maximum or minimum if unconstrained. When the constraints appear as linear equations and inequalities, the problem is called a *linear program* (LP). It may come as a surprise that a LP is more involved than a QP. The reason is that a LP is a nonsmooth problem, so the calculus criterion of zero gradient for optimality is no longer valid. The subject of linear programming often fills a full-semester senior or graduate-level course.

We consider a primal LP in the form:

$$minimize \quad \psi(\mathbf{x}) = \mathbf{c}^t \mathbf{x},$$

$$subject\ to \quad A\mathbf{x} \geq \mathbf{b}, \qquad (3.60)$$

$$\mathbf{x} \geq \mathbf{0}.$$

Here $\mathbf{x} \in \mathcal{R}^n$ is the unknown vector; $\mathbf{c} \neq \mathbf{0} \in \mathcal{R}^n$, $\mathbf{b} \in \mathcal{R}^m$, and $A$ are constant vectors and matrices, with $A$ $m \times n$. By simply replacing $\mathbf{c}$ by $-\mathbf{c}$, minimization is changed to maximization; hence we need to discuss only one of the two. The directions of the first $m$ inequalities are arranged to be $\geq$, and some of these are allowed to be equations; for simplicity, we will emphasize the case where all are inequalities and treat the case of some equations as an exception.

The next $n$ inequalities restrict all components of $\mathbf{x}$ to be nonnegative. This has a historical origin, for LP was first developed for managerial problems for which the variables are nonnegative. For scientific and engineering problems this convention presents an inconvenience which we shall replace in Section 3.8 by the requirement that the variables are *bounded*. The case where some of the $x_i$ are nonnegative and others are unrestricted is reduced to the case where all are nonnegative by the replacement of each unrestricted $x_i$ by $x_i' - x_i''$, where the new variables $x_i'$, $x_i''$ are nonnegative. For this section we stay with nonnegativity for the basic analysis of LP.

First we illustrate LP from a geometrical point of view:

*Example 1.* We consider the problem

$$minimize \quad x_1 + 2x_2 \quad subject\ to \quad 2x_1 + x_2 \geq 1, \quad x_1 \geq 0, \quad x_2 \geq 0.$$

We have $\mathbf{c} = (1,\ 2)^t$, $A = [2\ 1]$, and $\mathbf{b} = (1)$. The two nonnegativity constraints restrict the feasible solutions to the first quadrant of $\mathcal{R}^2$. The only other constraint represents a half-plane bounded by the line $2x_1 + x_2 = 1$. To satisfy all three constraints, the feasible solutions must lie in the intersection of the first quadrant and the half-plane that is, in the shaded region in Fig. 3.3, which we call the *feasible solution set* $S_f$. The figure also shows level curves of the *objective function* $\psi$, parallel straight lines. It is clear from these level curves that $\psi$ achieves its global minimum $\psi_{\min} = 2.5$ at the unique minimizer

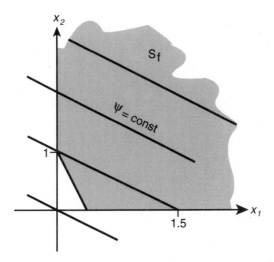

**Figure 3.3.** Linear programming problem in two dimensions.

$\mathbf{x}^* = (\frac{1}{2}, \ 0)^t$. We remark that, if we change "minimize" to "maximize," then no solution exists because $S_f$ is unbounded and $\psi$ is unbounded above on $S_f$.

For a general LP in the form of (3.60), the constraints together specify the *feasible solution set* $S_f$; as in Section 3.2, it is a closed convex subset of $\mathcal{R}^n$ and may be bounded or (as in Example 1) unbounded. When it is bounded, it is a convex polyhedron, with a finite number of extreme points (vertices) and the LP has a solution, as in Section 3.1, with every minimizer a relative boundary point of $S_f$ and at least one minimizer an extreme point. When $S_f$ is unbounded, the solution may fail to exist, as illustrated above; however, when it does exist, each minimizer is a relative boundary point of $S_f$ (last paragraph of Section 3.1). For example, if we change $\psi$ in Example 1 to $2x_1 + x_2$, then the minimum value is 1, taken along the line segment from (0.5, 0) to (0, 1), part of the relative boundary of $S_f$, including two extreme points of $S_f$.

When $S_f$ is unbounded, a solution fails to exist precisely when $S_f$ is "unbounded in the direction of $-\mathbf{c}$"; that is, when $\mathbf{c}^t\mathbf{x}$ is not bounded below. The vector $-\mathbf{c}$ for Example 1 is $(-1, -2)^t$; the set $S_f$ is bounded in that direction, so the minimum is achieved.

It can happen that $S_f$ is empty; then there is no solution, and the LP is considered to be ill-posed. The set $S_f$ can also reduce to a single point, so the problem is trivial.

In higher dimensions, $S_f$ is bounded by linear varieties, and the level sets of $\psi$ are parallel linear varieties. A higher-dimensional LP cannot be solved by graphing, but the geometrical concepts suggested above help in developing algorithms for computing solutions and detecting nonexistence or unboundedness.

## A Dual LP

The problem (3.60), assumed to have a minimizer $\mathbf{x}^*$, has a dual. We consider the case when no equations occur in the constraints $A\mathbf{x} \geq \mathbf{b}$. We then construct the Lagrangian function,

$$L(\mathbf{x}, \mathbf{y}) = \mathbf{c}^t\mathbf{x} - \mathbf{y}^t(A\mathbf{x} - \mathbf{b}),$$

where the dual variable $\mathbf{y}$, playing the role of Lagrangian multiplier, is also assumed nonnegative. For a fixed $\mathbf{y}$ we have the inequality

$$\mathbf{c}^t\mathbf{x}^* \geq \min_{\mathbf{x} \in S_f} L(\mathbf{x}, \mathbf{y}),$$

since a nonnegative second term is substracted from $\mathbf{c}^t\mathbf{x}$ in $L$. By rearranging terms, we obtain

$$\mathbf{c}^t\mathbf{x}^* \geq \min_{\mathbf{x} \in S_f}(\mathbf{c} - A^t\mathbf{y})^t\mathbf{x} + \mathbf{b}^t\mathbf{y} \geq \mathbf{b}^t\mathbf{y} = \phi(\mathbf{y}). \tag{3.61}$$

In order to achieve the last inequality in (3.61), which produces a lower bound function $\phi(\mathbf{y})$ for $\mathbf{c}^t\mathbf{x}^*$, we require that $\mathbf{c} - A^t\mathbf{y} \geq \mathbf{0}$ in addition to $\mathbf{y} \geq \mathbf{0}$. Hence $\phi(\mathbf{y})$ is a lower bound of $\psi(\mathbf{x})$ for an arbitrary admissible $\mathbf{y}$.

We seek the greatest lower bound by solving the dual problem,

$$\begin{aligned} maximize \quad & \phi(\mathbf{y}) = \mathbf{b}^t\mathbf{y} \\ subject\ to \quad & A^t\mathbf{y} \leq \mathbf{c}, \\ & \mathbf{y} \geq \mathbf{0}, \end{aligned} \tag{3.62}$$

whose solution (assumed to exist) is denoted by $\mathbf{y}^*$, the maximizer. We have established that

$$\mathbf{c}^t\mathbf{x}^* \geq \mathbf{b}^t\mathbf{y}^*. \tag{3.63}$$

Strict equality in (3.63) is needed in order to complete the duality theorem.

In order to prove the duality, we apply the Karush–Kuhn–Tucker conditions of Section 3.1 to the primal problem. To do so, we must write the constraints in the form $-x_i \leq 0, \ i = 1, \ldots, n, \ \mathbf{b} - A\mathbf{x} \leq \mathbf{0}$. We further must assume that at the minimizer $\mathbf{x}^*$ the gradients of the active constraints are linearly independent. This condition is normally satisfied; when it is not satisfied, one can show that a small perturbation of the primal problem will achieve the condition so that the duality can be proved. A limit process then shows that the duality is valid for the given primal problem.

The Karush–Kuhn–Tucker conditions (3.25) for our problem can be written as follows (Problem 3.46):

$$\mathbf{c} - (\lambda_1, \ldots, \lambda_n)^t - A^t(\lambda_{n+1}, \ldots, \lambda_{n+m})^t = \mathbf{0}. \tag{3.64}$$

Here all $\lambda_j \geq 0$ and only those $\lambda_j$ corresponding to constraints that are active at $\mathbf{x}^*$ are allowed to differ from 0. We denote the vector $(\lambda_1, \ldots, \lambda_n)^t$ by $\mathbf{g}$ and identify $\mathbf{y}^*$ with the vector $(\lambda_{n+1}, \ldots, \lambda_{n+m})^t$. We see at once that all components of $\mathbf{y}^*$ are nonnegative and, from (3.64), that $A^t\mathbf{y}^* \leq \mathbf{c}$ (Problem 3.49); thus $\mathbf{y}^*$ is in the feasible solution set for the dual problem. From (3.64) we find (Problem 3.47) that

$$\psi(\mathbf{x}^*) = \mathbf{c}^t\mathbf{x}^* = \mathbf{g}^t\mathbf{x}^* + \mathbf{y}^{*t}A\mathbf{x}^*.$$

But from the Karush–Kuhn–Tucker conditions a component $g_i$ of $\mathbf{g}$ differs from 0 only if the corresponding $x_i$ is 0. Thus the first term on the right of the last equation is $\mathbf{0}$. The second term can be written as

$$\mathbf{y}^{*t}(A\mathbf{x}^* - \mathbf{b}) + \mathbf{y}^{*t}\mathbf{b}.$$

Here again, by the Karush–Kuhn–Tucker conditions, each component $y_j$ of $\mathbf{y}$ can differ from 0 only if the corresponding component of $A\mathbf{x}^* - \mathbf{b}$ is 0. Thus the first term in the last display is $\mathbf{0}$, and the second term on the right in the previous display reduces to $\mathbf{y}^{*t}\mathbf{b} = \phi(\mathbf{y}^*)$. This shows that $\psi(\mathbf{x}^*) = \phi(\mathbf{y}^*)$, so $\mathbf{y}^*$ is indeed a maximizer for the dual problem and the duality is established.

We observe that in the course of the proof we have established that the components $x_i$ and $y_j$ of the minimizer and maximizer satisfy the "complementarity conditions":

$$y_j = 0 \quad \text{if } (A\mathbf{x})_j > b_j, \ y_j \geq 0 \quad \text{if } (A\mathbf{x})_j = b_j,$$
$$x_i = 0 \quad \text{if } (A^t\mathbf{y})_i < c_i, \ x_i \geq 0 \quad \text{if } (A^t\mathbf{y})_i = c_i. \tag{3.65}$$

(See Problem 3.50.)

**Example 2.** We seek the minimum of $\psi(x_1, x_2) = x_1 - x_2$ for $x_1 \geq 0$, $x_2 \geq 0$ and $x_1 + x_2 \leq 1$, $x_1 + 2x_2 \leq 2$. We find the minimizer to be $(0, 1)$ where $\psi = -1$. The dual problem is to maximize $\phi(y_1, y_2) = -y_1 - 2y_2$ for $y_1 \geq 0$, $y_2 \geq 0$ and $y_1 + y_2 + 1 \geq 0$, $y_1 + 2y_2 - 1 \geq 0$. The maximum is $-1$, attained along the line segment joining $(0, \frac{1}{2})$ to $(1, 0)$. In this example, *three* constraints are active at the minimizer for the primal problem. Thus the corresponding gradients cannot be linearly independent. We verify that perturbing the primal problem by choosing a small positive $\varepsilon$ and replacing the last constraint by the following: $(1 - \varepsilon)x_1 + 2x_2 \leq 2(1 - \varepsilon)$ leads to a primal problem for which the gradients are linearly independent at the minimizer; we also verify that the minimum for this new problem and the maximum for its dual (which are equal) vary continuously with $\varepsilon$ and approach $-1$ as $\varepsilon \to 0$. (See Problem 3.51.)

We have proved the duality under the assumption that no equations appear in the constraints. If some (or even all) of the constraints are equations

$(A\mathbf{x})_j = b_j$, then in the derivation of the dual one sees that the corresponding components of $\mathbf{y}$ are unrestricted in sign; in the complementarity conditions, the corresponding values of $j$ are omitted. (See Problem 3.52.)

REMARK. In Chapter 4 the duality theorem for LP is proved under general conditions as a consequence of the general theorems of Fenchel and Rockafellar. (See Sections 4.6 and 4.7.)

In the next section we describe a constructive procedure for seeking the minimum for the primal problem; the procedure is called the *simplex algorithm*. We now introduce a simplification that is helpful in the algorithm.

Our minimizer is sought at an extreme point of $S_f$, a boundary point of this set. It can be shown (see Problem 3.43) that each such point is specified by $n$ equations obtained by replacing $\geq$ by $=$ in $n$ of the constraints. In particular, a number $m'$, $0 \leq m' \leq n$ of the $m$ inequalities $A\mathbf{x} \geq \mathbf{b}$ will be active (become equations) at the solution. If we could identify the correct $m'$ active inequalities, we could ignore the remaining ones and restate (3.60) in the form

$$minimize \quad \psi(\mathbf{x}) = \mathbf{c}^t\mathbf{x},$$

$$subject\ to \quad A'\mathbf{x} = \mathbf{b}, \tag{3.60'}$$

$$\mathbf{x} \geq \mathbf{0}.$$

where $A'$ is $m' \times n$ and has rank $m' < n$. The LP in (3.60′) is called the *standard form* in most books on LP. However, there is a difficulty with this method of achieving the standard form: It is rarely possible to find the active constraints before the solution is known!

A standard form can be achieved in another way. An inequality $z \geq k$ can be replaced by an equation $z - s = k$ along with an inequality $s \geq 0$; here $s$ is the nonnegative surplus of $z$ over $k$. We call $s$ a *surplus variable*. We thus rewrite $A\mathbf{x} \geq \mathbf{b}$ as $A\mathbf{x} - \mathbf{s} = \mathbf{b}$, where $\mathbf{s} \geq \mathbf{0}$ is a vector of surplus variables. We may append $\mathbf{s}$ to $\mathbf{x}$ and still name the new variable $\mathbf{x}$, but it is now a vector in $\mathcal{R}^{m+n}$. With the vector $\mathbf{c}$ also expanded by adjoining $m$ zero components so that the surplus variables do not contribute to the value of $\psi$, the standard form (3.60′) is achieved with $A' = [A \mid -I]$, in block matrix notation (see Appendix A.20), where $I$ is the $m \times m$ identity matrix. The actual active inequality constraints at the optimal solution are those corresponding to zero surplus variables.

REMARK. In the exceptional case where the original constraints $A\mathbf{x} \geq \mathbf{b}$ include some equations, no surplus variables are needed for the equations.

We can illustrate this discussion by applying it to Example 1 above. The minimizer is specified by the two equations $2x_1 + x_2 = 1$, $x_2 = 0$. The LP can be rewritten as seeking the minimum of $x_1 + 2x_2 + 0x_3$, where $2x_1 + x_2 + x_3 = 1$, $x_1 \geq 0$, $x_2 \geq 0$, $x_3 \geq 0$. The feasible solution set is now a portion of a plane in the first octant and the minimum is achieved at the vertex $(0.5, 0, 0)$ at which the surplus variable $x_3$ is 0. It should be observed that by assigning to each point $(x_1, x_2)$ in the original set $S_f$ the point $(x_1, x_2, x_3)$ in the plane with the same values of $x_1$ and $x_2$ (and $x_3 = 1 - 2x_1 - x_2$), we obtain a one-to-one correspondence between the original feasible solution set and the new one. This reasoning holds in all cases: *By introducing surplus variables, we do not change the fundamental nature of the feasible solution set: a triangular, polygonal, polyhedral, ... region is transformed to a triangular, polygonal, polyhedral, ... region of the same dimension, same number of vertices, sides, . . . .*

In general, the feasible solution set for $(3.60')$ is the portion of a linear variety $A'\mathbf{x} = \mathbf{b}$ in the first orthant (all coordinates nonnegative) in $\mathcal{R}^n$; the minimum is found at an extreme point of this set, at which all $m'$ equations specifying the linear variety must be satisfied and $n - m'$ coordinates are 0.

***Example 3.*** We seek the minimum of $3 - x_1 - x_2$ for $-2x_1 - x_2 \geq -3$, $-x_1 - 2x_2 \geq -3$, $x_1 \geq 0$, $x_2 \geq 0$. The feasible solution set for the LP in this form is shown in Fig. 3.4; it is a region bounded by a quadrilateral. The minimum of 1 occurs at the extreme point $(1, 1)$. We convert the problem to one in standard form by subtracting surplus variables $x_3$, $x_4$ from the left sides of the first two inequalities and replacing the $\geq$ by $=$. After changing signs in the two new equations, we are led to the problem of minimizing $3 - x_1 - x_2$ for $2x_1 + x_2 + x_3 = 3$, $x_1 + 2x_2 + x_4 = 3$ and $x_i \geq 0$ for $i = 1, \ldots, 4$. We cannot graph this set in 4-dimensional space, but it is a 2-dimensional convex set in one-to-one correspondence with the set in Fig. 3.4; hence it must also be a set enclosed by a quadrilateral. The four vertices must satisfy the two linear equations and equations $x_i = 0$ for two choices of $i$. We find that there are six

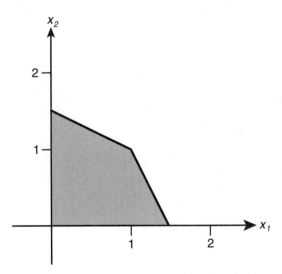

**Figure 3.4.** Feasible solution set for Example 3.

such points, but only four have nonnegative coordinates: (1, 1, 0, 0), (1.5, 0, 0, 1.5), (0. 1.5, 1.5, 0), and (0, 0, 3, 3). The minimum occurs at the first of these, at which both surplus variables are 0. (If the minimum had occurred at (0, 0, 3, 3), then neither surplus variable would be 0 at the point.) Verification of the details is left to Problem 3.45.

Example 3 indicates what must be done do find the minimizer for an LP in standard form. One concentrates on solutions $\mathbf{x} = (x_1, \ldots, x_n)^t$ of the equation $A'\mathbf{x} = \mathbf{b}$ such that $n - m'$ of the $x_i$ are 0 and the remaining $x_i$ are nonnegative. Each extreme point of the solution set is of this form. However, we must also ensure that the point is determined by $n$ independent conditions. This leads to the following definitions:

A *basic solution* of $A\mathbf{x} = \mathbf{b}$ is a solution that is obtained as follows: Select $m$ linearly independent columns of $A$ and form a nonsingular matrix $B$ having these as its column vectors, solve the equation $B\mathbf{v} = \mathbf{b}$ for $\mathbf{v}$, and then use the $m$ components of $\mathbf{v}$ and $n - m$ zeros to form $\mathbf{x}$ so that $A\mathbf{x} = \mathbf{b}$. If all components of $\mathbf{v}$ are nonnegative, then the basic solution $\mathbf{x}$ constructed from it is called a *basic feasible solution*. It can be shown that each basic feasible solution represents an extreme point of the feasible solution set and that all extreme points are obtained in this way. (See Hadley, 1962, sec. 3.10.)

For Example 3 in standard form, $A' = \begin{bmatrix} 2 & 1 & 1 & 0 \\ 1 & 2 & 0 & 1 \end{bmatrix}$ and $\mathbf{b} = (3, 3)^t$. If we form $B$ from the first two columns of $A$, then $\mathbf{v} = (1, 1)^t$, and we obtain the point (1, 1, 0, 0) again. Similarly the other three extreme points are obtained by forming $B$ from the other five pairs of columns of $A$ and rejecting the points having some negative coordinates.

### Karush–Kuhn–Tucker Conditions for LP in Standard Form

We consider an LP in standard form:

$$minimize \quad \psi(\mathbf{x}) = \mathbf{c}^t \mathbf{x}$$

$$subject\ to \quad A\mathbf{x} = \mathbf{b}, \tag{3.66}$$

$$\mathbf{x} \geq \mathbf{0}.$$

Because of the replacement of inequalities by equations in some of the constraints, the conditions now take the form

$$\mathbf{c} - \mathbf{g} - A^t(\mu_1, \ldots, \mu_m)^t = \mathbf{0}, \tag{3.67}$$

where $\mathbf{g} = (\lambda_1, \ldots, \lambda_n)^t$. Here only the $\lambda_i$ are necessarily nonnegative, and $\lambda_i = 0$ if $x_i > 0$. We are assuming, as before, that the gradients of the constraints which are active at the minimizer form a linearly independent set. From the previous discussion, the minimizer is a basic feasible solution of the equation

$A\mathbf{x} = \mathbf{b}$ and hence has $n - m$ zero components; there can be no more than $n - m$ zero components, for then one would have more than $n$ active constraints at the minimizer ($m$ coming from the equation $A\mathbf{x} = \mathbf{b}$, more than $n - m$ coming from the equations $x_i = 0$). We can assume that coordinates in $\mathcal{R}^n$ are numbered so that the first $n - m$ components of the minimizer $\mathbf{x}$ are positive and the remaining components are 0. Then the equation $A\mathbf{x} = \mathbf{b}$ can be written as $[B|R]\mathbf{x} = \mathbf{b}$, where $B$ is $m \times m$ and the minimizer is the vector $B^{-1}\mathbf{b}$ followed by $n - m$ zeros. If we denote by $\mathbf{c}_B$, $\mathbf{c}_R$, $\mathbf{g}_B$, $\mathbf{g}_R$ the vectors formed of the first $m$ and last $n - m$ components of $\mathbf{c}$ and $\mathbf{g}$, then (3.67) splits into two equations:

$$\mathbf{c}_B - \mathbf{g}_B - B^t(\mu_1, \ldots, \mu_m)^t = 0, \tag{3.68'}$$

$$\mathbf{c}_R - \mathbf{g}_R - R^t(\mu_1, \ldots, \mu_m)^t = 0. \tag{3.68''}$$

Here $\mathbf{g}_B = 0$ from the conditions on the $\lambda_i$. Since $B$ is nonsingular, we can eliminate the vector $(\mu_1, \ldots, \mu_m)$ from these two equations to obtain the following *necessary condition for a minimum of* $\psi$:

$$\mathbf{c}_R^t - \mathbf{c}_B^t B^{-1} R = \mathbf{g}_R^t = (\lambda_{n-m+1}, \ldots, \lambda_n) \geq 0. \tag{3.69}$$

(See Problem 3.53.) In the next section it will be seen how the simplex algorithm leads to a fulfillment of this condition at a minimizer. It should be stressed that the challenge is to find how to renumber coordinates in $\mathcal{R}^n$ so that $A$ takes on the form $[B \mid R]$ and all the conditions described are satisfied.

### Duality for LP in Standard Form

The discussion of duality given above can be repeated and $\mathbf{y}^*$ can be chosen as the vector $(\mu_1, \ldots, \mu_m)^t$. However, because of the replacement of some inequalities by equations, as remarked earlier, the components of $\mathbf{y}$ are unrestricted in sign in the dual problem. If we start with an LP (3.60) with inequalities as stated and then convert this to an LP in standard form by adding surplus variables, then we appear to have two different duals: one as in (3.62), in which all $y_j$ are nonnegative, and one like this except that the $y_j$ are unrestricted. However, one can verify from (3.67) that when the LP in standard form arises as in this case from (3.60) by introducing surplus variables, then the maximizer $\mathbf{y}^*$ of $\phi$ has only nonnegative components; in fact the process of introducing surplus variables has the effect of replacing the matrix $A$ by $A' = [A| - I]$ and the vector $\mathbf{c}$ by $\mathbf{c}'$ equal to $\mathbf{c}$ followed by $m$ zeros. Thus the condition $(A')^t\mathbf{y} \leq \mathbf{c}'$ for the LP in standard form is equivalent to the *two* conditions $A^t\mathbf{y} \leq \mathbf{c}$ and $\mathbf{y} \geq 0$. Accordingly the two dual problems are equivalent. (See Problem 3.54.)

CLOSING REMARK. Finally we observe that from (3.68') we obtain an expression for $\mathbf{y}^*$:

$$\mathbf{y}^* = (B^{-1})^t \mathbf{c}_B.$$

## 3.7  SIMPLEX ALGORITHM

We consider an LP in standard form:

$$minimize \quad \psi(\mathbf{x}) = \mathbf{c}^t \mathbf{x}$$
$$subject\ to \quad A\mathbf{x} = \mathbf{b}, \tag{3.70}$$
$$\mathbf{x} \geq \mathbf{0}.$$

Here, as in Section 3.6, $\mathbf{x} \in \mathcal{R}^n$, $A$ is $m \times n$ and has rank $m < n$, $\mathbf{b} \in \mathcal{R}^m$, and we assume that the feasible solution set $S_f$ is nonempty. We also assume that $\psi$ is bounded below on $S_f$, so the unbounded case is ruled out; however, our algorithm will reveal when we are in that case.

We now seek our minimum among the basic feasible solutions of the equation $A\mathbf{x} = \mathbf{b}$. The algorithm to be described has a simple geometrical interpretation. It is iterative, beginning with an extreme point $\mathbf{x}^{(0)}$. Its successor is then chosen among extreme points (or vertices) directly connected to $\mathbf{x}^{(0)}$ by an *edge* on the relative boundary of $S_f$ so as to achieve a maximum decrease in the value of $\mathbf{c}^t \mathbf{x}$. Thus the next vector $\mathbf{x}^{(1)}$ chosen, being basic and feasible, must satisfy $\mathbf{c}^t \mathbf{x}^{(1)} < \mathbf{c}^t \mathbf{x}^{(0)}$. Then $\mathbf{x}^{(1)}$ takes the place of $\mathbf{x}^{(0)}$, and the algorithm is repeated until no further decrease is possible. Since there are only a finite number of extreme points, the process stops after a finite number of steps and the minimizer is found. (A full justification requires more discussion. We have assumed that there is no unbounded solution, but trouble arises if there is degeneracy, as pointed out below. See Hadley, 1962, for a complete treatment.)

Having given this geometric view, we now fill in the algebraic details. The standard LP can be rewritten, after $m$ basis vectors are chosen to form the matrix $B$, in the form

$$minimize \quad \psi(\mathbf{v}, \mathbf{w}) = \mathbf{c}_B^t \mathbf{v} + \mathbf{c}_R^t \mathbf{w}$$
$$subject\ to \quad B\mathbf{v} + R\mathbf{w} = \mathbf{b},$$
$$\mathbf{v} \geq \mathbf{0}, \ \mathbf{w} \geq \mathbf{0},$$

where $R$ is an $m \times (n - m)$ matrix formed from those columns of $A$ not in $B$; $\mathbf{v}$ and $\mathbf{w}$ contain basic and nonbasic components of $\mathbf{x}$; $\mathbf{c}_B$ and $\mathbf{c}_R$ are formed from $\mathbf{c}$ by selecting first those components that correspond to the columns of $A$ used to form $B$, then those components corresponding to the columns of $A$ used to form $R$. The relation of $B$, $R$ to $A$; $\mathbf{v}$, $\mathbf{w}$ to $\mathbf{x}$; $\mathbf{c}_B$, $\mathbf{c}_R$ to $\mathbf{c}$ is defined by an integer row vector $\mathbf{N} = (N_1, \ldots, N_n)$ such that the $i$th column of $[B|R]$ is the $N_i$th column of $A$, the $i$th entry of $[\mathbf{v}^t|\mathbf{w}^t]$ is the $N_i$th component of $\mathbf{x}$, and the $i$th entry of $[\mathbf{c}_B^t|\mathbf{c}_R^t]$ is the $N_i$th component of $\mathbf{c}^t$. The process starts with a choice of this *index vector* $\mathbf{N}$, yielding a particular $B$, $R$, $\ldots$ as above. At each subsequent stage a permutation of this vector is made by interchanging the $I$th and $J$th components. Thus the matrices $B$ and $R$ change at each stage until the last one, at which, as will be seen, they are such that the conditions (3.69) are fulfilled.

The current basic feasible vector $\mathbf{x}^{(0)}$ corresponds to $\mathbf{w} = \mathbf{0}$ and $\mathbf{v} = B^{-1}\mathbf{b}$ which give the value $\psi(\mathbf{x}^{(0)}) = \mathbf{c}_B^t B^{-1}\mathbf{b}$. Trying to reduce the value of $\psi$, we let one component of $\mathbf{w}$ increase from 0. Which component should be chosen? We first express $\mathbf{v}$ in terms of $\mathbf{w}$ to get $\mathbf{v} = B^{-1}\mathbf{b} - B^{-1}R\mathbf{w}$; then we derive

$$\psi(\mathbf{v}(\mathbf{w}), \mathbf{w}) = \mathbf{c}_B^t(B^{-1}\mathbf{b} - B^{-1}R\mathbf{w}) + \mathbf{c}_R^t\mathbf{w} = \psi(\mathbf{x}^{(0)}) + \mathbf{p}^t\mathbf{w},$$

where $\mathbf{p}^t = \mathbf{c}_R^t - \mathbf{c}_B^t B^{-1}R$. We can write $R\mathbf{w} = w_1\mathbf{r}_1 + \cdots + w_{n-m}\mathbf{r}_{n-m}$, where the $\mathbf{r}_j$ are the columns of $R$; a component of $\mathbf{w}$ for which the corresponding column vector $\mathbf{r}_j = \mathbf{0}$ cannot help. (We exclude the case where $R$ is a zero matrix, where the problem is trivial.)

Therefore, we choose a $J$ such that $p_J = \min_j\{p_j \mid \mathbf{r}_j \neq \mathbf{0}\}$. If $p_J \geq 0$ (all components of $\mathbf{p}$ are nonnegative), no reduction of $\psi$ is possible since all components of $\mathbf{w}$ must be nonnegative to produce a feasible vector $\mathbf{x}$. In this case, the current solution $\mathbf{x}^{(0)}$ is the minimizer and

$$\mathbf{c}_R^t - \mathbf{c}_B^t B^{-1}R \geq \mathbf{0},$$

so condition (3.69) holds. Otherwise ($p_J < 0$), we let the corresponding $w_J$ increase from 0. This leads to

$$B\mathbf{v}' + \mathbf{r}_J w_J = \mathbf{b} = B\mathbf{v}, \tag{3.71}$$

where $\mathbf{v}'$ is the new solution after making $w_J > 0$. The columns $\mathbf{b}_i$ of $B$ (unrelated to $\mathbf{b}$) form a basis for $\mathcal{R}^m$ (Section 1.6); hence

$$\mathbf{r}_J = \sum_{i=1}^m \gamma_{iJ}\mathbf{b}_i \tag{3.72}$$

for appropriate coefficients $\gamma_{iJ}$.

Substituting (3.72) into (3.71) produces

$$\sum_{i=1}^m (v_i' + \gamma_{iJ}w_J)\mathbf{b}_i = \sum_{i=1}^m v_i\mathbf{b}_i,$$

and equating component to component in the above expression, we obtain

$$v_i' = v_i - \gamma_{iJ}w_J, \qquad i = 1, 2, \ldots, m. \tag{3.73}$$

As $w_J$ increases from 0, the right-hand sides of the $m$ equations in (3.73) will decrease or increase depending on the signs of each $\gamma_{iJ}$. These coefficients cannot be all zero, since $\mathbf{r}_J$ cannot be a zero vector. If $\gamma_{iJ} \leq 0$ for all $i$, some components of $\mathbf{v}'$ will increase indefinitely while the solution remains feasible. Thus we are in the unbounded case, and no solution is possible. If this does not occur, we continue the process as before. If a coefficient $\gamma_{iJ}$ is positive, the corresponding component of $v_i'$ decreases linearly from a positive value $v_i$ as $w_J$ increases; this component eventually becomes negative to render the solution infeasible. If more than one coefficient is positive, the first $v_i'$ reaching zero determines the maximum value of $w_J$ that keeps the new solution feasible. We choose

$$w_J = \min_i \left\{ \frac{v_i}{\gamma_{iJ}} \mid \gamma_{iJ} > 0 \right\}. \tag{3.74}$$

This choice also makes the new solution basic, since there are at most $m$ nonnegative components in $\mathbf{x}^{(1)}$. Let the minimizing index in (3.74) be $i = I$. We then make the permutation interchanging $I$ and $m + J$ in the index vector, corresponding to the interchange of the columns $\mathbf{b}_I$ and $\mathbf{r}_J$. Since $\gamma_{IJ} > 0$, the new matrix $B$ is nonsingular (Problem 3.48). We can then write *in the renumbered coordinates*:

$$\mathbf{x}^{(1)} = (v_1', \ldots, v_{I-1}', w_J, v_{I+1}', \ldots, v_m', 0, \ldots, 0)^t,$$

which is the replacement for $\mathbf{x}^{(0)}$. The equation $\psi(\mathbf{x}^{(1)}) = \psi(\mathbf{x}^{(0)}) + p_J w_J^*$ produces a reduction (descent) of the function value in the minimizing sequence. The procedure described above can now be repeated. It is possible that the basic variable $\mathbf{v}$ may have a zero component to produce a minimum ratio $v_i/\gamma_{iJ} = 0$. Then the incoming and outgoing basic vectors are both associated with a zero component and produce no change in $\psi$. This is the *degenerate case* in which the simplex algorithm cycles among feasible basic vectors producing no change in the values of $\psi$. This cycling is extremely rare; it can be removed by a small perturbation of a zero component of the basic variable. If a finite solution exists, the simplex algorithm will reach a feasible basic vector that minimizes $\psi$.

   Some LPs arise naturally from managerial problems in business and industry. Minimizing cost and maximizing profit is the goal of every factory manager who is also facing limits on raw materials, energy, facility, labor force, and a host of other constraints. This is how the subject of LP began. Many other problems share this mathematical formulation, such as transportation, network, government, and military operations. (For example, see Pierre, 1969, pp. 239–263; Dantzig, 1963.)

   A typical LP usually involves several hundred variables or more. We present a small but realistic example:

***Example 1.*** A small auto company has two plants A and B: A for three types of power trains and B for three sizes of bodies. Together they produce compact (C), midsize (M) and large (L) cars. Let $x_i$, $i = 1, 2, 3$ be the respective units of C, M, and L cars to be produced each month. The facility and manpower are organized as work units. At plant A, 8, 8, 9 units are required for the power train of each respective C, M, L car, while at plant B, the units are 8, 9, 10 for the respective bodies. The total monthly work units available are 10,000 and 11,000 for plants A and B, respectively. We may write these two constraints as

$$8x_1 + 8x_2 + 9x_3 \leq 10,000, \qquad 8x_1 + 9x_2 + 10x_3 \leq 11,000.$$

Since one cannot produce a negative number of cars, we also need the constraints

$$x_1 \geq 0, \qquad x_2 \geq 0, \qquad x_3 \geq 0.$$

The profit for each C, M, and L car is 5000, 5400, and 6125, respectively, in 1992 dollars. One seeks to maximize the total profit

$$\phi(\mathbf{x}) = 5000x_1 + 5400x_2 + 6125x_3,$$

subject to the above constraints. Here the primal LP is a maximization, and the inequality $A\mathbf{x} \leq \mathbf{b}$ has direction opposite to that of (3.60). We will follow through the simplex algorithm, slightly modified to accomodate these differences. To achieve standard form (3.60′), we add nonnegative *slack* variables (similar to surplus variables) $x_4$ and $x_5$ to the left-hand side of $A\mathbf{x} \leq \mathbf{b}$ to obtain $A'\mathbf{x}' = \mathbf{b}$, where

$$A' = \begin{bmatrix} 8 & 8 & 9 & 1 & 0 \\ 8 & 9 & 10 & 0 & 1 \end{bmatrix}, \qquad \mathbf{b} = \begin{pmatrix} 10,000 \\ 11,000 \end{pmatrix},$$

$$\mathbf{c}' = (5000, 5400, 6125, 0, 0)^t, \quad \text{and} \quad \mathbf{x}' = (x_1, x_2, x_3, x_4, x_5)^t.$$

We may choose the $\mathbf{x}^{(0)} = (0, 0, 0, 10,000, 11,000)^t$ as the starting point, since it is a basic feasible solution. We have $\phi_0 = 0$. This technique for starting is often used when the elements of $\mathbf{b}$ are arranged to yield positive slack (or surplus) variables. Using the chosen initial index vector $(4, 5, 1, 2, 3)$,

$$\mathbf{v}^t = (10,000, 11,000), \qquad B = \begin{bmatrix} 1 & 0 \\ 0 & 1 \end{bmatrix}$$

$$R = \begin{bmatrix} 8 & 8 & 9 \\ 8 & 9 & 10 \end{bmatrix}, \quad \mathbf{c}_B^t = (0, 0), \quad \mathbf{c}_R^t = (5000, 5400, 6125),$$

we compute

$$\mathbf{p}^t = \mathbf{c}_R^t - \mathbf{c}_B^t B^{-1} R = (5000, 5400, 6125).$$

Since we are maximizing, we choose $p_J = p_{max} = 6125$, so $J = 3$, and

$$\mathbf{r}_J = \begin{pmatrix} 9 \\ 10 \end{pmatrix} = 9 \begin{pmatrix} 1 \\ 0 \end{pmatrix} + 10 \begin{pmatrix} 0 \\ 1 \end{pmatrix},$$

so $\gamma_{1J} = 9$ and $\gamma_{2J} = 10$. From these we can find

$$w_J = \min\{10,000/9, 11,000/10\} = 1100,$$

so $I = 2$. We also obtain

$$v_1' = v_1 - \gamma_{1J}w_J = 10,000 - 9900 = 100,$$

$$v_2' = v_2 - \gamma_{2J}w_J = 11,000 - 11,000 = 0.$$

We interchange the second and fifth columns of $[B \mid R]$ and have a new index vector $(4, 3, 1, 2, 5)$ and new $\mathbf{v}^t = (100, 11,000)$. This leads to

$$\mathbf{x}^{(1)} = (0, 0, 1100, 100, 0)^t, \qquad \phi_1 = 0 + 6125(1100) = 6,737,500,$$

where $\phi_1$ is the profit from producing 1100 L cars per month and no other cars. It is not optimal.

We continue the algorithm with this new feasible basic vector. We have a new $\mathbf{v}^t$ as above and new

$$B = \begin{bmatrix} 1 & 9 \\ 0 & 10 \end{bmatrix} \quad \text{and} \quad R = \begin{bmatrix} 8 & 8 & 0 \\ 8 & 9 & 1 \end{bmatrix}.$$

With $\mathbf{c}_B^t = (0, 6125)$ and $\mathbf{c}_R^t = (5000, 5400, 0)$ we obtain

$$\mathbf{p}^t = \mathbf{c}_R^t - \mathbf{c}_B^t B^{-1} R = (100, -112.5, -612.5).$$

Now $p_J = p_{max} = 100$, with $J = 1$ and

$$\mathbf{r}_J = \begin{pmatrix} 8 \\ 8 \end{pmatrix} = 0.8 \begin{pmatrix} 1 \\ 0 \end{pmatrix} + 0.8 \begin{pmatrix} 9 \\ 10 \end{pmatrix},$$

so $\gamma_{1J} = \gamma_{2J} = 0.8$. We find $w_J = \min\{100/0.8, 1100/0.8\} = 125$, so $I = 1$. We again obtain

$$v_1' = v_1 - 0.8(125) = 100 - 100 = 0,$$

$$v_2' = v_2 - 0.8(125) = 1100 - 100 = 1000.$$

Thus we interchange the first and third columns, to obtain a new index vector $(1, 3, 4, 2, 5)$ and new $\mathbf{v}^t = (125, 1000)$. This corresponds to

$$\mathbf{x}^{(2)} = (125, 0, 1000, 0, 0)^t, \qquad \phi_2 = 6,737,500 + 100(125) = 6,750,000,$$

where $\phi_2$ is the profit from producing 1000 L cars and 125 C cars. Using the new index vector, we find that $\mathbf{p} = (-125, -100, -500)^t$. Since all components of $\mathbf{p}$ are now negative, no further increase of profit is possible. The solution $\mathbf{x}^{(2)}$ and the profit $\phi_2$ are optimal.

The dual LP for this problem is

$$minimize \quad \psi(\mathbf{y}) = 10{,}000y_1 + 11{,}000y_2$$

$$subject\ to \quad 8y_1 + 8y_2 \geq 5000,$$

$$8y_1 + 9y_2 \geq 5400,$$

$$9y_1 + 10y_2 \geq 6125,$$

$$y_1 \geq 0,\ y_2 \geq 0.$$

Using the last basic matrix

$$B^* = \begin{bmatrix} 8 & 9 \\ 8 & 10 \end{bmatrix}, \quad we\ solve \quad (B^*)^t\mathbf{y} = \mathbf{c}_B = \begin{pmatrix} 5000 \\ 6125 \end{pmatrix}$$

to obtain $\mathbf{y}^* = (125, 500)^t$ and $\psi^* = 6{,}750{,}000$, which are, respectively, the minimizer and minimum objective function value of the dual problem.

Since we know which constraints are active at the optimum, we can check our results by examining the standard form obtained by ignoring the inactive inequality constraints:

$$maximize \quad \phi(\mathbf{x}) = 5000x_1 + 5400x_2 + 6125x_3$$

$$subject\ to \quad 8x_1 + 8x_2 + 9x_3 = 10{,}000,$$

$$8x_1 + 9x_2 + 10x_3 = 11{,}000,$$

$$x_1 \geq 0,\ x_2 \geq 0,\ x_3 \geq 0.$$

This is equivalent to the original LP with inequality constraints, since both inequalities are active at the optimal solution. The simplex algorithm for this smaller problem again yields $\mathbf{x}^* = (125, 0, 1000)^t$ and $\mathbf{y}^* = (125, 500)^t$. Techniques of identifying active inequality constraints, sometimes iteratively, are important in optimization. Slack and surplus variables can be considered as tools to achieve this identification. A zero slack or surplus variable makes the original inequality active.

The meaning of each dual variable for this example is the cost plus expected profit per work unit in a plant. The second inequality in the dual problem gives $8y_1^* + 9y_2^* = 5500$, which is greater than the actual cost plus expected profit of $5400 that can be made on each M car. This is why M cars are not produced.

This small example did not take the market into account and ignored the customers who want M cars. The $125:1000$ ratio of C cars and L cars is determined purely from considerations of profit and of work unit distributions at the plants. This is probably a true reflection of the decision-making process of U.S. auto industries in the last few decades. Now the internal competition, fuel economy, environment, and family sizes may all play a role in decision making. Introducing all such constraints can lead to a large LP of a few thousand variables; this can be solved easily on modern computers. For a student of LP, this example provides a complete demonstration of the simplex algorithm for primal and dual solutions.

REMARKS. (a) There are a number of ways to obtain an initial basic feasible solution. For an LP in standard form (3.70) one can introduce a new LP:

$$minimize \quad \chi(\mathbf{u}) = \mathbf{w}^t \mathbf{u}$$

$$subject\ to \quad A\mathbf{x} + \mathbf{z}^t \mathbf{u} = \mathbf{b}, \tag{3.75}$$

$$\mathbf{x} \geq \mathbf{0},\ \mathbf{u} \geq \mathbf{0}.$$

Here $w_i = 1$ for $i = 1, \ldots, m$ and $z_i = sign(b_i) = \pm 1$ (with $+$ when $b_i \geq 0$ and $-$ otherwise) for $i = 1, \ldots, m$. If $\mathbf{x}$ is a feasible solution of (3.70), then $\mathbf{u} = \mathbf{0}$ is a feasible solution of this LP, and this choice of $\mathbf{u}$ minimizes $\chi(\mathbf{u})$. Conversely, if a pair $\mathbf{x}, \mathbf{u}$ is a minimizer of the LP (3.75) (and the original LP (3.70) has a feasible solution), then $\mathbf{u}$ must be $\mathbf{0}$ and $\mathbf{x}$ must be a feasible solution of (3.70). By the choice of $\mathbf{z}$, one can apply the simplex method to (3.75) with initial basic feasible solution: $\mathbf{x} = \mathbf{0}$, $u_i = |b_i|$ for $i = 1, \ldots, m$. The simplex method produces a minimizer of $\chi$ which is a basic feasible solution of (3.75), so the $\mathbf{x}$ found is a feasible solution of (3.70); in fact one verifies at once that it is a basic feasible solution of (3.70). The procedure described is also applicable if some or all of the $x_i$ are unrestricted; as pointed out at the beginning of Section 3.6, this case can always be reduced to one of nonnegative variables. The procedure also applies to finding a feasible solution of a QP as in (3.40). By introducing surplus variables $v_i = x_{n+i}$, one can reduce the side conditions to the form $A\mathbf{x} = \mathbf{b}$, $v_i \geq 0$.

(b) The task of writing a software that provides an initial iterate and takes care of all possible nonconvergent, nonexistent, and degenerate cases is quite tedious. Since many good LP codes exist (see Moré and Wright, 1993, ch. 5), we need not start from scratch.

(c) The key issue in LP is computational efficiency. Otherwise, one could examine all basic solutions of $A\mathbf{x} = \mathbf{b}$, a finite number, at most $N = n!/[m!(n-m)!]$, of vectors from which to find the optimizer. But the factorial growth of $N$ makes this approach impractical and prohibitive for large $n$. Current research in LP include sparse matrix updating and interior path (nonsimplex) methods.

(d) Standard LP is developed for nonnegative yet unbounded variables. Several a posteriori methods are available to adapt the simplex method to the case of arbitrarily signed and bounded variables, but they all pay a price of increased computer memory requirement, added complexity of coding, and much reduced efficiency. In the next section we present another version of LP to handle such problems. The dual of the new LP leads naturally to a nonsimplex method.

## PROBLEMS

**3.39.** Find the maximum or minimum if it exists, state what the maximizers or minimizers are, and if there is no global maximum or minimum, explain why this occurs.

(a) $x_1 \geq 0$, $x_2 \geq 0$, $-3x_1 - x_2 \geq -1$, maximize $\phi(x_1, x_2) = 7x_1 + x_2$.

(b) $x_1 \geq 0$, $x_2 \geq 0$, $x_1 + 2x_2 \geq 3$, minimize $\psi(x_1, x_2) = x_1 - x_2$.

(c) $x_1 \geq 0$, $x_2 \geq 0$, $x_1 - x_2 \geq 0$, $x_2 \geq 1$, minimize $\psi(x_1, x_2) = x_1 + x_2$.

(d) $x_1 \geq 0$, $x_2 \geq 0$, $x_1 + x_2 \geq 5$, $-3x_1 - x_2 \geq -2$, minimize $\psi(x_1, x_2) = 2x_1 + x_2$.

(e) $x_1 \geq 0$, $x_2 \geq 0$, $x_3 \geq 0$, $-x_1 - x_2 - x_3 \geq -1$, maximize
$\phi(x_1, x_2, x_3) = x_1 + x_2 + x_3$.

(f) $x_1 \geq 0$, $x_2 \geq 0$, $x_3 \geq 0$, $-x_1 - x_2 - x_3 \geq -1$, $\quad$ $3x_1 + x_2 + x_3 \geq 1$, minimize $\psi(x_1, x_2, x_3) = 7x_1 + x_2$.

**3.40. (a)** $\Rightarrow$ **(f)** By introduction of surplus variables, replace each of the LP problems of Problem 3.39 by problems in standard form (3.60′).

*****3.41.** Consider an LP (3.60) in $\mathcal{R}^2$ for which the set $S_f$ is unbounded but a minimizer exists as in Example 1 in Section 3.6. Show that one minimizer is an extreme point of $S_f$. [Hint: The known minimizer cannot be a relative interior point of $S_f$, since $\nabla \psi = \mathbf{c} \neq \mathbf{0}$. Hence it is a relative boundary point of $S_f$. If it is not at a vertex of $S_f$, then by linearity $\psi$ must be constant along the whole line, which must pass through a vertex of $S_f$, an extreme point, which is a minimizer; if it is a vertex, then again it is an extreme point.]

**3.42. (a)** Give an example of an LP (3.60) in $\mathcal{R}^2$ for which $S_f$ is empty.

**(b)** Give an example of an LP (3.60) in $\mathcal{R}^2$ for which $S_f$ is a single point.

*****3.43.** Give a geometric argument to show why for $n = 2$ or $n = 3$ each extreme point of a convex set in $\mathcal{R}^n$ defined by a finite number of linear inequalities is the solution of $n$ linear equations in $n$ unknowns with non-singular coefficient matrix, each equation being obtained by replacing one of the inequality signs by an $=$ sign.

**3.44.** Consider an LP in $\mathcal{R}^2$ with $S_f$ defined by the inequalities $x_1 \geq 0$, $x_2 \geq 0$, and $-x_1 - x_2 \geq -1$. Introduce the surplus variable $x_3$ to obtain the feasible solution set $S_f'$ for the LP in standard form, defined by the inequalities $x_1 \geq 0$, $x_2 \geq 0$, $x_3 \geq 0$, and the equation $x_1 + x_2 + x_3 = 1$. Show that the two sets are related by a one-to-one correspondence pairing the point $(x_1, x_2)$ in $S_f$ with the point $(x_1, x_2, x_3)$ such that $x_3 = 1 - x_1 - x_2$ in $S_f'$; show that in each case the former point can be interpreted as the perpendicular projection of the latter point on the $x_1 x_2$-plane.

**3.45.** Carry out the details of the analysis in Example 3 in Section 3.6 to find the four vertices of the quadrilateral in $\mathcal{R}^4$.

**3.46.** Verify that the Karush–Kuhn–Tucker conditions (3.26) for the problem (3.60) can be written in the form (3.64), with the $\lambda_j$ as described.

**3.47.** From (3.64) deduce that $\psi(\mathbf{x}^*) = \mathbf{g}^t \mathbf{x}^* + \mathbf{y}^{*t} A \mathbf{x}^*$, where $\mathbf{g}$ and $\mathbf{y}^*$ are as described.

**3.48.** Let $B$ be a nonsingular $m \times m$ matrix with columns $\mathbf{b}_i, i = 1, \ldots, m$. Let $\mathbf{r}_J$ be a linear combination of these columns, as defined by (3.72), where $\gamma_{IJ} \neq 0$. Show that replacement of $\mathbf{b}_I$ by $\mathbf{r}_J$ in $B$ results in a nonsingular matrix.

**3.49.** Show from the steps in the proof of duality in Section 3.6 that $\mathbf{c}^t - \mathbf{y}^{*t} A \geq \mathbf{0}$.

**3.50.** Show that by the duality theorem for (3.60), the optimum solutions $\mathbf{x}$, $\mathbf{y}$ satisfy the complementarity conditions (3.65). [Hint: The reasoning in Section 3.6 shows that for all $\mathbf{x}$ and $\mathbf{y}$ in their respective feasible solution sets, $\psi(\mathbf{x}) \geq L(\mathbf{x}, \mathbf{y}) \geq \phi(\mathbf{y})$. Hence the optimum solutions satisfy this statement with $\geq$ replaced by $=$. The first equation then gives $\mathbf{y}^t(A\mathbf{x} - \mathbf{b}) = 0$. Write this as a sum, and use the fact that in each term of the sum both factors are nonnegative to deduce the first line of (3.65). The second line is deduced in the same way from the second equation.]

**3.51.** Verify the assertions in Example 2 in Section 3.6.

**3.52.** From the proof of duality in Section 3.6 show that, when some of the constraints are equations: $(A\mathbf{x})_j = b_j$; then in the dual problem the corresponding components of $\mathbf{y}$ are unrestricted in sign, and in the complementarity conditions the corresponding values of $j$ are omitted.

**3.53.** Deduce (3.69) from (3.68') and (3.68'') as suggested in the text.

**3.54.** Verify the equivalence of the two duals considered in the next to last paragraph of Section 3.6.

**3.55.** Work through Example 1 in Section 3.7, and verify all the steps as applications of the simplex algorithm.


## 3.8  LP WITH BOUNDED VARIABLES

### p-Norms, Hölder Inequality

In Section 1.14 a norm in $\mathcal{R}^n$ is defined by the conditions (1.149), which we repeat:

$$\|\mathbf{x}\| > 0, \quad \text{except for} \quad \|\mathbf{0}\| = 0, \quad \|c\mathbf{x}\| = |c| \|\mathbf{x}\|, \quad \|\mathbf{x} + \mathbf{y}\| \le \|\mathbf{x}\| + \|\mathbf{y}\|. \quad (3.80)$$

Thus far we have encountered the Euclidean norm and the closely related $A$-norm. Also at the end of Section 1.14 another norm is mentioned for $\mathcal{R}^2$, which one can generalize to $\mathcal{R}^n$. This is the *1-norm*:

$$\|\mathbf{x}\|_1 = \sum_{i=1}^{n} |x_i|. \quad (3.81)$$

We now define a whole class of norms in $\mathcal{R}^n$, the $p$-norms, for $1 < p < \infty$:

$$\|\mathbf{x}\|_p = \left( \sum_{i=1}^{n} |x_i|^p \right)^{1/p}. \quad (3.82)$$

The first two norm properties in (3.80) can be immediately verified for the $p$-norm (Problem 3.56); for the last property (the "triangle inequality"), see Section 4.9. We observe that for $p = 1$ the $p$-norm reduces to the 1-norm (3.81), and for $p = 2$ it reduces to the Euclidean norm. Furthermore one can also make a definition for the case $p = \infty$ and obtain an $\infty$-norm:

$$\|\mathbf{x}\|_\infty = \max_{1 \le i \le n} |x_i|. \quad (3.83)$$

For this norm the properties (3.80) are easily verified (Problem 3.57).

The Hölder inequality is a relationship between the $p$-norm and the $q$-norm, for

$$\frac{1}{p} + \frac{1}{q} = 1;$$

here we allow $p = 1$ and $q = \infty$. The inequality asserts that

$$\sum_{i=1}^{n} |x_i y_i| \le \|\mathbf{x}\|_p \|\mathbf{y}\|_q. \quad (3.84)$$

For a proof, see Section 4.9. The inequality is also valid in the case $p = 1$, $q = \infty$ (Problem 3.58). We remark that for $p = q = 2$, the Hölder inequality (3.84) reduces to the *Schwarz inequality* (1.61').

We now return to a consideration of linear programming (LP) and proceed to replace the constraint $\mathbf{x} \geq 0$ by constraints that require all variables to be bounded. Since a nonnegative variable is not natural in science (e.g., tension and compression are negatives of each other) and quite often a variable is bounded above and below for physical reasons, we replace the constraint $\mathbf{x} \geq 0$ by $l_i \leq x_i \leq u_i$, $i = 1, 2, \ldots, n$, where $l_i$ (usually negative) and $u_i$ (usually positive) are constants; such constraints appear naturally in engineering in accordance with the properties of materials or the limits on devices. One should observe that, if all $l_i$ are $-1$ and all $u_i$ are $+1$, then the set of inequalities can be written as $\|\mathbf{x}\|_\infty \leq 1$.

We further replace the inequality $A\mathbf{x} \geq \mathbf{b}$ by an *equation* $A\mathbf{x} = \mathbf{b}$. Such an equation often comes naturally from conservation laws of physics. Otherwise, inequalities can always be replaced by equations, by introduction of *surplus* and *slack* variables (see Sections 3.6 and 3.7). These extra variables will then be bounded below by 0; they are also bounded above because they are expressible in terms of bounded variables through the equations; for example, if $x_1 - 3x_2 + s = 5$, where $s$ is a surplus variable and say $0 \leq x_1 \leq 7$, $0 \leq x_2 \leq 11$, then $s = 5 - x_1 + 3x_2$ implies that $s \leq 38$, and so finally $0 \leq s \leq 38$.

We present the new LP as a maximization, since many problems arise naturally in this form:

$$\textit{maximize} \quad \phi(\mathbf{x}) = \mathbf{c}^t \mathbf{x}$$

$$\textit{subject to} \quad A\mathbf{x} = \mathbf{b},$$

$$l_i \leq x_i \leq u_i, \qquad i = 1, 2, \ldots, n,$$

where matrix $A$ is $m \times n$, $m < n$, $A$ has rank $m$, $\mathbf{c} \in \mathcal{R}^n$, $\mathbf{c} \neq \mathbf{0}$, and $\mathbf{b} \in \mathcal{R}^m$. By shifting the origin and scaling (a linear transformation), we can symmetrize and normalize $\mathbf{x}$ to achieve desired values for the $l_i$ and $u_i$. By making them $-1$ and 1, as above, we achieve a new form for our LP, which we can write (without changing notation) as follows:

$$\textit{maximize} \quad \phi(\mathbf{x}) = \mathbf{c}^t \mathbf{x}$$

$$\textit{subject to} \quad A\mathbf{x} = \mathbf{b}, \tag{3.85}$$

$$\|\mathbf{x}\|_\infty \leq 1.$$

We call this problem the *standard* LP1 (1 stands for normalized bounds). The choice of a standard form is a matter of convenience. Since $S_f$ (if nonempty) is a bounded, closed convex set defined by linear equations and inequalities, the global maximum of the linear function $\phi$ is attained at some extreme point of $S_f$.

It is convenient to be able to ensure that $S_f$ is not empty. A necessary and sufficient condition that $S_f$ be nonempty is that the minimum distance from the origin to the linear variety $A\mathbf{x} = \mathbf{b}$ in the infinity norm be at most 1. (Finding this minimum distance is itself a problem in mathematical programming. A sufficient condition that is easier to implement is obtained by replacing the infinity norm by the Euclidean norm, since $\|\mathbf{x}\|_2 \leq 1$ implies that $\|\mathbf{x}\|_\infty \leq 1$, as in Problem 3.59. By squaring the Euclidean norm one is led to the problem of Section 3.4 with equality constraints: in particular, to (3.40') with $\psi(\mathbf{x}) = \mathbf{x}^t\mathbf{x}$. As in Problem 3.38, in the present notation, the minimum value of $\psi$ is $\mathbf{b}^t(AA^t)^{-1}\mathbf{b}$. Hence, if this last quantity is less than 1, the set $S_f$ is not empty.)

In the following discussion, we will assume that $S_f$ is not empty and that it includes interior points of the hypercube $\|\mathbf{x}\|_\infty \leq 1$ so that $S_f$ is itself an $(n-m)$-dimensional set, a convex polyhedron of that dimension. This is ensured by replacing $\leq$ by $<$ in the appropriate inequalities above: for example by ensuring that the Euclidean distance from the origin to the linear variety $A\mathbf{x} = \mathbf{b}$ is less than 1.

***Example 1.*** We give an example from mechanics in the form of LP1. A three-bar truss carries a load $P$ which is to be maximized. The truss and the load are shown with node and bar numbers in Fig. 3.5, where $\theta$ and $\beta$ are given angles and $t_1$, $t_2$, and $t_3$ denote tensions in the corresponding bars, which are fixed at nodes 1, 2, and 3. The equilibrium equations which balance the tensions with the load are

$$(-\cos\beta)t_1 \quad -t_2 \quad (-\cos\beta)t_3 \quad +(\cos\theta)P \quad = 0,$$
$$(\sin\beta)t_1 \qquad\qquad (-\sin\beta)t_3 \quad +(\sin\theta)P \quad = 0.$$

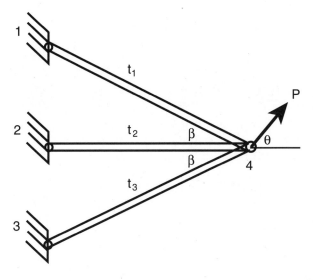

**Figure 3.5.** A three-bar truss under load.

We use the convention that a negative tension is compression. The inequalities $|t_i| \leq 1$, $i = 1, 2, 3$, represent the strength limits of the bars. Since $P \geq 0$ is unbounded above, we assume a large number $U$ (e.g., $U > 2$) to be an unattainable upper bound. We define a new variable $t_4 = (2P - U)/U$ (a linear transformation); then $|t_4| \leq 1$. By letting $\mathbf{c} = (0, 0, 0, U/2)^t$ and $\mathbf{x} = (t_1, t_2, t_3, t_4)^t$, we obtain a problem in the form of LP1 with a $2 \times 4$ matrix $A$ (of rank 2 except for very special $\beta$) and a vector $\mathbf{b} = (U/2)(-\cos\theta, -\sin\theta)^t$. The problem seeks the maximum equilibrium load under the given limits of bar strength. Such problems are called limit analysis problems in structural mechanics. We here consider only the special case $\beta = 45°$, $\theta = 0°$. This simple problem can be solved by a few steps of trial and error to give $t_1 = t_2 = t_3 = 1$ and $P_{\max} = 1 + \sqrt{2}$, so $\phi_{\max} = 1 + \sqrt{2} - (U/2)$.

A modified simplex method for general LP1 is presented in Section 3.11 below.

## Duality

We return to the general LP1 (3.85) whose feasible solution set $S_f$ is defined by the conditions $A\mathbf{x} = \mathbf{b}$, $\|\mathbf{x}\|_\infty \leq 1$. The problem has a dual, which we find by a familiar procedure. We let

$$L(\mathbf{x}, \mathbf{y}) = \mathbf{c}^t\mathbf{x} - \mathbf{y}^t(A\mathbf{x} - \mathbf{b}). \qquad (3.86)$$

Then by the Hölder inequality (3.84),

$$L(\mathbf{x}, \mathbf{y}) = (\mathbf{c} - A^t\mathbf{y})^t\mathbf{x} + \mathbf{b}^t\mathbf{y}$$

$$\leq \|\mathbf{c} - A^t\mathbf{y}\|_1 \|\mathbf{x}\|_\infty + \mathbf{b}^t\mathbf{y} \leq \|\mathbf{c} - A^t\mathbf{y}\|_1 + \mathbf{b}^t\mathbf{y} = \psi(\mathbf{y}).$$

Here $\mathbf{y} \in \mathcal{R}^m$ is unrestricted. For $\mathbf{x} \in S_f$, $L(\mathbf{x}, \mathbf{y}) = \mathbf{c}^t\mathbf{x} = \phi(\mathbf{x})$. Thus for all $\mathbf{y}$, $\phi(\mathbf{x}) \leq \psi(\mathbf{y})$. We show below that equality holds for appropriate $\mathbf{x}$ in $S_f$ and $\mathbf{y}$ in $\mathcal{R}^m$ so that the problem:

$$minimize \quad \psi(\mathbf{y}) = \|\mathbf{c} - A^t\mathbf{y}\|_1 + \mathbf{b}^t\mathbf{y} \quad \text{in} \quad \mathcal{R}^m \qquad (3.87)$$

is dual to the problem (3.85). From the rules of Section 3.2, we see that $\psi$ is convex (Problem 3.61). Hence the dual problem is one of minimizing a convex function without constraints. Such problems are considered in Section 3.11.

***Example 2.*** Consider the function $\psi$ for Example 1. The "dual variable" is a vector $\mathbf{y} = (u, v)^t$ in $\mathcal{R}^2$ with the meaning of the velocity vector of node 4 when the truss is under the maximum load and is about to collapse. For the case $\beta = 45°$, $\theta = 0$, we have

$$\psi(u, v) = \frac{1}{\sqrt{2}}|u - v| + |u| + \frac{1}{\sqrt{2}}|u + v| + \frac{U}{2}(|1 - u| - u).$$

By observation, this function takes its minimum value

$$\psi_{\min} = 1 + \sqrt{2} - \left(\frac{U}{2}\right) = \phi_{\max}$$

for $u = 1$, $-1 \leq v \leq 1$. In this case the dual minimizer is not unique.

*Proof of duality.* We seek $\mathbf{x}$ and $\mathbf{y}$ such that $\phi(\mathbf{x}) = \psi(\mathbf{y})$. As in Section 3.7 we write the equation $A\mathbf{x} = \mathbf{b}$ as

$$B\mathbf{v} + R\mathbf{w} = \mathbf{b}, \tag{3.88}$$

where we have permuted the columns of $A$ so that the first $m$ columns form the nonsingular matrix $B$; we denote these columns by $\mathbf{b}_1, \ldots, \mathbf{b}_m$ as before; also we again denote the columns of $R$ by $\mathbf{r}_1, \ldots, \mathbf{r}_{n-m}$ and define the vectors $\mathbf{c}_B$ and $\mathbf{c}_R$ so that $\mathbf{c}^t \mathbf{x} = \mathbf{c}_B^t \mathbf{v} + \mathbf{c}_R^t \mathbf{w}$. As for LP in Section 3.6 we can determine the form of the extreme points of $S_f$: They have $n - m$ coordinates $w_j$ equal to $\pm 1$ and the remaining coordinates $v_i$ in the interval $[-1, 1]$; in each case they provide a solution of (3.88) as described above. Now let $\phi$ attains its maximum at such a point with $\mathbf{v}$ and $\mathbf{w}$ as described. We further assume that $-1 < v_i < 1$ for $i = 1, \ldots, m$; otherwise, we are dealing with a degenerate solution. As in Section 3.7 we calculate

$$\phi_{\max} = \mathbf{c}_B^t B^{-1} \mathbf{b} + (\mathbf{c}_R^t - \mathbf{c}_B^t B^{-1} R)\mathbf{w}.$$

We can write this as

$$\phi_{\max} = \mathbf{c}_B^t B^{-1} \mathbf{b} + h_1 w_1 + \cdots + h_{n-m} w_m,$$

where

$$h_j = c_{R,j} - \mathbf{c}_B^t B^{-1} \mathbf{r}_j.$$

Next we write

$$\psi(\mathbf{y}) = \|\mathbf{c} - A^t \mathbf{y}\|_1 + \mathbf{b}^t \mathbf{y}$$

$$= \sum_{i=1}^{m} |c_{B,i} - \mathbf{b}_i^t \mathbf{y}|$$

$$+ \sum_{j=1}^{n-m} |c_{R,j} - \mathbf{r}_j^t \mathbf{y}| + \mathbf{b}^t \mathbf{y}.$$

We now choose $\mathbf{y}$ to make the first sum vanish, that is, so that $\mathbf{c}_B - B^t \mathbf{y} = \mathbf{0}$ (exactly as for LP!). We claim that, with this choice of $\mathbf{y}$, $\psi(\mathbf{y}) = \phi_{\max}$. We first verify (Problem 3.61) that with this choice of $\mathbf{y}$,

$$\psi(\mathbf{y}) = |h_1| + \cdots + |h_{n-m}| + \mathbf{c}_B^t B^{-1} \mathbf{b}.$$

Hence the desired equality will follow if we can show that all $w_j h_j$ are nonnegative; that is, $w_j = 1$ if $h_j > 0$, $w_j = -1$ if $h_j < 0$.

To establish these conditions, we apply the Karush–Kuhn–Tucker conditions (3.26) at the maximum point $(v_1, \ldots, w_1, \ldots)$. The conditions are applicable, since, by the restriction imposed above, there are $n$ active constraints at the point which determine the $v_i$ and $w_j$ uniquely. From the components of the gradients involving the $v_i$, we obtain a relation

$$\mathbf{c}_B^t - (\mu_1, \ldots, \mu_m)B = \mathbf{0}.$$

From the components involving the $w_j$, we obtain a relation

$$\mathbf{c}_R^t - (\mu_1, \ldots, \mu_m)R = (\lambda_1 w_1, \ldots, \lambda_{n-m} w_{n-m}),$$

where $\lambda_j \geq 0$ if $w_j = 1$, and $\lambda_j \leq 0$ if $w_j = -1$. From these equations we deduce that $\lambda_j w_j = h_j$ and we are done. (See Problem 3.61.) ∎

REMARK. As in the proof of duality for LP in Section 3.6, a continuity argument shows that the duality remains valid even if some $v_i = \pm 1$ at the maximizer of $\phi$.

### Relation between $\mathbf{x}^*$ and $\mathbf{y}^*$

If the maximizer $\mathbf{x}^*$ is known, then $B$, $R$, $c_B$, $c_R$ are known, and hence $\mathbf{y}^*$ is found from the defining equation $c_B - B^t \mathbf{y}^* = \mathbf{0}$. If the dual problem has a unique solution $\mathbf{y}^*$, then it must be the solution just considered. Now the maximizer $\mathbf{x}^*$ of $\phi$ satisfies the equations

$$\mathbf{c}^t \mathbf{x}^* = \mathbf{c}^t \mathbf{x}^* - \mathbf{y}^{*t}(A\mathbf{x}^* - \mathbf{b}) = (\mathbf{c} - A^t \mathbf{y}^*)^t \mathbf{x}^* + \mathbf{b}^t \mathbf{y}^* = \sum_{i=1}^{n} g_i x_i^* + \mathbf{b}^t \mathbf{y}^*,$$

where $g_i = (\mathbf{c}^t - A^t \mathbf{y}^*)_i$ for $i = 1, \ldots, n$. From the defining equation for $\mathbf{y}^*$, we conclude that $m$ of the coefficients $g_i$ are 0; those which are 0 determine $B$ and $R$. The remaining terms in the sum are precisely the terms of the sum $h_1 w_1 + \cdots + h_{n-m} w_{n-m}$. The coefficients $h_j$ are simply the corresponding $g_i$, which are known. In the proof above we saw that each of these terms is nonnegative. Hence, if none is zero, we can find each $w_j$, chosen as 1 or $-1$. With the $w_j$ known, the $\mathbf{v}$ part of $\mathbf{x}^*$ can be found from the equation $\mathbf{v} = B^{-1}(\mathbf{b} - R\mathbf{w})$. This recovery of the solution of the primal problem requires only one matrix computation.

Since the LP1 in (3.85) can be converted to a standard LP, the simplex algorithm can be applied to it. In fact the algorithm can be modified to solve (3.85) directly without conversion. But the simplex method is not always the preferred method, since it requires a feasible starting point and its rate of convergence depends on the choice of this starting point. We will see in Section 3.11

that there is a good nonsimplex method for solving the dual problem. For the cases in which this method is successful and economical, it is a good way to obtain both solutions, primal and dual. As our discussion indicates, there may be complications: Because of degeneracy there may be a nonunique dual solution $\mathbf{y}^*$; one cannot expect a computer procedure to produce $g_i$'s that are exactly 0. One responds to both difficulties by selecting the $m$ $g_i$ that are smallest in absolute value. It turns out that one still obtains a correct $B$. This in turn produces a correct $\mathbf{x}$ from the equation $B\mathbf{v} = \mathbf{b} - R\mathbf{w}$; typically some small inaccuracy in $\mathbf{y}$ does not affect the $w_j$, since they are chosen as $\pm 1$.

***Example 3.*** Maximize $x_1 + x_2$ subject to $x_1 - x_2 = b$, $|x_1| \leq 1$ and $|x_2| \leq 1$. This small problem has $m = 1$, $n = 2$. It is easy to derive the dual: Minimize $|1 - y| + |1 + y| + by$. For $b = 0.5$ we find that $y^* = -1$ and $\psi(-1) = 1.5$. To recover the primal solution, we compute $g_1 = 1 - (-1) = 2$, $g_2 = 1 + (-1) = 0$. Hence $x_2 = v$, $x_1 = w = 1$, and $B = [-1]$. We solve the $1 \times 1$ matrix equation, $[-1]v = b - [1]w$, to obtain $x_2 = v = -0.5 + 1 = 0.5$. As a check, $\phi(1, 0.5) = 1 + 0.5 = 1.5 = \psi(-1)$.

For the case $b = 0$, the dual solution is not unique (a case of degeneracy); all $\mathbf{y}^*$ in the interval $[-1, 1]$ lead to the same $\psi_{\min} = 2$. For $y^* = \pm 1$, one of the $(g_1, g_2)$ will be zero. The recovery procedure works normally and produces $x_1 = 1$, $x_2 = 1$, and $\phi_{\max} = 2$. But for $y^* = 0$, no zero $g_i$ can be found $(g_1 = 1, g_2 = 1)$. We may arbitrarily choose $w = x_2 = 1$, which leads to $[1]v = x_2 = 1$ and $\phi_{\max} = 2$.

REMARK. From the materials presented up to this point in this chapter, we might now expect to proceed to functions and constraints involving higher-degree polynomials or more general nonlinearity. But general nonlinear optimization problems are very difficult, and the field of *nonlinear programming* (NP) is much less developed compared with QP and LP. We will follow an intermediate itinerary through a special class of nonlinear optimization problems where the objective function and constraints are *convex*. Many problems in the physical sciences and engineering naturally embody the property of convexity. *Convex programming* (CP) includes many of the most important optimization problems arising in applications. The classes LP and QP are subclasses of CP; therefore, methods developed in CP apply to LP and QP.

## PROBLEMS

**3.56.** For the $p$-norm defined by (3.82) for $1 < p < \infty$, verify the first two properties in (3.80).

**3.57.** For the $\infty$-norm defined by (3.83) verify the properties (3.80).

**3.58.** Verify the Hölder inequality (3.84) for the case $p = 1, q = \infty$.

**3.59.** Show that for $\mathbf{x}$ in $\mathcal{R}^n$, $\|\mathbf{x}\|_\infty \leq \|\mathbf{x}\|_2$, and show the geometric meaning of this inequality for $n = 2$ by graphing the sets $\|\mathbf{x}\|_\infty \leq 1$ and $\|\mathbf{x}\|_2 \leq 1$ in the $x_1 x_2$-plane.

**3.60.** Consider Example 1 in Section 3.8 for the case $\beta = 45°$, $\theta = 0$.

    **(a)** Use trial and error to seek the maximum of the load $P$ satisfying the stated equilibrium equations, where $|t_i| \leq 1$ for $i = 1, 2, 3$.

    **(b)** Use the substitution suggested to write the problem in the form (3.85) and from (3.87) obtain the dual as stated in Example 2 of Section 3.8. Verify that (for $U > 2$) $\psi$ takes on the value $\phi_{\max}$ as stated.

**3.61.** In Section 3.8 a proof is given that the dual of the program (3.85) is the program (3.87).

    **(a)** Show that the function $\psi$ defined by (3.87) is convex. [Hint: See Section 3.1.]

    **(b)** In the notations of that proof, verify that

$$\psi(\mathbf{y}) = |h_1| + \cdots + |h_{n-m}| + \mathbf{c}_B^t B^{-1} \mathbf{b}.$$

    **(c)** In the notations of the proof, verify that $\lambda_j w_j = h_j$ and hence that $w_j = 1$ if $h_j > 0$, $w_j = -1$ if $h_j < 0$.

**3.62.** Verify all the details in Example 3 of Section 3.8.

## 3.9 CONVEX FUNCTIONS AND CONVEX PROGRAMMING

### Minimizing a Convex Function Over Its Domain

A convex function is always defined over a convex set $E$ (Section 3.1). Unless $E$ is the whole space $\mathcal{R}^n$, this domain set $E$ is an implied constraint. Therefore unconstrained minimization of a convex function assumes that the function is defined over the whole space. Commonly the convex set $E$ is described by *functional constraints*; that is, by constraints in the form of equations or inequalities which must be satisfied by certain functions of the coordinates. Such constraints are illustrated by all the programming problems considered earlier in this chapter.

In this section we present some theoretical results and methods for minimizing a convex function $\psi(\mathbf{x})$ defined over a closed convex set $E$ in $\mathcal{R}^n$. The minima may occur in the relative interior or on the relative boundary of $E$. By Section 3.1, if $\psi$ is strictly convex and has a local minimum on $E$, the problem

$$\min_{\mathbf{x} \in E} \psi(\mathbf{x}) \tag{3.90}$$

always has a unique minimizer, and it provides the global minimum of $\psi$. Furthermore, the minimizer provides the only local minimum of $\psi$. This is a very attractive property, since an iterative method that produces a decrease in the function value in each step can be expected to converge to the minimizer. One should think of walking down the slope of a landscape that is bowl-shaped. The simplex method for LP can be considered as such an iterative method that does converge. Various descent methods will be presented in this section.

If $\psi$ is convex, but not strictly convex, then as in Section 3.1 each local minimum of $\psi$ is a global minimum. The minimizer of $\psi$, if there is one, may not be unique. However, the set of minimizers has another nice property: The set $M$ of all minimizers forms a convex subset of $E$ on which $\psi$ is constant. (See Section 1.9.) Thus the graph of $\psi$ has a flat bottom over the minimizing set. A descent method may converge to one point in $M$; if so, one may examine the neighborhood for more solutions if desired.

Further theoretical results can be derived when the convex function is once or twice differentiable. Such smoothness conditions are needed for a number of theorems in optimization. It is generally true that nonsmooth optimization problems are more difficult and require special techniques (Sections 3.6 and 3.7). We turn now to the smooth problems and assume that $E$ is a closed region in $\mathcal{R}^n$. We denote the gradient vector $\nabla\psi$ by $\mathbf{g}(\mathbf{x})$ and the Hessian matrix $(\psi_{x_i x_j})$ by $H(\mathbf{x})$. We assume that these are defined and continuous in an open region containing $E$.

A necessary condition for $\psi$ to have a local minimum at an interior point $\mathbf{x}$ of $E$ is that at $\mathbf{x}$ all directional derivatives of $\psi$ are nonnegative; that is,

$$\mathbf{g}^t \mathbf{d} \geq 0 \tag{3.91}$$

for all unit vectors $\mathbf{d}$. If $\mathbf{x}$ is a boundary point of $E$, at which the boundary is smooth, there is a similar necessary condition: (3.91) must hold for all $\mathbf{d}$ that point to the interior of $E$ or are tangent to the boundary; this is the essential idea in the proof of the Karush–Kuhn–Tucker conditions in Section 2.8. For an interior point, the condition is equivalent to the condition that $\mathbf{g} = \mathbf{0}$ at the point; as in Section 1.9, if also $H(\mathbf{x})$ is positive definite, then the critical point does provide a local minimum for $\psi$. Finding such a critical point is equivalent to solving $n$ nonlinear equations. A standard method for finding the solution is *Newton's iteration:*

$$\mathbf{x}^{(k+1)} = \mathbf{x}^{(k)} - \left[ H(\mathbf{x}^{(k)}) \right]^{-1} \mathbf{g}(\mathbf{x}^{(k)}), \qquad k = 0, 1, \ldots,$$

which converges if the starting point is sufficiently close to the critical point and the Hessian matrix $H$ is positive definite in a neighborhood of the point. The Newton method is known to lead to a quadratic rate of convergence. (See Isaacson and Keller, 1966, ch. 3.)

***Example 1.*** We seek the minimum of the function $\psi(x, y) = x^2 + y^2 + e^{-x-y}$ of Example 1 in Section 1.11. As pointed out in Section 1.11, the function has a positive definite Hessian matrix in $\mathcal{R}^2$ and hence strictly convex in $\mathcal{R}^2$. We find that Newton's iteration can be written in the form $\mathbf{x}^{(k+1)} = \Phi(\mathbf{x}^{(k)})$, where

$$\Phi((x, y)^t) = \frac{x + y + 1}{2(e^{x+y} + 1)}(1, \ 1)^t.$$

Thus, if we start at $(x_0, \ y_0)^t = (1, \ 1)^t$, then we obtain successive iterates

$$(0.17880, \ 0.17880)^t, \quad (0.27935, \ 0.27935)^t, \quad (0.28357, \ 0.28357)^t, \ldots$$

The last one shown agrees with the results found in Section 1.11. (See Problem 3.63.)

If $H$ is positive definite in $E$, then $\psi$ is strictly convex (Section 1.7), and we know that there is at most one local minimum, which provides the global minimum of $\psi$. If $H$ is only positive semidefinite on $E$, then $\psi$ is convex on $E$ and, as in Section 1.7, we are sure that every local minimum of $\psi$ is a global minimum. If $E$ is bounded, such a local minimum must exist, but it may not be unique.

### Method of Steepest Descent

The method is intuitively natural. One imagines that one is walking blind-folded on the slope of a bowl-shaped valley, trying to reach the bottom. At a given starting point, one feels a direction of steepest slope and proceeds downhill in that direction until no more descent is felt. Then one starts over from the point reached in another steepest descent direction. By the convexity of the valley's shape function, one should eventually attain the bottom.

Again we discuss the case where the minimizer is an interior point of $E$. Let $\mathbf{g}_k$ be the gradient vector at the current point $\mathbf{x}^{(k)}$. The next point is reached by the formula

$$\mathbf{x}^{(k+1)} = \mathbf{x}^{(k)} - \alpha\mathbf{g}_k, \tag{3.92}$$

where $\alpha$ is a variable to be chosen as the value of $t$ which minimizes the function: $\bar{\psi}(t) = \psi(\mathbf{x}^{(k)} - t\mathbf{g}_k)$ over the interval for which the function is defined. This is a 1-dimensional (or line) search. We find (Problem 3.64) $\bar{\psi}'(0) = -\mathbf{g}_k^t\mathbf{g}_k = -\|\mathbf{g}_k\|^2 \leq 0$. For small $t$, $\psi(\mathbf{x}^{(k)} - t\mathbf{g}_k) < \psi(\mathbf{x}^{(k)})$ if $\mathbf{g}_k \neq 0$. This also implies that

$$\psi(\mathbf{x}^{(k+1)}) = \min_{t>0} \psi(\mathbf{x}^{(k)} - t\mathbf{g}_k) < \psi(\mathbf{x}^{(k)}), \quad \mathbf{g}_k \neq 0,$$

and we have a descent step. Under appropriate conditions the algorithm will converge to a point $\mathbf{x}^*$ at which $\mathbf{g}(\mathbf{x}^*) = 0$ and $\psi(\mathbf{x}^*) = \psi_{\min}$. (See Fletcher, 1987, sec. 2.5.)

The method of steepest descent is simple and requires only the first derivatives of the functions to be evaluated. In practice, this method is normally slower than Newton's method. The rates for various modified and quasi-Newton methods fall between the two.

The method of steepest descent also applies to nonconvex functions as a procedure for seeking local minima. Newton's method can be applied generally to seek critical points of a function.

There are various known ways of carrying out the line search (see Fletcher, 1987, sec. 2.6).

***Example 2.*** We apply the method of steepest descent to the problem of Example 1 above. We observe that along each straight line $x = x_0 + at$, $y = y_0 + bt$, the function $\psi$ becomes a strictly convex function of the form

$$f(t) = p + qt + rt^2 + se^{ut}.$$

Further $f$ has limit $\infty$ as $|t| \to \infty$, so $f$ has a unique minimum. We can find the minimizer by finding the unique critical point of $f$, the solution of $f'(t) = 0$. This solution can be found by Newton's iteration and we verify that this follows the formula:

$$t_k = \Phi(t_{k-1}), \quad \text{where} \quad \Phi(t) = \frac{use^{ut}(ut - 1) - q}{2r + su^2 e^{ut}}.$$

The gradient vector $\mathbf{g}$ is $(2x - e^{-x-y}, 2y - e^{-x-y})^t$. If we start at the vector $(1, 1)^t$, then the line directed by the negative gradient passes through the known minimizer $(0.28357, 0.28357)^t$; then minimizing $f(t)$ for this line, by Newton's method, yields the solution sought. If we start at a point not on this line, more than one choice of direction is needed. (See Problem 3.65.)

### Steepest Descent by Solving Ordinary Differential Equations

Closely related to the method of steepest descent is the method described in Sections 1.11 and 2.4: one simply finds a solution $\mathbf{x} = \mathbf{x}(t)$ of the differential equation

$$\frac{d\mathbf{x}}{dt} = -\mathbf{g}(\mathbf{x}).$$

As in the sections mentioned, under appropriate hypotheses the solution approaches a minimizer as the limit, as $t \to \infty$.

### Conjugate Gradient (CG) Method

This is an extension of a method with the same name commonly used for solving linear equations (or for minimizing quadratic functions). The original method can, in principal, be adapted to minimizing a nonquadratic function $\psi$, but it would then require computing the Hessian matrix $H(\mathbf{x})$ repeatedly; for this reason this procedure is rarely used.

We present another version of a nonquadratic CG method, using only the gradient vectors $\mathbf{g}_k$ and the conjugate direction vectors $\mathbf{d}_k$:

0.  Start with an arbitrary $\mathbf{x}^{(0)}$.
1.  Set $k = 0$, $\mathbf{g}_0 = \nabla\psi(\mathbf{x}^{(0)})$, $\mathbf{d}_0 = -\mathbf{g}_0$.
2.  For $k = 1$ to $n$:
    $\mathbf{x}^{(k)} = \mathbf{x}^{(k-1)} + \alpha\mathbf{d}_{k-1}$ where $\alpha$ minimizes $\psi$ along the direction $\mathbf{d}_{k-1}$
    $\mathbf{g}_k = \nabla\psi(\mathbf{x}^{(k)})$
    $\beta = [\mathbf{g}_k^t\mathbf{g}_k]/[\mathbf{g}_{k-1}^t\mathbf{g}_{k-1}]$
    $\mathbf{d}_k = -\mathbf{g}_k + \beta\mathbf{d}_{k-1}$
    Next $k$.
3.  If $\|\mathbf{g}_n\| \geq \varepsilon$, $\mathbf{x}^{(0)} = \mathbf{x}^{(n)}$ go to step 1.
4.  End.

For a quadratic function with Hessian matrix $A$, the method has a simple geometric meaning (see Hestenes, 1975, ch. 1): One modifies the steepest descent method by changing the direction at each stage to produce a sequence of directions $\mathbf{d}_k$ which form an orthogonal system with respect to the inner product associated with the $A$-norm—that is, so that $\mathbf{d}_i^t A\mathbf{d}_j = 0$ for $i \neq j$. In general, two directions determined by vectors $\mathbf{p}, \mathbf{q}$ are said to be *conjugate* (with respect to the matrix $A$) when $\mathbf{p}^t A\mathbf{q} = 0$. In the quadratic case the procedure terminates after at most $n$ iterations in step 2, so step 3 is not needed. (The method is discussed in Fletcher 1987, ch. 4.)

### Penalty Method

Now let us bring in the constraints. Consider a *convex program* (CP) in the form

$$\begin{aligned} \textit{minimize} \quad & \psi(\mathbf{x}) \\ \textit{subject to} \quad & \mathbf{x} \in S_f \subseteq \mathcal{R}^n. \end{aligned} \tag{3.93}$$

The penalty method replaces (3.93) by an unconstrained problem:

$$\textit{minimize} \quad \psi(\mathbf{x}) + c\rho(\mathbf{x}).$$

Here $c$ is a positive constant and $\rho(\mathbf{x})$ is a penalty function constructed to have the properties that $\rho$ is convex in $\mathcal{R}^n$, is positive outside of $S_f$, and has a flat bottom: $\rho(\mathbf{x}) = 0$ for all $\mathbf{x} \in S_f$. Examples of penalty functions are given in Fletcher (1987, ch. 12). The idea of the method is to make $c$ very large so that an infeasible point adds a large positive value to the function to be minimized. The minimizing process naturally keeps the iterates feasible or close to $S_f$.

There is a common problem associated with all penalty functions. If $c$ is too large, the value of $c\rho$ swamps that of $\psi$, and every point in $S_f$ becomes an approximate minimizer. In practice, the method diverges or converges very slowly. A remedy is to use a sequence of increasing values of $c$, with a small initial value. Initially the minimum of $\psi + c\rho$ is given by a point close to the minimizer of $\psi$ in $\mathcal{R}^n$, which may not be feasible. Subsequent larger values of $c$ force the minimizer of the function with penalty term toward the feasible region. As $c \to \infty$, if the value of $c\rho$ at the minimizer approaches 0, then the method produces an exact solution of the original problem (3.93).

**Example 3.** We give a simple example: Find the shortest distance from the origin of $\mathcal{R}^2$ to the line $ax + by + c = 0$, which defines the feasible set $S_f$. We write

$$minimize \quad x^2 + y^2 + \mu(ax + by + c)^2,$$

where $\mu$ is now the penalty parameter. We can solve this problem exactly to obtain

$$x(\mu) = -\frac{ac\mu}{1 + (a^2 + b^2)\mu}, \qquad y(\mu) = -\frac{bc\mu}{1 + (a^2 + b^2)\mu}.$$

As $\mu \to \infty$, we obtain $x^* = -ac/(a^2 + b^2)$ and $y^* = -bc/(a^2 + b^2)$, the exact solution.

Choice of a proper penalty function and a proper sequence of penalty parameters determines the success or failure of the method. Simplicity is the major attraction of the method.

We now consider a CP in the form:

$$minimize \quad \psi(\mathbf{x})$$
$$subject\ to \quad A\mathbf{x} = \mathbf{b}, \tag{3.94}$$
$$\mathbf{x} \in \mathcal{R}^n.$$

Here $A$ is an $m \times n$ matrix and is assumed to have rank $m < n$ and $\psi$ is assumed to be a smooth function which is strictly convex on the convex set $S_f$.

For the problem (3.94), there are many possible choices of a penalty function $\rho$. We can use $\|A\mathbf{x} - \mathbf{b}\|_2^2$, as suggested by the example given above. Or we can use

$$\rho(\mathbf{x}) = \sum_{i=1}^{m} \max\{0,\ a_{i1}x_1 + \cdots + a_{in}x_n - b_i\}.$$

Construction of a penalty function is not an exact science, but convergence of iterative methods with certain penalty functions has been analyzed (Fletcher, 1987, ch. 12).

Penalty methods are like the Lagrange method with only one multiplier. They lump all constraints into a single real functional constraint and thereby gain simplicity. They do not have the elegance of the Lagrange method, which is often accompanied by duality theorems.

### Duality of Convex Programming

In the following sections we give some further examples of duals of convex programs. As pointed out in Section 3.2, knowledge of a dual program may provide a useful additional way of seeking the minimizer or maximizer of a given program. This idea is also illustrated in the discussion to follow. General methods for seeking duals of convex programs are given in Chapter 4.

### PROBLEMS

**3.63.** Verify the results found in Example 1 in Section 3.9.

**3.64.** Let $f(t) = \psi(\mathbf{a} - t\mathbf{b})$, where $\mathbf{a}$ is a constant vector in $\mathcal{R}^n$, $\psi(\mathbf{x})$ is differentiable in a neighborhood of $\mathbf{a}$ and $\mathbf{b} = \nabla\psi(\mathbf{a})$. Show that $f'(0) = -\|\mathbf{b}\|^2$.

**3.65.** Consider Example 2 in Section 3.9.

    **(a)** Verify the results given in the text.

    **(b)** Seek the minimizer by the same method but start at $(1, 2)^t$.

**3.66.** Let $\psi(\mathbf{x})$ be the quadratic function $x^2 + 3y^2 + 3z^2 + 2xy + 4yz + 4x + 6y + 2z$ of $(x, y, z)$.

    **(a)** Show that $\psi$ is strictly convex and find its minimizer.

    **(b)** Follow the method of conjugate gradients starting at $(0, 0, 0)$ to find the minimizer. Verify that the vectors $\mathbf{d}_0, \mathbf{d}_1, \mathbf{d}_2$ are pairwise orthogonal with respect to the inner product associated with the $A$-norm, where $A$ is the Hessian matrix of $\psi$. Thus these vectors have conjugate directions.

**3.67.** Apply the method of conjugate gradients to seek the minimizer of the function $\psi$ of Example 1 in Section 3.9, starting at $(1, 0)$. Carry out the loop $k = 1, 2$ just once to obtain $\mathbf{d}_1, \mathbf{x}^{(1)}, \mathbf{x}^{(2)}$. To save effort, use software (such as *fzero* in MATLAB) to find the values of $\alpha$.

**3.68.** Verify the details in Example 3 in Section 3.9.

### 3.10 THE FERMAT–WEBER PROBLEM AND A DUAL PROBLEM

The Fermat–Weber problem is a classical problem of plane geometry; it is also an example of the *facility location problem*, of much practical importance.

For $k \geq 2$, one is given $k$ fixed distinct points $(x_1, y_1), \ldots, (x_k, y_k)$ in the plane. One considers these points as given locations of consumption (e.g., households in a village that need water). A single point $(x, y)$ (e.g., a well) is to be located from which to distribute the supply. One assumes that the lines of distribution (e.g., footpaths or pipelines), connecting each point of consumption to the point of supply, are straight lines. One seeks the best location in the plane for the point of supply by choosing it to minimize the total length of the supply lines. The statement

$$minimize \quad \psi(x, y) = \sum_{i=1}^{k} \sqrt{(x - x_i)^2 + (y - y_i)^2} \qquad (3.100)$$

defines the Fermat–Weber problem in a plane. (No constraint is given, so one is seeking the minimum of $\psi(x, y)$ in $\mathcal{R}^2$.)

Many problems in engineering, economics, management and other sciences can be formulated as a version of this well-known problem of Euclidean geometry or of a more general problem in $\mathcal{R}^n$:

$$minimize \quad \sum_{i=1}^{k} \sqrt{\mathbf{x}^t A_i \mathbf{x} + \mathbf{b}_i^t \mathbf{x} + c_i}. \qquad (3.101)$$

Here each $A_i$ is assumed to be a positive definite symmetric matrix of order $n$; the $\mathbf{b}_i$ and $c_i$ are vector and scalar constants. It is assumed that in each term the quadratic function under the square root sign has a nonnegative minimum (found as in Section 3.3), so each term is defined in $\mathcal{R}^n$.

The simple-looking problem (3.100) does not have an explicit solution except in some very special cases. That a global minimum exists is intuitively obvious and easily proved (Problem 3.69). A numerical solution can be obtained only by an iterative process. The function $\psi$ is a member of the class of functions discussed in the next to last paragraph of Section 3.1. Hence, except when the given points are on a line, the function is strictly convex, so the minimum occurs at a unique point $(x^*, y^*)$. When all the given points are on a line, the function remains convex, and the minimizer may not be unique. This is illustrated by the case $k = 2$, where all points of the segment between $(x_1, y_1)$ and $(x_2, y_2)$ minimize the function.

Similar remarks apply to the more general problem (3.101). The function has a global minimum by the rule for functions with limit $\infty$ for $\|\mathbf{x}\| \to \infty$ (Section 1.12) (Problem 3.70). The function is convex, and except in certain special cases, it is strictly convex and the minimizer is unique.

There is one more complication to consider when problems of this type are solved numerically. The function in (3.100) is not differentiable at the fixed points $\mathbf{x}_1, \ldots, \mathbf{x}_k$. Evaluation of a gradient at or near one of these points will cause floating point overflow and failure of the computational procedure. Methods based on gradients and/or higher derivatives are very common in optimization, as illustrated in Section 3.9.

Much research has been conducted on the Fermat–Weber problem and many good algorithms for it are now available; one algorithm is given in Section 3.11 below. Here the consideration of two simple examples will reveal certain features of the problem. If we assign a unit mass to each fixed point, the center of mass seems to be a logical choice for location of the optimal point $(x^*, y^*)$ for (3.100). It is obviously correct for the case of three points at the vertices of an equilateral triangle or the case of more than three points arranged in a symmetrical configuration. But we already know in the case $k = 2$ that points other than the midpoint are also optimal.

***Example 1.***   For $k = 3$ let the points be $(0, 0)$, $(1, 0)$ and $(0, 2)$. The unique optimal solution is found to be

$$(x^*, y^*) = (0.3045036, 0.2545693),$$

with seven digits of accuracy. The reader should verify the result by numerical experimentation (Problem 3.71). This optimal point is surprisingly far from the mass center $(0.3333333, 0.6666667)$.

***Example 2.***   For $k = 3$ let the three points be $(0, 1)$, $(0, -1)$, and $(q, 0)$, where $0 \leq q < \infty$. By symmetry, we conclude that $y^* = 0$. The function $x^*(q)$ is graphed in Fig. 3.6.

Figure 3.6 illustrates an important aspect of the solution: When the third point $(q, 0)$ is far from the other two, the optimal solution is hardly affected by the location of the third point. One finds generally that a cluster of points has more power to attract the optimal point to their neighborhood. Thus the well for a village water supply system should be located near the concentration of houses or on the property of a specific house among them (Problem 3.72).

If one has no algorithm available for attacking this problem, then a brute force search method will suffice to produce optimal solutions for small $k$. The experience of a naive search method even for the case $k = 3$, $n = 2$ will definitely reinforce one's opinion that an efficient and stable method is absolutely necessary for solving the Fermat-Weber problems for larger choices of $k$; this conclusion is all the more true for generalizations to higher dimensions.

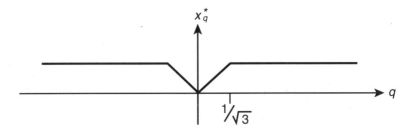

**Figure 3.6.** Solution of a Fermat–Weber problem depending on a parameter.

A slightly more general version of the Fermat-Weber problem (3.100) inserts a positive constant as multiplier (weight) of each term. The behavior of the solution will be similar to that of (3.100) except that a point with greater weight exerts more attraction on the optimal point.

## Dual Problems

In the LP and QP problems presented earlier, duality theorems provided a basis for alternative computational methods. The property of duality appears also in convex problems. We will now describe a second program that turns out to be a dual problem of (3.100).

We consider the mechanics problem of finding the equilibrium point of $k$ attached strings pulled by weights. Let $k$ strings be given, labeled by index $i$, $i = 1, \ldots, k$. Let the $i$th one be threaded through a hole located at the point $\mathbf{x}_i$ in a horizontal plate as shown in Fig. 3.7, and let all the strings be tied together at one end $P$ in the plane of the plate. Each end below the plate is connected to a unit weight $W$ as shown in Fig. 3.7.

We assume that there is no friction, so the strings move freely on the plate and through the holes. Because of gravity the strings will be taut under tension exerted by the weights. When equilibrium is reached, the point $P$ assumes the location $(x^*, y^*)$, called the equilibrium point. Let the forces exerted on the point $P$ by the strings be denoted by $\mathbf{f}_1, \mathbf{f}_2, \ldots, \mathbf{f}_k$, each a vector in the plane. They satisfy the equilibrium equation:

$$\sum_{i=1}^{n} \mathbf{f}_i = \mathbf{0}, \tag{3.102}$$

and the magnitude of each force is equal to unity if the point $P$ is not located at a hole. If $P$ is at a hole, say at $\mathbf{x}_I$, the force exerted by the $I$th string on $P$ is

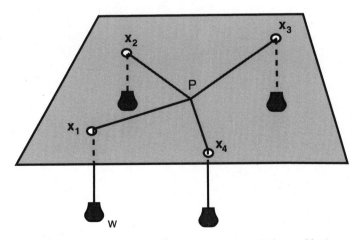

**Figure 3.7.** Mechanics model for a dual Fermat–Weber problem.

not determined by the weight but rather by the equilibrium equation (3.102), and it satisfies the inequality $\|\mathbf{f}_l\| \leq 1$. Without predetermining the location of the equilibrium point, whether it coincides with a hole or not, the forces $\{\mathbf{f}_i, \ i = 1, 2, \ldots, k\}$ can be determined by the following constrained maximization program:

$$maximize \quad \phi(\mathbf{f}_1, \ldots, \mathbf{f}_k) = \sum_{i=1}^{k} \mathbf{x}_i^t \mathbf{f}_i$$

$$subject \ to \quad \sum_{i=1}^{k} \mathbf{f}_i = \mathbf{0}, \tag{3.103}$$

$$\|\mathbf{f}_i\|_2 \leq 1, \qquad i = 1, 2, \ldots, k.$$

[We can arrive at (3.103) by the following physical reasoning: The first constraint is to be expected for an equilibrium state. Next observe that by this constraint we can replace $\mathbf{x}_i$ by $\mathbf{x}_i - \mathbf{x}^*$ in the function $\phi$. The forces should be unit vectors that are positive scalars times these vectors, and such choices of the forces clearly maximize $\phi$ under the given constraints. This argument has to be modified when $\mathbf{x}^*$ is one of the $\mathbf{x}_i$.]

Since the objective function in (3.103) is linear and the constraint set is convex (Problem 3.73), (3.103) is again a CP. Since the constraint set is also closed and bounded, the function has a global maximum.

Given the same set of points (data) for (3.100) and (3.103), the two seemingly quite unrelated problems have the same optimal value of their respective objective functions, and an optimal location of (3.100) determines a maximizer for (3.103). We now prove this duality theorem, considering two cases: Case 1. The function $\psi$ has a minimizer $\mathbf{x}^*$ which is not one of the given points $\mathbf{x}_i$. Case 2. The function $\psi$ is minimized at $\mathbf{x}_k$.

*Proof for Case 1.*   By assumption,

$$\psi_{\min} = \sum_{i=1}^{k} \|\mathbf{x}_i - \mathbf{x}^*\|_2. \tag{3.104}$$

We may write $\mathbf{x}_i = \mathbf{x}^* + \mathbf{u}_i$, $i = 1, 2, \ldots, k$, where $\mathbf{u}_i$ is the position vector of $\mathbf{x}_i$ relative to $(x^*, y^*)$. For each feasible point for (3.103), the objective function $\phi$ has the value

$$\sum \mathbf{x}_i^t \mathbf{f}_i = \sum (\mathbf{x}^* + \mathbf{u}_i)^t \mathbf{f}_i = \sum \mathbf{u}_i^t \mathbf{f}_i \leq \sum \|\mathbf{u}_i\|_2 = \psi_{\min}, \tag{3.105}$$

where the summations are all from $i = 1$ to $i = k$. The third expression follows from the second by splitting the second into two terms and factoring the constant vector $\mathbf{x}^*$ out of the first term. The inequality comes from the Schwarz inequality (1.61′). We show that the $\mathbf{f}_i$ can be chosen for equality to hold throughout in (3.105), so duality is valid.

To that end we observe that in this case the function $\psi$ is differentiable at $\mathbf{x}^*$, so its gradient is $\mathbf{0}$ at the point. We find that

$$\nabla\psi = \sum_{i=1}^{k} \frac{\mathbf{u}_i}{\|\mathbf{u}_i\|_2}. \tag{3.106}$$

Hence, if we set $\mathbf{f}_i = \mathbf{u}_i/\|\mathbf{u}_i\|_2$ for $i = 1,\, ,\ldots, k$, the constraints in (3.103) are satisfied, and the calculation (3.105) shows that $\phi(\mathbf{f}_1, \ldots, \mathbf{f}_k) = \psi_{\min}$. Thus duality holds.  ∎

*Proof for Case 2.*   Here the last term in the sum (3.100) is not differentiable at $\mathbf{x}^* = \mathbf{x}_k$. However, this term has *directional derivative* equal to 1 along each direction from $\mathbf{x}_k$. Since $\psi$ is minimized at $\mathbf{x}_k$, its directional derivative along every such direction at $\mathbf{x}_k$ must be nonnegative. If $\mathbf{v}$ is a unit vector, we conclude that at $\mathbf{x}_k$

$$\mathbf{v}^t \sum_{i=1}^{k-1} \frac{\mathbf{u}_i}{\|\mathbf{u}_i\|_2} + 1 \geq 0.$$

Since $\mathbf{v}$ is arbitrary, we conclude (Problem 3.75) that the vector

$$\sum_{i=1}^{k-1} \frac{\mathbf{u}_i}{\|\mathbf{u}_i\|_2}$$

has 2-norm at most 1. Thus, if we again define $\mathbf{f}_i$ as $\mathbf{u}_i/\|\mathbf{u}_i\|_2$ for $i = 1, \ldots, k-1$, we can choose $\mathbf{f}_k$ to equal $-(\mathbf{f}_1 + \cdots + \mathbf{f}_{k-1})$, and then all constraints in (3.103) are satisfied. The calculation (3.105) can be repeated, with $\mathbf{u}_k = \mathbf{0}$, to show (Problem 3.76) that $\phi(\mathbf{f}_1, \ldots, \mathbf{f}_k) = \psi_{\min}$.  ∎

Such related pairs of problems, illustrated by the primal problem (3.100) and its dual (3.103), are omnipresent in the fields of convex programming and calculus of variations. The problem having a more natural origin is usually called primal, but the roles can be interchanged. We have shown that a dual of the Fermat–Weber problem (3.100) of geometry is a problem having a mechanics interpretation.

REMARK. The results of this section are generalized in the following section and, to a much greater extent, in Section 4.10.


## PROBLEMS

**3.69.** Show that the function $\psi(x, y)$ defined by (3.100) has limit $\infty$ as $\|\mathbf{x}\| \to \infty$, and hence the function has a minimizer (see Section 1.12).

**\*3.70.** Show that the function to be minimized in (3.101) has a minimum under the hypotheses stated. [Hint: Proceed as in Problem 3.69 with the aid of (1.85).]

**3.71.** Try to verify the result of Example 1 in Section 3.10 by computing the values of the function $\psi$ for points near the asserted minimizer.

**3.72.** Use intuitive reasoning to justify the assertion that if $k$ is large and $k - 1$ of the points $(x_i, y_i)$ are located inside a circle while the $k$th point is far from the circle, then the minimizer of the function $\psi$ in (3.100) should be inside the circle.

**3.73.** Show that the feasibility set $S_f$ for the program (3.103) is convex.

**3.74.** Verify the expression (3.106) for the gradient of the function (3.100) at a minimizer $\mathbf{x}^*$, where $\mathbf{u}_i = \mathbf{x}_i - \mathbf{x}^*$ for each $i$.

**3.75.** Let the vector $\mathbf{z}$ in $\mathcal{R}^2$ be such that $\mathbf{v}^t\mathbf{z} + 1 \geq 0$ for every unit vector $\mathbf{v}$. Show that $\|\mathbf{z}\|_2 \leq 1$.

**3.76.** With the aid of the result of Problem 3.75, verify the details of the proof for case 2 at the end of Section 3.10.

## 3.11  A DUALITY RELATION IN HIGHER DIMENSIONS

We will present in this section a more general version of the duality relation of the previous section. It has applications to a wide class of problems in plasticity. (For example, see Yang 1993.)

### Block Matrices

We will take advantage of the block matrix representation described in the Appendix, A.20. In particular, we will use a matrix $A$ that is formed of a single row of blocks:

$$A = (A_1, \ldots, A_k), \quad \text{where} \quad A_i \text{ is } m \times p, \quad i = 1, \ldots, k, \tag{3.110}$$

so $A$ is $m \times kp$. We will also use column vectors $\mathbf{x}$ and $\mathbf{c}$ in $\mathcal{R}^{kp}$ which are formed of $k$ vertical blocks—that is, column vectors—in $\mathcal{R}^p$. We can write

$$\mathbf{x} = (\mathbf{x}_1^t, \mathbf{x}_2^t, \ldots, \mathbf{x}_k^t)^t, \tag{3.111}$$

where $\mathbf{x}_1, \ldots, \mathbf{x}_k$ are all in $\mathcal{R}^p$. We can write $\mathbf{c}$ in analogous fashion. We call the vectors $\mathbf{x}_i$ and $\mathbf{c}_i$ *subvectors* of $\mathbf{x}$ and $\mathbf{c}$, respectively.

Such a representation is common in the finite-difference and finite-element methods for vector-valued functions. For instance, the $k$ force vectors in (3.103), each having two components $(X_i, Y_i)$, can be represented by a single vector

$$\mathbf{f} = (X_1, Y_1, X_2, Y_2, \ldots, X_k, k)^t \in \mathcal{R}^{2k};$$

here the subvectors are 2-dimensional. Similarly the set of fixed points can be represented by a constant vector

$$\mathbf{c} = (x_1, y_1, x_2, y_2, \ldots, x_k, y_k)^t \in \mathcal{R}^{2k}.$$

## A Program in CP and Its Dual

Let $A$ be a constant $m \times kp$ matrix, $k \geq 2$, written in block-matrix form as in (3.110); let $m \leq p$, and let each $A_i$ have rank $m$. Let $\mathbf{x}$ be a variable vector in $\mathcal{R}^{kp}$ and $\mathbf{c}$ a constant nonzero vector in the same space, written in block form in terms of subvectors in $\mathcal{R}^p$ as in (3.111).

We introduce the primal program:

$$\begin{aligned} maximize \quad & \phi(\mathbf{x}) = \mathbf{c}^t \mathbf{x} \\ subject\ to \quad & A\mathbf{x} = \mathbf{0}, \\ & \|\mathbf{x}_i\|_2 \leq 1, \qquad i = 1, 2, \ldots, k. \end{aligned} \qquad (3.112)$$

We assert that it has a dual:

$$minimize \quad \psi(\mathbf{y}) = \sum_{i=1}^{k} \|\mathbf{c}_i - A_i^t \mathbf{y}\|_2 \quad in \quad \mathcal{R}^m, \qquad (3.113)$$

where $\mathbf{c}_i$ is the $i$th subvector of $\mathbf{c}$ and $\mathbf{y} \in \mathcal{R}^m$ is the dual variable. We make the assumption that no two terms of the sum in (3.113) equal zero for the same $\mathbf{y}$.

A special case of (3.112) and (3.113) is the pair (3.103) and (3.100) with

$$\mathbf{c}_i = \begin{pmatrix} x_i \\ y_i \end{pmatrix}, \quad \mathbf{x}_i = \begin{pmatrix} X_i \\ Y_i \end{pmatrix}, \quad A_i = \begin{bmatrix} 1 & 0 \\ 0 & 1 \end{bmatrix}, \quad \mathbf{y} = \begin{pmatrix} x \\ y \end{pmatrix}. \qquad (3.114)$$

The function $\phi$ is defined on a bounded, closed, convex set; it is linear and hence concave. By continuity, it attains its maximum, but the maximizer need not be unique (Problem 3.78).

The function $\psi$, a generalization of (3.100), is convex in $\mathcal{R}^m$, as follows from the results of Section 3.1. It attains a global minimum, though the minimizer need not be unique. One can justify existence of a minimizer by the following reasoning: In $\mathcal{R}^{kp}$ a valid norm is

$$\|\mathbf{x}\| = \sum_{i=1}^{k} \|\mathbf{x}_i\|_2 \qquad (3.115)$$

(Problem 3.79). In terms of this norm one introduces distance as usual. Then the minimum sought is the shortest distance from $\mathbf{c}$ to the set $E$ consisting of all $\mathbf{x}$ in $\mathcal{R}^{kp}$ for which $\mathbf{x}_i = A_i^t \mathbf{y}$ for some $\mathbf{y}$ in $\mathcal{R}^m$. The set $E$ is a subspace of $\mathcal{R}^{kp}$ and is hence a closed set; as in Section 1.12 there is a point of $E$ closest to the given point $\mathbf{c}$. Thus the global minimum is attained. In many cases the function $\psi$ is strictly convex, so the minimizer is unique; one can determine these cases with the aid of the rules in Section 3.1.

To prove the duality of (3.112) and (3.113), we begin with the primal problem and replace the constraint $A\mathbf{x} = \mathbf{0}$ by the "weak form" of this condition: namely

$$\mathbf{y}^t A\mathbf{x} = 0 \qquad \text{for all } \mathbf{y} \in \mathcal{R}^m. \tag{3.116}$$

Equation (3.116), called a virtual work relation in mechanics literature, is equivalent to $A\mathbf{x} = \mathbf{0}$, since $\mathbf{y}$ can be arbitrarily chosen in the full space $\mathcal{R}^m$. Because of this equivalence, we can write the objective function $\phi(\mathbf{x})$ in various ways:

$$\phi(\mathbf{x}) = \mathbf{c}^t \mathbf{x} = \mathbf{c}^t \mathbf{x} - \mathbf{y}^t A\mathbf{x} = (\mathbf{c} - A^t \mathbf{y})^t \mathbf{x} \qquad \text{for all } \mathbf{y} \in \mathcal{R}^m.$$

The last expression can written out as

$$\sum_{i=1}^{k} (\mathbf{c}_i - A_i^t \mathbf{y})^t \mathbf{x}_i.$$

Therefore, from the Schwarz inequality (1.61′) and the last constraint in (3.112), we conclude that

$$\phi(\mathbf{x}) \leq \sum_{i=1}^{n} \|\mathbf{c}_i - A_i^t \mathbf{y}\|_2 = \psi(\mathbf{y}).$$

This inequality is valid for every $\mathbf{x} \in \mathcal{R}^{kp}$ that satisfies the constraints in (3.112) and for every $\mathbf{y}$ in $\mathcal{R}^m$.

To complete the proof of duality, we show that $\phi(\mathbf{x}^*) = \psi(\mathbf{y}^*)$ for appropriate $\mathbf{x}^*$ and $\mathbf{y}^*$. As in Section 3.10 we consider two cases. In the first case, there is a minimizer for (3.113) at which all terms in (3.113) are differentiable; that is, each term in (3.113) is positive at $\mathbf{y}^*$. Then, as in Section 3.10 the gradient of $\psi$ is $\mathbf{0}$ at the point. After applying rules of calculus (Section 2.1), we find that

$$\nabla \psi = - \sum_{i=1}^{k} A_i \mathbf{x}_i,$$

where

$$\mathbf{x}_i = (\mathbf{c}_i - A_i^t \mathbf{y}) / \|\mathbf{c}_i - A_i^t \mathbf{y}\|_2, \qquad i = 1, \ldots, k.$$

If we now let $\mathbf{y} = \mathbf{y}^*$ in these equations and let $\mathbf{x}^*$ be the corresponding $\mathbf{x}$, then the condition that the gradient is $\mathbf{0}$ at $\mathbf{y}^*$ implies that this $\mathbf{x}$ satisfies the constraints in (3.112). Finally, we verify (Problem 3.80) that $\phi(\mathbf{x}^*) = \psi(\mathbf{y}^*)$. The proof is complete for the case considered.

In the second case, we assume that no minimizer for $\psi$ is a point of differentiability. Thus it occurs at a $\mathbf{y}^*$ at which at least one term in (3.113) is 0. By hypothesis, only one term in the sum is 0 at $\mathbf{y}^*$. We can assume that that term is the last one; thus all the other terms are differentiable at $\mathbf{y}^*$. As in Section 3.10 we then consider the directional derivative of $\psi$ at $\mathbf{y}^*$ in the direction of a unit vector $\mathbf{u}$ of $\mathcal{R}^m$. The sum of the first $k - 1$ terms has a gradient that can be calculated as above, and this sum has a corresponding directional derivative at $\mathbf{y}^*$ equal to the inner product of the gradient vector with $\mathbf{u}$. For the last term we let $\mathbf{y}$ vary on a ray in the direction of $\mathbf{u}$ starting at $\mathbf{y}^*$ : $\mathbf{y} = \mathbf{y}^* + s\mathbf{u}$, with $s > 0$, and find the directional derivative to be $\|A_k^t\mathbf{u}\|_2$. Since $\mathbf{y}^*$ provides a local minimum for $\psi$, the directional derivative of $\psi$ along the ray chosen must be nonnegative. Hence, with the $\mathbf{x}_i^*$ chosen as above for $i = 1, \ldots, k - 1$ and the notation

$$z = -\sum_{i=1}^{k-1} A_i\mathbf{x}_i^*,$$

we conclude that

$$\mathbf{z}^t\mathbf{u} + \|A_k^t\mathbf{u}\|_2 \geq 0.$$

This inequality is also valid with $\mathbf{u}$ replaced by its negative. Therefore, for all unit vectors $\mathbf{u}$ in $\mathcal{R}^n$,

$$|\mathbf{z}^t\mathbf{u}| \leq \|A_k^t\mathbf{u}\|_2.$$

It is now a matter of algebra to show that the last assertion implies that a vector $\mathbf{x}_k^*$ of 2-norm at most 1 exists such that

$$\mathbf{z} - A_k\mathbf{x}_k^* = \mathbf{0}.$$

(See Problem 3.82). Thus $\mathbf{x}^*$ has been found, satisfying the constraints in (3.112). Furthermore we verify that $\phi(\mathbf{x}^*) = \psi(\mathbf{y}^*)$, taking advantage of the fact that the last term in (3.113) is 0 for $\mathbf{y} = \mathbf{y}^*$. Thus duality is proved in this case also.

### The 1-Norm Minimization

In the case $p = 1$ we have $\|\mathbf{x}_i\|_2 = |x_i|$, $A_i^t = \mathbf{a}_i^t \in \mathcal{R}^m$ (a row vector) and the sum from $i = 1$ to $k$ of the terms $\|\mathbf{c}_i - A_i^t\mathbf{y}\|_2 = |c_i - \mathbf{a}_i^t\mathbf{y}|$, where $\mathbf{a}_i$ is the ith column of $A$, equals $\|\mathbf{c} - A^t\mathbf{y}\|_1$. The pair (3.112) and (3.113) reduce to the primal problem

$$maximize \quad \mathbf{c}^t\mathbf{x}$$

$$subject\ to \quad A\mathbf{x} = \mathbf{0}, \tag{3.117}$$

$$\|\mathbf{x}\|_\infty \leq 1,$$

and its dual

$$minimize \quad \|\mathbf{c} - A^t\mathbf{y}\|_1. \tag{3.118}$$

These are the same as the pair (3.36) and (3.38) in LP1, with $\mathbf{b} = \mathbf{0}$. These optimization problems appear frequently in plasticity, where the vectors $\mathbf{x}$, $\mathbf{y}$, and $\mathbf{c}$ have physical and geometrical meaning. When these global vectors have block structures as in (3.111), the subvectors represent pointwise physical and geometrical vector- (or tensor-)valued quantities. A norm of any type on the subvectors can be interpreted as a "subspace norm." By adding such norms as in (3.115), one obtains a valid norm in the whole vector space. The process of adding the norms on the subspace is analogous to that for forming the 1-norm of a vector in $\mathcal{R}^n$. Therefore we also refer to the problem in (3.113) as a 1-norm minimization. While (3.117) is a problem of pure 1-norm minimization, the Fermat–Weber problem (3.100) is a 1-norm minimization with 2-norm subspace measure. An example of the pure 1-norm minimization is provided by a problem which we term the Fermat–Weber problem in a city. Consider a retail chain with $k$ stores scattered in a city having a rectangular grid of streets. A single warehouse is to be located with minimum total travel distance to all stores. (We assume that there are no one-way, dead-end or overpass streets.) Formulation of this problem is left as an exercise (Problem 3.84).

One can interpret the minimization problems discussed here in another way: namely as problems of minimizing a function

$$\psi(\mathbf{y}) = \rho(\mathbf{c} - A^t\mathbf{y}) + \mathbf{b}^t\mathbf{y},$$

where $\rho$ is a suitable norm in $\mathcal{R}^m$. This problem is considered in full generality in Section 4.10, and a dual is obtained. This general result includes all the duality theorems of this chapter.

## Numerical Methods

The nonsmoothness of the objective function $\psi$ in (3.87), (3.113), and (3.118) is a source of difficulty for many computational methods of optimization. We mention here two combined smoothing and successive approximations algorithms that have been shown to be effective for these problems. They were motivated by the desire to make these nonsmooth functions differentiable everywhere and the algorithms numerically stable. Their success has led to new mathematical analysis and further refinements of the algorithms. References are given at the end of this section.

For the problem (3.113) we let $\varepsilon > 0$ be a smoothing parameter. A perturbed objective function of (3.113) is defined as the function

$$\psi_\varepsilon(\mathbf{y}) = \sum_{i=1}^{k} \sqrt{\|\mathbf{c}_i - A_i^t \mathbf{y}\|_2^2 + \varepsilon^2}.$$

This new function is differentiable everywhere, and it is strictly convex in $\mathcal{R}^m$, as follows from the results of Section 3.1. It has a global minimum in $\mathcal{R}^m$, since the function has limit $\infty$ for $\mathbf{x} \to \infty$. Therefore it has a unique minimizer. Furthermore the function converges to the original objective function as $\varepsilon \to 0$. The optimality condition for minimizing this perturbed objective function leads naturally to a fixed point iteration:

$$\mathbf{y}^{(j+1)} = \left[ \sum_{i=1}^{k} w_i^{(j)} A_i A_i^t \right]^{-1} \left( \sum_{i=1}^{k} w_i^{(j)} A_i \mathbf{c}_i \right), \qquad j = 0, 1, 2, \ldots$$

Here the $w_i^{(j)} = 1/\sqrt{\|\mathbf{c}_i - A_i^t \mathbf{y}^{(j)}\|_2^2 + \varepsilon^2}$ are weights that change from iteration to iteration. An initial $\mathbf{y}^{(0)}$ is assumed chosen.

This iteration has been found to be robust. This conclusion can be deduced from the following theorem: *From any starting point $\mathbf{y}^{(0)} \in \mathcal{R}^m$, $\mathbf{y}^{(1)}$ is always in the convex hull* (Problem 3.16) *of the data set* $\{A_i \mathbf{c}_i \mid i = 1, 2, \ldots, n\}$. For engineering accuracy, it has been found that only a few iterations are needed.

We next consider the dual (3.82) of the standard LP1 (3.80). This is the problem

$$minimize \quad \psi(\mathbf{y}) = \|\mathbf{c} - A^t \mathbf{y}\|_1 + \mathbf{b}^t \mathbf{y} = \sum_{i=1}^{n} |c_i - \mathbf{a}_i^t \mathbf{y}| + \mathbf{b}^t \mathbf{y},$$

where $A$ is an $m \times n$ matrix of rank $m$, $m < n$, $\mathbf{c} \in \mathcal{R}^n$, and $\mathbf{b} \in \mathcal{R}^m$; the vectors $\mathbf{a}_i$ are the columns of $A$. It is assumed that the equation $A\mathbf{x} = \mathbf{b}$ has a solution $\mathbf{x}$ with $|x_i| < 1$ for $i = 1, \ldots, n$.

The function to be minimized is convex and unconstrained, continuous but not differentiable everywhere. Again we introduce a positive $\varepsilon$ as a smoothing parameter and introduce the modified objective function

$$\psi(\mathbf{y}, \varepsilon) = \sum_{i=1}^{n} \sqrt{(c_i - \mathbf{a}_i^t \mathbf{y})^2 + \varepsilon^2} + \mathbf{b}^t \mathbf{y}.$$

This function has the following properties: It is strictly convex, by the results of Section 3.1; it is everywhere differentiable; for each $\varepsilon$, $\psi(\mathbf{y}, \varepsilon) > \psi(\mathbf{y})$ for all $\mathbf{y}$; for each $\mathbf{y}$,

$$\lim_{\varepsilon \to 0} \psi(\mathbf{y}, \varepsilon) = \psi(\mathbf{y}).$$

For a given $\varepsilon$, $\psi(\mathbf{y}, \varepsilon)$ has a global minimum and therefore has a unique minimizer. The minimizer is the unique critical point of the function, and it can be found by solving $m$ nonlinear equations. Methods of Section 3.9 can be used. However, for the special type of nonlinearity in this function, it is preferable to use the following fixed point iteration procedure:

$$\mathbf{y}^{(k+1)} = T(\mathbf{y}^{(k)}), \qquad k = 0, 1, 2, \ldots,$$

$$T(\mathbf{y}) = [A^t M(\mathbf{y}, \varepsilon)A]^{-1} A^t M(\mathbf{y}, \varepsilon)\mathbf{c},$$

where $M$ is a diagonal matrix of order $m$ with entries

$$\mu_i(\mathbf{y}, \varepsilon) = \frac{1}{\sqrt{(c_i - \mathbf{a}_i^t \mathbf{y})^2 + \varepsilon^2}}.$$

The matrix $M$ is positive definite, as is $A^t M A$. The iteration converges to the unique minimizer $\mathbf{y}^*$ with arbitrary starting point.

As we saw in Section 3.8, the solution of the primal problem (3.80) can be recovered from the solution of the dual problem. Therefore the method presented here provides an alternative to a simplex method for (3.80). The new method has two major advantages over the simplex method: No special initial point (like a basic feasible solution for the simplex method) is needed; it is simpler to program, and the rate of convergence is comparable to or better than that of the simplex method, especially for very large problems.

For the numerical methods discussed here, refer to the two papers by Yang listed at the end of the chapter.

## PROBLEMS

**3.77.** Verify that the pair of programs (3.103) and (3.100) forms a special case of the pair (3.112), (3.113), where the $\mathbf{c}_i$, $\mathbf{x}_i$, $A_i$ and $\mathbf{y}$ are chosen as in (3.114).

**\*3.78.** Give an example of the program (3.112), with $m = k = p = 2$ and $\mathbf{c} \neq \mathbf{0}$, such that the maximizer is not unique.

**3.79.** Verify that (3.115) is a valid norm in $\mathcal{R}^{kp}$, with the notations as given preceding (3.115) (see the beginning of Section 3.8).

**3.80.** Verify the details of the proof of duality for the first case considered in Section 3.11.

**3.81.** Let the last term in the sum in (3.113) equal 0 for $\mathbf{y} = \mathbf{y}^*$. Show that this term has directional derivative at $\mathbf{y}^*$, in the direction of the unit vector $\mathbf{u}$, equal to $\|A_k^t \mathbf{u}\|_2$.

**3.82.** Let $A$ be an $m \times n$ matrix of rank $m \leq n$. Let $\mathbf{z}$ be a vector of $\mathcal{R}^m$ such that $|\mathbf{z}^t\mathbf{u}| \leq \|A^t\mathbf{u}\|$ for all unit vectors $\mathbf{u}$ in $\mathcal{R}^m$.

   **(a)** Show that $|\mathbf{z}^t\mathbf{v}| \leq \|A^t\mathbf{v}\|$ for *all* vectors $\mathbf{v}$ in $\mathcal{R}^m$.

   **(b)** Let $\mathbf{q} = A^t(AA^t)^{-1}\mathbf{z}$. (By Problem 1.82, $AA^t$ is nonsingular.) Show that $A\mathbf{q} = \mathbf{z}$.

   **(c)** Write $\mathbf{q}^t\mathbf{q}$ as $\mathbf{z}^t\mathbf{v}$ for appropriate $\mathbf{v}$, and conclude from **(a)** that $\|\mathbf{q}\|^2 \leq \|\mathbf{q}\|$ and hence $\|\mathbf{q}\| \leq 1$.

**3.83.** With the aid of the results of Problems 3.81 and 3.82, verify the details of the proof of duality for the second case as given in Section 3.11.

**3.84.** Formulate as a mathematical program the Fermat–Weber problem in a city, as described in the lines following (3.118) (See Witzgall, 1964, pp. 11 ff., for a discussion of this problem.)

## REFERENCES

Dantzig, G. (1963). *Linear Programming and Extensions*, Princeton University Press, Princeton.

Fletcher, R. M. (1971). "General quadratic programming alogorithm," in *Journal of the Institute of Mathematics and Applications*, 7, 76–91.

Fletcher, R. M. (1987). *Practical Methods of Optimization*, Wiley, New York.

Gill, P. E., and Murray, W. (1978). "Numerically stable methods for quadratic programming," in *Mathematical Programming*, 14, 349–372.

Hadley, G. (1962). *Linear Programming*, Addison-Wesley, Reading, Mass.

Hestenes, M. (1975). *Optimization Theory, The Finite Dimensional Case*, R. E. Krieger, Huntington, NY.

Isaacson, E., and Keller, H. B. (1966). *Analysis of Numerical Methods*, Wiley, New York.

Kuhn, H. W., and Tucker, A. W. (1951). "Nonlinear programming," in *Proceedings of the Second Berkeley Symposium on Mathematical Statistics and Probability*, 481–492, Univ. of California Press, Berkeley.

Moré, J. J., and Wright, S. J. (1993). *Optimization Software Guide*, Society for Industrial and Applied Mathematics, Philadelphia.

O'Leary, D. P., and Yang, W. H. (1978). "Elastoplastic torsion by quadratic programming," in *Computer Methods in Applied Mechanics and Engineering*, 16, 361–368.

Pierre, D. A. (1969). *Optimization Theory with Applications*, Wiley, New York.

Webster, R. (1994). *Convexity*, Oxford University Press, Oxford.

Witzgall, C. (1964). "Optimal location of a central facility, mathematical models and concepts," National Bureau of Standards Report 8388, Washington, DC.

Yang, W. H. (1991). "A duality theorem for plane torsion," in *International Journal of Solids and Structures*, 27, 1981–1989.

Yang, W. H. (1993). "Large deformation of structures by sequential limit analysis," in *International Journal of Solids and Structures*, 30, 1001–1013.

# 4

# Fenchel–Rockafellar Duality Theory

## 4.1 GENERALIZED DIRECTIONAL DERIVATIVE

Let $f$ be a function of $n$ variables defined in a neighborhood of a point $\mathbf{x}$ of $\mathcal{R}^n$. In the calculus one defines the directional derivative of $f$ at $\mathbf{x}$ in the direction of a nonzero vector $\mathbf{y}$ of $\mathcal{R}^n$ as

$$\lim_{h \to 0+} \frac{f(\mathbf{x} + h\mathbf{y}) - f(\mathbf{x})}{h\|\mathbf{y}\|}$$

if the limit exists. One then shows that for a differentiable function $f$, this directional derivative is the component in the direction $\mathbf{y}$ of the gradient of $f$ at $\mathbf{x}$. In particular, for $f(x, y)$, the directional derivative in the direction of the $x$-axis is $\partial f / \partial x$. (By our general convention, $\|\mathbf{y}\|$ is the Euclidean norm of $\mathbf{y}$.)

We now define the generalized directional derivative of $f$ at $\mathbf{x}$ in the direction of the vector $\mathbf{y}$ as

$$f'(\mathbf{x}; \mathbf{y}) = \lim_{h \to 0+} \frac{f(\mathbf{x} + h\mathbf{y}) - f(\mathbf{x})}{h} \tag{4.10}$$

provided that the limit exists. This is clearly equal to $\|\mathbf{y}\|$ times the previously defined directional derivative. If $\mathbf{y}$ happens to be a unit vector, the two definitions agree. Otherwise, the directional derivative depends only on the *direction* of $\mathbf{y}$, whereas the generalized directional derivative depends on the *length and direction* of $\mathbf{y}$; it is defined, with value 0, for $\mathbf{y} = \mathbf{0}$.

***Example 1.*** Let $f$ be linear, $f(\mathbf{x}) = \mathbf{c}^t\mathbf{x}$. Then $f'(\mathbf{x}; \mathbf{y}) = \mathbf{c}^t\mathbf{y}$ independent of $\mathbf{x}$. (See Problem 4.1.)

***Example 2.*** Let $f$ be quadratic, $f(\mathbf{x}) = \mathbf{x}^t A\mathbf{x}$, where $A$ is a symmetric matrix. Then $f'(\mathbf{x}; \mathbf{y}) = 2\mathbf{x}^t A\mathbf{y}$, a "bilinear function" of $\mathbf{x}$ and $\mathbf{y}$ (Problem 4.2).

***Example 3.*** Let $f(\mathbf{x}) = \|\mathbf{x}\|$. Then $f'(\mathbf{x}; \mathbf{y}) = (1/\|\mathbf{x}\|)\mathbf{x}^t\mathbf{y}$ for $\mathbf{x} \neq \mathbf{0}$ and $f'(\mathbf{0}; \mathbf{y}) = \|\mathbf{y}\|$ (Problem 4.3).

Let $f'(\mathbf{x}; \mathbf{y})$ exist for a particular $\mathbf{x}$ and $\mathbf{y}$. Then $f'(\mathbf{x}; t\mathbf{y})$ exists for all $t \geq 0$ and

$$f'(\mathbf{x}; t\mathbf{y}) = tf'(\mathbf{x}; \mathbf{y}) \quad \text{for } t \geq 0. \tag{4.11}$$

We say that the generalized directional derivative is *positively homogeneous*. (See Problem 4.4.)

Now let the function $f$ be defined on the $n$-dimensional convex set $E$, and let $f$ be a convex function. Let $\mathbf{x}$ be a fixed interior point of $E$. Then (1) *the generalized directional derivative of $f$ at $\mathbf{x}$ in direction $\mathbf{y}$ exists for all $\mathbf{y} \in \mathcal{R}^n$ and* (2) $f'(\mathbf{x}; \mathbf{y})$ *is a convex function of $\mathbf{y}$.* Furthermore (3) $\mathbf{x}$ *is a minimizer of $f$ if and only if $f'(\mathbf{x}; \mathbf{y}) \geq 0$ for all $\mathbf{y}$.*

*Proof.* 1. The existence of the generalized directional derivative follows from the existence of directional derivatives of convex functions (Section 1.4).

2. To prove the convexity, we observe that by the convexity of $f$,

$$f(\mathbf{x} + h(\mathbf{y} + \mathbf{z})) = f\left(\tfrac{1}{2}(\mathbf{x} + 2h\mathbf{y}) + \tfrac{1}{2}(\mathbf{x} + 2h\mathbf{z})\right)$$
$$\leq \tfrac{1}{2} f(\mathbf{x} + 2h\mathbf{y}) + \tfrac{1}{2} f(\mathbf{x} + 2h\mathbf{z}),$$

and therefore

$$\frac{f(\mathbf{x} + h(\mathbf{y} + \mathbf{z})) - f(\mathbf{x})}{h} \leq \frac{f(\mathbf{x} + 2h\mathbf{y}) - f(\mathbf{x})}{2h} + \frac{f(\mathbf{x} + 2h\mathbf{z}) - f(\mathbf{x})}{2h}.$$

We let $h \to 0+$ and conclude that

$$f'(\mathbf{x}; \mathbf{y} + \mathbf{z}) \leq f'(\mathbf{x}; \mathbf{y}) + f'(\mathbf{x}; \mathbf{z}).$$

We replace $\mathbf{y}$ by $\alpha\mathbf{y}$ and $\mathbf{z}$ by $(1 - \alpha)\mathbf{z}$, where $0 \leq \alpha \leq 1$, and obtain the inequality

$$f'(\mathbf{x}; \alpha\mathbf{y} + (1 - \alpha)\mathbf{z}) \leq f'(\mathbf{x}; \alpha\mathbf{y}) + f'(\mathbf{x}; (1 - \alpha)\mathbf{z}).$$

Therefore, by positive homogeneity,

$$f'(\mathbf{x}; \alpha\mathbf{y} + (1 - \alpha)\mathbf{z}) \leq \alpha f'(\mathbf{x}; \mathbf{y}) + (1 - \alpha)f'(\mathbf{x}; \mathbf{z}),$$

and the convexity is proved.

3. Let $\mathbf{x}$ be a minimizer of $f$. If, for some $\mathbf{y}$, $f'(\mathbf{x}; \mathbf{y}) < 0$, then for $h$ sufficiently small and positive, one must have $f(\mathbf{x} + h\mathbf{y}) - f(\mathbf{x}) < 0$, which contradicts the assumption that $\mathbf{x}$ is a minimizer of $f$. Hence $f'(\mathbf{x}; \mathbf{y}) \geq 0$ for all $\mathbf{y}$.

Next let $f'(\mathbf{x}; \mathbf{y}) \geq 0$ for all $\mathbf{y}$. If $\mathbf{x}$ is not a minimizer of $f$, then for some $\mathbf{x}_1 \in E$, $f(\mathbf{x}_1) < f(\mathbf{x})$. Let $\mathbf{y} = \mathbf{x}_1 - \mathbf{x}$. Then the function $g(h) = f(\mathbf{x} + h\mathbf{y})$ is a convex function of $h$ for $0 \leq h \leq 1$ (by the convexity of $E$ and the convexity of $f$). Also $g(1) = f(\mathbf{x}_1) < g(0) = f(\mathbf{x})$. Since $g$ is convex, its graph lies below the chord for the interval $0 \leq h \leq 1$. Hence the slope of the chord for the interval $[0, h]$ is less than or equal to that for the interval $[0, 1]$, which is the negative constant $c = f(\mathbf{x}_1) - f(\mathbf{x})$. Accordingly, for $0 < h < 1$,

$$\frac{f(\mathbf{x} + h\mathbf{y}) - f(\mathbf{x})}{h} \leq c < 0.$$

If we let $h \to 0+$, we conclude that $f'(\mathbf{x}; \mathbf{y}) \leq c < 0$, which contradicts the assumption that $f'(\mathbf{x}; \mathbf{y}) \geq 0$ for all $\mathbf{y}$.                                   ■

We will refer to the rule 3 as the *minimum principle for convex functions*. We remark that the results of this section can be applied to the case when $E$ has dimension less than $n$. When $E$ is $d$-dimensional, with $d < n$, one can choose coordinates so that $E$ lies in the $d$-dimensional subspace of $\mathcal{R}^n$ formed of the points whose coordinates $x_{d+1}, \ldots, x_n$ are all 0, and this subspace can be regarded as $\mathcal{R}^d$. The term "interior" must then be replaced by "relative interior," and $\mathbf{y}$ must be restricted to $\mathcal{R}^d$.


## PROBLEMS

**4.1.** Verify the results of Example 1 in Section 4.1.

**4.2.** Verify the results of Example 2 in Section 4.1.

**4.3.** Verify the results of Example 3 in Section 4.1.

**4.4.** **(a)** Let $f$ be a function having a generalized directional derivative at a point $\mathbf{x}$. Show that (4.11) holds and hence the generalized directional derivative in direction $\mathbf{y}$ is positively homogeneous in $\mathbf{y}$.

  **(b)** Show by example that, for a function $f$ having a generalized directional derivative at $\mathbf{x}$, the equation in (4.11) may fail to hold for negative $t$.

**4.5.** Apply the minimum principle for convex functions to show that the function $f(\mathbf{x}) = \|\mathbf{x}\|$ has a minimum only at $\mathbf{x} = \mathbf{0}$.


## 4.2. LOCAL STRUCTURE OF THE BOUNDARY OF A CONVEX SET

All examples of convex sets encountered in earlier chapters have had simple relative boundaries: polygonal, polyhedral, smooth curves or surfaces. We now show that near each relative boundary point of a convex set, the relative boundary can be represented as the graph of a continuous convex function. It suffices to consider only $n$-dimensional convex sets in $\mathcal{R}^n$ and the actual boundaries of such sets. For a $d$-dimensional convex set in $\mathcal{R}^n$, with $d<n$, one can choose coordinates so that the set lies in the set of points with coordinates $(x_1, \ldots, x_d, 0, \ldots, 0)$, hence in effect as a $d$-dimensional set in $\mathcal{R}^d$.

## Motivation

Let $y=f(x)$, $-\delta<x<\delta$, be a convex function with $f(0)=0$. As in Section 1.2, $f$ is necessarily continuous. We assume that for some positive constant $k$, $-k<f(x)<k$ for $-\delta<x<\delta$. (See Fig. 4.1.) Then the set

$$E_0 = \{(x, y) \mid -\delta<x<\delta, \; f(x)<y<k\} \tag{4.20}$$

is a convex set (Problem 4.6) whose boundary includes the graph of $f$. Our goal is to show that for every 2-dimensional convex set $E$ in the plane having a boundary point, one can choose coordinates in the plane with origin at the boundary point, positive constants $k$, $\delta$ and a convex function $f$, so that the portion of the interior of $E$ in the rectangular region

$$R = \{(x, y) \mid -\delta<x<\delta, \; -k<y<k\}$$

consists of the corresponding set $E_0$ and the portion of the boundary of $E$ in $R$ consists of the graph of $f$. We also wish to establish the analogous theorem for $n$-dimensional convex sets in $\mathcal{R}^n$ for all $n \geq 2$.

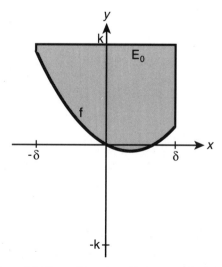

**Figure 4.1.** Part of boundary of convex set in the plane.

For the $n$-dimensional case it will be convenient to denote a vector

$$(x_1, \ldots, x_{n-1}, x_n)^t$$

of $\mathcal{R}^n$ by $(\mathbf{u}, v)$, where $\mathbf{u}$ is regarded as the vector $(x_1, \ldots, x_{n-1})^t$ of $\mathcal{R}^{n-1}$ and $v = x_n$.

**Theorem.** *Let $n \geq 2$. Let $E$ be an $n$-dimensional convex set in $\mathcal{R}^n$, and let $\mathbf{x}_0$ be a boundary point of $E$. Then one can choose coordinates in $\mathcal{R}^n$ with origin at $\mathbf{x}_0$ and positive constants $\delta$, $k$ such that, in the notation given above, the portion of the interior of $E$ in the set*

$$R = \{(\mathbf{u}, v) \mid \|\mathbf{u}\| < \delta, \ -k < v < k\}$$

*consists of the set*

$$E_0 = \{(\mathbf{u}, v) \mid \|\mathbf{u}\| < \delta, \ f(\mathbf{u}) < v < k\}$$

*and the portion of the boundary of $E$ in $R$ consists of the graph of the function $f$, where $f$ is a convex function defined for $\|\mathbf{u}\| < \delta$ such that $-k < f(\mathbf{u}) < k$ for each $\mathbf{u}$.*

*Proof.* Since $E$ is $n$-dimensional, it has interior points (Section 3.1). Let $\mathbf{x}_1$ be an interior point so that $\mathbf{x}_1 \neq \mathbf{x}_0$. We now choose coordinates in $\mathcal{R}^n$, with $\mathbf{x}_0$ as origin, so that $\mathbf{x}_1$ has coordinates $(0, \ldots, 0, k)$ with $k > 0$. We choose $\delta > 0$ so small that the set

$$V_\delta = \{(\mathbf{u}, v) \mid \|\mathbf{u}\| < \delta, \ v = k\}$$

is contained in the interior of $E$ and define $R$ as above. We also define

$$W_\delta = \{(\mathbf{u}, v) \mid \|\mathbf{u}\| < \delta, \ v = -k\}.$$

We now observe that the intersection of the $n$-dimensional convex set $E$ with each straight line $L$ in $\mathcal{R}^n$ containing an interior point $\mathbf{a}$ of $E$ is a convex subset of $L$ and is thus an "interval" $I$ on $L$. The relative interior of $I$ consists of interior points of $E$. Each endpoint of $I$ (if $I$ has endpoints) is a boundary point of $E$; the remaining points of $L$ are not in $E$ and not boundary points of $E$. We refer to this statement as the *line-intersection rule*. A proof of the rule is given at the end of this section.

No point $(\mathbf{u}, -k)$ of $W_\delta$ is in $E$ or a boundary point of $E$. This follows from the line-intersection rule applied to the straight line $L$ through the point $(\mathbf{u}, -k)$ and the origin. (See Problem 4.7.)

We next apply the line-intersection rule to each parallel to the $x_n$-axis in $\mathcal{R}^n$ through a point $(\mathbf{u}, k)$ of $V_\delta$. For each $\mathbf{u}$ we obtain an interval on the corresponding line. By the properties of the sets $V_\delta$, $W_\delta$ given above, there is a uniquely determined function $f(\mathbf{u})$ such that one endpoint of the interval is a point $(\mathbf{u}, f(\mathbf{u}))$ with $-k < f(\mathbf{u}) < k$. From its definition the function $f$ has the property that the point $(\mathbf{u}, v)$ of $R$ is in the interior of $E$ for $v > f(\mathbf{u})$, is a boundary point of $E$ for $v = f(\mathbf{u})$, or is not in $E$ or its boundary for $v < f(\mathbf{u})$.

We now assert that the function $f$ is convex. It is defined on the convex set: $\|\mathbf{u}\| < \delta$. For $\mathbf{u}_1, \mathbf{u}_2$ in this set and $v_1 = f(\mathbf{u}_1)$, $v_2 = f(\mathbf{u}_2)$, the points $\mathbf{x}_i = (\mathbf{u}_i, v_i)$ are boundary points of $E$, with $v_i < k$ ($i = 1, 2$). Hence we can choose numbers $v_i' > v_i$ arbitrarily close to the $v_i$ such that the points $\mathbf{x}_i' = (\mathbf{u}_i, v_i')$ are in $E$ and in $R$. Since $E$ and $R$ are convex, for each $t$ in the interval $0 \leq t \leq 1$, the point $\mathbf{x}' = (1 - t)\mathbf{x}_1' + t\mathbf{x}_2'$ is in $E$ and in $R$. Therefore

$$(1 - t)v_1' + tv_2' \geq f((1 - t)\mathbf{u}_1 + t\mathbf{u}_2).$$

We now let the $v_i' \to v_i$ and conclude that

$$(1 - t)f(\mathbf{u}_1) + tf(\mathbf{u}_2) \geq f((1 - t)\mathbf{u}_1 + t\mathbf{u}_2).$$

Thus $f$ is convex and hence continuous.

As noted above, the points of $R$ above the graph of the function $f$ are interior to $E$, those on the graph are boundary points of $E$, and those below the graph are neither interior points nor boundary points (and are hence interior to the complement of $E$ in $\mathcal{R}^n$). Thus our theorem is proved.  ■

*Proof of the Line-Intersection Rule.*   Since $\mathbf{a}$ is interior to $E$, for some $\varepsilon > 0$ the $\varepsilon$-neighborhood of $\mathbf{a}$, which we denote by $U_\varepsilon$, is contained in the interior of $E$. It follows that $I$ contains an open interval of length $2\varepsilon$, each point of which is an interior point of $E$. If $I$ has an endpoint $\mathbf{c}$, then every neighborhood of $\mathbf{c}$ contains points in $E$ and points not in $E$, and so $\mathbf{c}$ is a boundary point of $E$.

Now let $\mathbf{b}$ be a point of $L$ which is either a point of $E$ or a boundary point of $E$. We now show that *all points of the line segment joining $\mathbf{b}$ to $\mathbf{a}$, except perhaps for $\mathbf{b}$, are in the interior of $E$.* From this assertion it follows that all points of the interval $I$, except for endpoints of $I$, are interior to $E$, and points of $L$ that are not on $I$ and not endpoints of $I$ are neither in $E$ nor boundary points of $E$. Thus the line-intersection rule would be completely established.

To prove our assertion, we observe that there is a sequence of points $\mathbf{b}_m$ of points of $E$ converging to $\mathbf{b}$. This is clear if $\mathbf{b}$ is a boundary point of $E$, and if $\mathbf{b}$ is in $E$, we can choose all $\mathbf{b}_m$ equal to $\mathbf{b}$. Let $\mathbf{d}$ be a point, other than $\mathbf{b}$, of the line segment joining $\mathbf{b}$ to $\mathbf{a}$. We can write $\mathbf{d} = t\mathbf{a} + (1 - t)\mathbf{b}$ with $0 \leq t < 1$. Now choose $M$ so large that

$$\|\mathbf{b}_m - \mathbf{b}\| < \frac{t}{1 - t}\varepsilon \qquad \text{for } m > M.$$

By the convexity of $E$, all points

$$\mathbf{c}_m = t(\mathbf{a} + \mathbf{r}) + (1 - t)\mathbf{b}_m$$

with $\|\mathbf{r}\| < \varepsilon$ are points of $E$. We can write

$$\mathbf{c}_m = t\mathbf{a} + (1 - t)\mathbf{b} + t\mathbf{r} + (1 - t)(\mathbf{b}_m - \mathbf{b}) = \mathbf{d} + t\mathbf{r} + (1 - t)(\mathbf{b}_m - \mathbf{b}).$$

Here, for $m > M$, the last term has norm at most $t\varepsilon$, and the next to the last term varies over all vectors of norm at most $t\varepsilon$. Hence a value $\mathbf{r}_0$ of $\mathbf{r}$ can be chosen, with norm less than $\varepsilon$, so that $t\mathbf{r}_0$ equals the negative of the last term. We can thus write

$$\mathbf{c}_m = \mathbf{d} + t(\mathbf{r} - \mathbf{r}_0).$$

For each choice of $\mathbf{r}$ with norm less than $\varepsilon$, we know that $\mathbf{c}_m$ is in $E$. In particular, for all $\mathbf{r}$ with

$$\|\mathbf{r} - \mathbf{r}_0\| < \varepsilon - \|\mathbf{r}_0\|;$$

$\mathbf{c}_m$ is in $E$. This statement shows that $\mathbf{d}$ has a neighborhood in $E$; accordingly $\mathbf{d}$ is in the interior of $E$, and our proof is complete. ∎

## PROBLEMS

**4.6.** Show that under the assumptions stated, the set $E_0$ specified in (4.20) is convex and that its boundary includes the graph of the function $f$.

**4.7.** Let the sets $V_\delta$ and $W_\delta$ be defined as in Section 4.2 so that each point of $V_\delta$ is an interior point of the convex set $E$. Let $(\mathbf{u}, -k)$ be a particular point of $W_\delta$. Follow the suggestion in the text to show that $(\mathbf{u}, -k)$ is not in $E$ and is not a boundary point of $E$.

**4.8.** Verify each step of the proof of the line-intersection rule given at the end of Section 4.2.

## 4.3  SUPPORTING HYPERPLANE, SEPARATING HYPERPLANE

A linear variety of dimension $n - 1$ in $\mathcal{R}^n$ is called a *hyperplane*. It can be represented by a linear equation

$$\mathbf{y}^t \mathbf{x} = w, \tag{4.30}$$

where the nonzero vector $\mathbf{y}$ is normal to the hyperplane and $w$ is a scalar; if $\|\mathbf{y}\| = 1$ and so $\mathbf{y}$ is a unit vector, then $|w|$ is the distance from the origin $\mathbf{0}$ to the hyperplane (Problem 4.9). The hyperplane determines two *half-spaces* in $\mathcal{R}^n$: one in which $\mathbf{y}^t \mathbf{x} \le w$; one in which $\mathbf{y}^t \mathbf{x} \ge w$. Each of these half-spaces is an $n$-dimensional convex set whose boundary (and relative boundary) is the hyperplane.

The hyperplane (4.30) is said to *support* a convex set $E$ in $\mathcal{R}^n$ if $E$ lies wholly in one of the two half-spaces. If $E$ is supported by the hyperplane but is not contained in the hyperplane, and the relative boundary of $E$ contains a point $\mathbf{x}_0$

in the hyperplane, then the hyperplane is said to be a *supporting hyperplane* for $E$ at $\mathbf{x}_0$. (See Fig. 4.2.)

If $E_1$ and $E_2$ are convex sets in $\mathcal{R}^n$ that are contained in opposite half-spaces determined by the hyperplane (4.30), then the hyperplane is said to *separate* $E_1$, $E_2$, or to be a *separating hyperplane* for these two convex sets. The separation is said to be *proper* if the hyperplane does not contain both sets. (See Fig. 4.3.)

***Example 1.*** Let $A$, $B$ be convex sets containing distinct points $\mathbf{a} \in A$, $\mathbf{b} \in B$ such that $\|\mathbf{a} - \mathbf{b}\|$ minimizes the distance between the sets:

$$\|\mathbf{a} - \mathbf{b}\| = \min \|\mathbf{a}' - \mathbf{b}'\|, \qquad \mathbf{a}' \in A, \ \mathbf{b}' \in B.$$

Then the hyperplane (4.30) which is the perpendicular bisector of the line segment from $\mathbf{a}$ to $\mathbf{b}$ separates $A$, $B$. Its equation can be written in the form (4.30) with $\mathbf{y} = \mathbf{b} - \mathbf{a}$, $w = (\frac{1}{2})(\|\mathbf{b}\|^2 - \|\mathbf{a}\|^2)$. (See Problem 4.10.) If, instead, $w$ is taken as $\mathbf{b}^t \mathbf{a} - \|\mathbf{a}\|^2$ and $A$ is $n$-dimensional, then the hyperplane (4.30) is a supporting hyperplane for $A$ at $\mathbf{a}$. (Problem 4.10(c).)

REMARK. The definitions given above continue to apply in the case $n = 1$. In $\mathcal{R}^1$ a hyperplane reduces to a point, defined by an equation $yx = w$, or $x = x_0 = w/y$. A 1-dimensional convex set $E$ in $\mathcal{R}^1$ is an interval, perhaps infinite, and its relative boundary is formed of the one or two endpoints if there are any. Each endpoint is a supporting hyperplane for $E$. One can even consider a 0-dimensional convex set in $\mathcal{R}^1$. This is a set consisting of one point; as in Section 3.1 it has no relative boundary. It has no supporting hyperplane. In the discussion to follow we will tacitly assume that $n \geq 2$; the results generally remain valid for $n = 1$ and for convex sets of dimension 0 or 1, but we will omit a discussion of these cases.

## Supporting Hyperplane at a Point at Which the Boundary Is Smooth

Let $E$ be an $n$-dimensional convex set in $\mathcal{R}^n$ with boundary point $\mathbf{x}_0$. In Section 4.2 it is shown that coordinates can be chosen in $\mathcal{R}^n$ with origin at $\mathbf{x}_0$

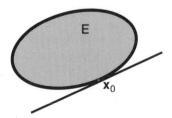

**Figure 4.2.** Supporting hyperplane.

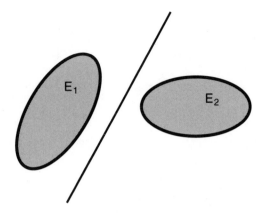

**Figure 4.3.** Hyperplane properly separating two convex sets.

such that, in a neighborhood $R$ of the origin, the boundary of $E$ is represented as the graph of a continuous convex function $x_n = f(x_1, \ldots, x_{n-1})$ and the part of $E$ in $R$ lies above this graph. If $f$ is differentiable, then a supporting hyperplane for $E$ at the origin is the *tangent hyperplane* to the graph of $f$ at the origin:

$$x_{n-1} = \sum_{i=1}^{n-1} g_i x_i, \tag{4.31}$$

where $\mathbf{g} \in \mathcal{R}^{n-1}$ is the gradient vector of $f$ at the origin. Indeed, as in Problem 1.55 the tangent hyperplane lies below the graph of $f$, hence below the part of $E$ in $R$. If a point $\mathbf{x}_1$ of $E$ were *strictly below* this tangent hyperplane, then the line segment joining an interior point of $E$ to the origin would contain points of $E$ in $R$ strictly below the tangent hyperplane and hence strictly below the graph of $f$, which is not possible. (See Problem 4.11.) Hence $E$ lies wholly on one side of the tangent hyperplane at the origin, and this is a supporting hyperplane. In this case there is only one supporting hyperplane at the point $\mathbf{x}_0$ (see Problem 1.26(d)).

When $E$ is of dimension $d$ at least 2 but less than $n$, again a supporting hyperplane can be found at each point on a smooth portion of the relative boundary. Here we can assume that $E$ is contained in a linear variety of dimension $d < n$, and for proper choice of coordinates, $E$ lies in the subspace of $\mathcal{R}^n$ for which the coordinates $x_i$ are 0 for $i = d+1, \ldots, n$. Then we can assume that the *relative* boundary of $E$ is given as above by an equation $x_d = f(x_1, \ldots, x_{d-1})$ in this subspace, where $f$ has properties as in the preceding paragraph. The equation analogous to (4.31) gives a tangent hyperplane in $\mathcal{R}^d$. An arbitrary hyperplane of $\mathcal{R}^n$ intersecting the $d$-dimensional subspace in this hyperplane is then a supporting hyperplane for $E$ in $\mathcal{R}^n$.

***Example 2.*** A supporting hyperplane at (1, 1, 0) for the 2-dimensional convex set $E$ bounded by the ellipse $x^2 + 4y^2 = 5$, $z = 0$ in $\mathcal{R}^3$ is an arbitrary hyperplane meeting the $xy$-plane in the tangent line $2x + 8y = 10$, $z = 0$ to the relative boundary of $E$ at (1, 1). For example, the hyperplane $2x + 8y + 7z = 10$ is such a supporting hyperplane at the point. (See Fig. 4.4.)

### A Supporting Hyperplane Exists at Each Relative Boundary Point x₀ of a Convex Set

As the discussion just given shows, it is sufficient to consider the case when $E$ is $n$-dimensional and the relative boundary is the actual boundary. In this case, as in the case of a smooth boundary, each hyperplane through the point for which $E$ is included in one of the two corresponding half-spaces is a supporting hyperplane for $E$ at the point.

We now assume that coordinates and the region $R$ are chosen as in Section 4.2, so that the part of $E$ in $R$ lies above the graph of the convex function $f$, and we seek a supporting hyperplane for $E$ at the origin. Since $f$ is convex, the generalized directional derivative $f'(\mathbf{0}; \mathbf{y})$ exists (Section 4.1). We denote this convex function in $\mathcal{R}^{n-1}$ by $g(\mathbf{y})$. Since $g$ is continuous, the graph of $g$ is a closed set in $\mathcal{R}^n$. We let $A$ be the set of all points $(\mathbf{u}, v)$ of $\mathcal{R}^n$ for which $v \geq g(\mathbf{u})$; this is the *epigraph* of $g$, and as shown in Section 4.4, it is a convex set. Here it is also $n$-dimensional and is a closed set (Problem 4.12). We choose a point $\mathbf{b} = (\mathbf{u}_0, v_0)$ of $\mathcal{R}^n$ which is strictly below the graph of $g$: for example, the point with $\mathbf{u}_0 = \mathbf{0}$ and $v_0 = -1$ (recall that $g(\mathbf{0}) = 0$). There is a unique point $\mathbf{a}$ of the epigraph $A$ that is nearest to $\mathbf{b}$; by the choice of $\mathbf{b}$, $\mathbf{a}$ must be a point of the graph of $g$ (Problem 4.13). As in Example 1 of Section 4.3, the hyperplane through $\mathbf{a}$ with normal $\mathbf{b} - \mathbf{a}$ is a supporting hyperplane for $A$ at $\mathbf{a}$. If $\mathbf{a}$ is the origin of $\mathcal{R}^n$, then this hyperplane has the equation $v = 0$. Otherwise, the hyperplane contains the straight line through $\mathbf{a}$ and the origin. Indeed, by the positive homegeneity of $g$, the graph of $g$, and hence the epigraph $A$, must contain this straight line, so $\mathbf{a}$ must be the point of this line nearest to $\mathbf{b}$; by geometry, $\mathbf{a}$ must be perpendicular

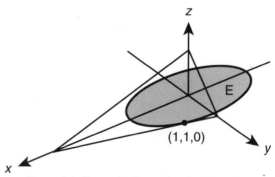

**Figure 4.4.** Supporting hyperplane for Example 1.

to $\mathbf{b} - \mathbf{a}$, so the supporting hyperplane must pass through the origin and contains the straight line in question. (See Fig. 4.5.) Thus the supporting hyperplane must in any case have an equation of the form $v = \mathbf{c}^t \mathbf{u}$ for an appropriate vector $\mathbf{c}$ of $\mathcal{R}^{n-1}$, and we can assert that

$$g(\mathbf{u}) = f'(\mathbf{0}; \mathbf{u}) \geq \mathbf{c}^t \mathbf{u} \tag{4.32}$$

for all $\mathbf{u}$.

We now let $\phi(\mathbf{u}) = f(\mathbf{u}) - \mathbf{c}^t \mathbf{u}$. This is a convex function for $\|\mathbf{u}\| < \delta$. Furthermore its generalized directional derivative at $\mathbf{u} = \mathbf{0}$ in the direction $\mathbf{y}$ is

$$\phi'(\mathbf{0}; \mathbf{y}) = g(\mathbf{y}) - \mathbf{c}^t \mathbf{y}.$$

(See Example 1 in Section 4.1.) By (4.32) this function is nonnegative for all $\mathbf{y}$. Therefore, by the minimum principle for convex functions (Section 4.1), $\phi$ has its minimum at $\mathbf{u} = \mathbf{0}$, where $\phi = 0$. Accordingly $f(\mathbf{u}) \geq \mathbf{c}^t \mathbf{u}$ for $\|\mathbf{u}\| < \delta$, and the argument given above for the case of a smooth boundary shows that the hyperplane $v = \mathbf{c}^t \mathbf{u}$ is a supporting hyperplane for $E$ at the origin. ∎

For nonsmooth boundaries the supporting hyperplane is generally not unique. For example, let $E$ be a triangular region in the plane. Then at each vertex the supporting hyperplanes are all straight lines through the vertex not entering the interior of the triangle. At each boundary point not a vertex, the supporting hyperplane is the straight line containing the edge through the point. (See Fig. 4.6.)

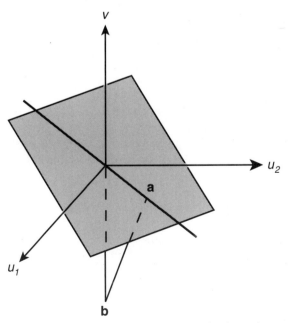

**Figure 4.5.** Proof of existence of a supporting hyperplane.

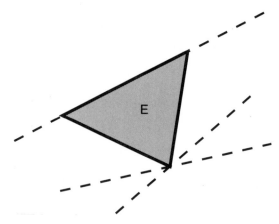

**Figure 4.6.** Supporting hyperplanes for a triangular region in the plane.

### Separation Theorems

We first remark that the theorem on existence of a supporting hyperplane can be interpreted as a separation theorem, namely that *the convex set consisting of the single point* $x_0$ *can be properly separated by a hyperplane from a convex set E of which it is a relative boundary point.* The supporting hyperplane found does not contain $E$ but does contain $x_0$, and $E$ lies in one of the two half-spaces determined by the hyperplane.

From this result we deduce the following theorem:

**Theorem.** *If $E_1$ and $E_2$ are nonempty convex sets in $\mathcal{R}^n$ that have no common points, then $E_1$ and $E_2$ can be properly separated by a hyperplane.*

*Proof.* Let $E$ be the set of all vectors of the form $x_1 - x_2$, where $x_1 \in E_1$ and $x_2 \in E_2$. Then $E$ is a convex set (Problem 4.14). Since $E_1$ and $E_2$ have no common points, $E$ does not contain the point $\mathbf{0}$. If $\mathbf{0}$ is not a relative boundary point of $E$, then $\mathbf{0}$ does not belong to the closed set formed of $E$ and its relative boundary points, and hence, as in Example 1 above, there is a hyperplane separating $\mathbf{0}$ from $E$. If $\mathbf{0}$ is a relative boundary point of $E$, then a supporting hyperplane for $E$ at $\mathbf{0}$ is the separating hyperplane sought. So in either case there is a hyperplane separating $\mathbf{0}$ from $E$. Let its equation by $y^t x = w$. (See Fig. 4.7.) Then the expression $y^t x - w$ must have opposite signs at $\mathbf{0}$ and at the points of $E$, for example, negative or 0 at $\mathbf{0}$ and positive or 0 on $E$. In that case, $w \geq 0$ and $y^t(x_1 - x_2) - w \geq 0$ for each point $x_1 - x_2$ of $E$. Since the hyperplane does not contain $E$, there must be some points $x_1'$ in $E_1$, $x_2'$ in $E_2$ for which

$$y^t(x_1' - x_2') - w > 0. \tag{4.33}$$

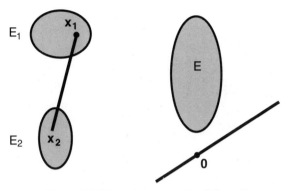

**Figure 4.7.** Hyperplane separating **0** from $E$.

Now we have for all $\mathbf{x}_1 \in E_1$ and $\mathbf{x}_2 \in E_2$,

$$\mathbf{y}^t\mathbf{x}_1 \geq \mathbf{y}^t\mathbf{x}_2 + w \geq \mathbf{y}^t\mathbf{x}_2.$$

Hence $\mathbf{y}^t\mathbf{x}$ is bounded below on $E_1$ and bounded above on $E_2$, and if $w_1$ is the greatest lower bound of $\mathbf{y}^t\mathbf{x}$ on $E_1$ and $w_2$ is the least upper bound of $\mathbf{y}^t\mathbf{x}$ on $E_2$, then $w_1 \geq w_2$. Thus $\mathbf{y}^t\mathbf{x} \geq w_1$ on $E_1$ and $\mathbf{y}^t\mathbf{x} \leq w_2 \leq w_1$ on $E_2$. Hence $\mathbf{y}^t\mathbf{x} - w_1 \geq 0$ on $E_1$, and $\mathbf{y}^t\mathbf{x} - w_1 \leq 0$ on $E_2$; that is, the hyperplane $\mathbf{y}^t\mathbf{x} = w_1$ has $E_1$, $E_2$ in opposite half-spaces. It cannot contain both $E_1$ and $E_2$, for then $\mathbf{y}^t\mathbf{x}_1' = w_1$, $\mathbf{y}^t\mathbf{x}_2' = w_1$ or $\mathbf{y}^t(\mathbf{x}_1' - \mathbf{x}_2') = 0$. This contradicts (4.33), since $w \geq 0$. Accordingly, the hyperplane properly separates $E_1$ and $E_2$. ∎

As a consequence of this theorem we can now establish a basic result:

**Separation Theorem.** *Two convex sets in $\mathcal{R}^n$ with no common relative interior points can be properly separated by a hyperplane.*

*Proof.* The relative interiors of the two sets form two convex sets with no common points (Problem 4.15). Hence by the theorem just proved, there is a hyperplane that properly separates these interiors. The relative boundary points of each set lie in the same half-space determined by the hyperplane as the relative interior and so the hyperplane separates the given convex sets. If the hyperplane contained both sets, it would have to contain their relative interiors, which are nonempty. This cannot be, since the hyperplane properly separates the relative interiors. Thus the hyperplane properly separates the given convex sets. ∎

REMARK. There is also a converse theorem: *If two convex sets can be properly separated, then they have no common relative interior points.* In particular, there is no supporting hyperplane to a convex set at a relative interior point of the set.

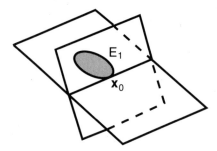

**Figure 4.8.** Proof of converse of Separation Theorem.

Indeed, let $E_1$ and $E_2$ be convex sets in $\mathcal{R}^n$ that are properly separated by the hyperplane (4.30), and let $\mathbf{x}_0$ be a relative interior point of both $E_1$ and $E_2$. It then follows that $\mathbf{x}_0$ lies in the hyperplane (4.30). If $E_1$ is not contained in the hyperplane and $E_1$ has dimension $d_1$, then $E_1$ lies in a linear variety of dimension $d_1$; $\mathbf{x}_0$ lies in the intersection of this linear variety with the hyperplane and must hence be a *relative boundary point* of $E_1$, contrary to assumption (see Fig. 4.8, which illustrates the case $n = 3$, $d_1 = 2$). Thus $E_1$ is contained in the hyperplane, and similarly $E_2$ is contained in the hyperplane. But that means that the separation cannot be proper. ∎

## PROBLEMS

**4.9.** As in plane and solid analytic geometry, the *normal form* of the equation of a hyperplane in $\mathcal{R}^n$ is $\mathbf{y}^t\mathbf{x} = w$, where $\mathbf{y}$ is a unit vector, normal to the hyperplane, and $w$ is the distance from the origin to the hyperplane. Establish this rule by using the method of Lagrange multipliers to minimize the distance squared from the origin to the hyperplane.

**4.10.** Let $A$, $B$ be convex sets in $\mathcal{R}^n$, and let $\mathbf{a} \in A$, $\mathbf{b} \in B$ be distinct points such that $\|\mathbf{a} - \mathbf{b}\|$ minimizes the distance between $A$ and $B$.

    **(a)** Show that the hyperplane which is the perpendicular bisector of the segment joining $\mathbf{a}$ to $\mathbf{b}$ has the equation $\mathbf{y}^t\mathbf{x} = w$ with $\mathbf{y} = \mathbf{b} - \mathbf{a}$, $w = \left(\frac{1}{2}\right)(\|\mathbf{b}\|^2 - \|\mathbf{a}\|^2)$.

    **(b)** Show that the hyperplane of part (a) separates $A$ and $B$.

    **(c)** Show that if $A$ is $n$-dimensional and in the equation of the hyperplane of part (a) $w$ is replaced by $\mathbf{b}^t\mathbf{a} - \|\mathbf{a}\|^2$, then the new equation is that of a supporting hyperplane for $A$ at $\mathbf{a}$.

**4.11.** In Section 4.3 it is shown that in the case of a point at which the boundary of the $n$-dimensional convex set $E$ is smooth, coordinates can be chosen with origin at the point such that the tangent hyperplane to the boundary at the origin lies below the graph of the boundary function $f$ in $R$. Show that this implies that no point $\mathbf{x}_1$ of the convex set $E$ can lie strictly below this tangent hyperplane. [Hint: If such a point $\mathbf{x}_1$ exists, then there must be an interior point $\mathbf{x}_2$ of $E$ lying strictly below the tangent hyperplane. Apply the line-intersection rule of Section 4.2 to the straight line joining $\mathbf{x}_2$ to the origin.]

**4.12.** Let $g$ be a convex function in $\mathcal{R}^{n-1}$, and let $A$ be the epigraph of $g$, a convex subset of $\mathcal{R}^n$. Show that $A$ is a closed set. [Hint: Use the continuity of $g$.]

**4.13.** In the notations of Problem 4.12, let $\mathbf{b}$ be a point of $\mathcal{R}^n$ that is strictly below the graph of $g$, and let $\mathbf{a}$ be the point of $A$ nearest to $\mathbf{b}$. Show that $\mathbf{a}$ lies on the graph of $g$.

**4.14.** Let $E_1$ and $E_2$ be nonempty convex sets in $\mathcal{R}^n$. Let $E$ be the set of all vectors of the form $\mathbf{x}_1 - \mathbf{x}_2$, where $\mathbf{x}_1 \in E_1$ and $\mathbf{x}_2 \in E_2$. Show that $E$ is a convex set.

**4.15.** **(a)** Show that the interior of an $n$-dimensional convex set $E$ in $\mathcal{R}^n$ is a convex set. [Hint: Let $\mathbf{a}$ and $\mathbf{b}$ be distinct interior points of $E$. One must show that these two points can be joined by a line segment that consists of interior points of $E$. Choose $\mathbf{a}' \in E$ and $\mathbf{b}' \in E$ such that the line segment joining these two points contains $\mathbf{a}$ and $\mathbf{b}$ in its relative interior. Now apply the line-intersection rule.]

**(b)** Apply the result of part **(a)** to show that the relative interior of a $d$-dimensional convex set in $\mathcal{R}^n$ is a convex set.

## 4.4  NEW DEFINITION OF CONVEX FUNCTION, EPIGRAPH, HYPOGRAPH

Following Fenchel, we now define a convex function $f$ in $\mathcal{R}^n$ to be a function having finite values on a nonempty convex set $K(f)$ and satisfying the usual convexity condition on this set; furthermore $f$ has the value $+\infty$ on all other points of $\mathcal{R}^n$. Similarly a concave function $g$ in $\mathcal{R}^n$ satisfies the usual concavity condition on a convex set $K(g)$ and has the value $-\infty$ elsewhere in $\mathcal{R}^n$. The obvious operations on $+\infty$ and $-\infty$ are allowed and lead to no difficulty. Thus the sum of two convex functions $f_1$, $f_2$ is a convex function provided that the sets $K(f_1)$, $K(f_2)$ have common points and that $K(f_1 + f_2) = K(f_1) \cap K(f_2)$. If $f$ is convex and $c$ is a constant, not 0, then $cf$ is convex if $c > 0$ and is concave if $c < 0$. One of course avoids expressions such as $\infty - \infty$ and $0 \cdot \infty$.

It will be convenient to denote by $E^I$ the relative interior of the convex set $E$. In particular, we will refer to $K(f)^I$ for a convex $f$.

The *graph* of a function is a familiar concept. For a convex function $f$ in $\mathcal{R}^n$, the graph is a set in $\mathcal{R}^{n+1}$, consisting of all points $(x_1, \ldots, x_n, z) = (\mathbf{x}, z)$

such that $z = f(\mathbf{x})$. Here we allow only finite values, so $\mathbf{x}$ varies over $K(f)$. By the *epigraph* of $f$, we mean the set $[f]$ of all points of $\mathcal{R}^{n+1}$ that lie above the graph; that is, all $(\mathbf{x}, z)$ for which $z$ is finite and $z \geq f(\mathbf{x})$. Similarly the *hypograph* of a concave function $g$ in $\mathcal{R}^n$ consists of all $(\mathbf{x}, z)$ for which $z \leq g(\mathbf{x})$ (see Fig. 4.9); we denote the hypograph by $[g]$. (The context will always make clear whether it is an epigraph or hypograph.)

Each epigraph is a convex set in $\mathcal{R}^{n+1}$. In fact the epigraph $[f]$ of convex $f$ is simply the sublevel set $f(\mathbf{x}) - z \leq 0$ of the convex function $f(\mathbf{x}) - z$ of $\mathbf{x}$ and $z$. Hence as in Section 1.9, $[f]$ is a convex set. Similarly the hypograph $[g]$ is a convex set; it is a sublevel set of the *convex* function $z - g(\mathbf{x})$.

If $K(f)$ is $d$-dimensional, then $[f]$ has dimension $d + 1$. For example, if $E = K(f)$ is a closed region in $\mathcal{R}^n$, so $E$ has dimension $n$ and $f$ is continuous on $E$, then $[f]$ is a closed region in $\mathcal{R}^{n+1}$. Its interior consists of all $(\mathbf{x}, z)$ for which $\mathbf{x}$ is interior to $E$ and $z > f(\mathbf{x})$, for each such point has a neighborhood contained in $[f]$: namely the set of all $(\mathbf{x}', z')$ for which $\mathbf{x}'$ lies in a $\delta$-neighborhood of $\mathbf{x}$ contained in $E$ and $z'$ lies in a sufficiently small open interval containing $z$. Similarly one verifies that each point $(\mathbf{x}, z)$ of $[f]$, for which $\mathbf{x}$ is a boundary point of $E$ and/or $z = f(\mathbf{x})$, is a boundary point of $[f]$.

By similar reasoning, one verifies that in general $[f]$ has dimension $d + 1$ and that the relative interior of $[f]$ consists of the points $(\mathbf{x}, z)$ for which $\mathbf{x}$ is a relative interior point of $K(f)$ and $z > f(\mathbf{x})$, whereas the relative boundary of $[f]$ includes all $(\mathbf{x}, z)$ for which $\mathbf{x}$ is a relative interior point of $K(f)$ and $z = f(\mathbf{x})$.

The relative boundary of $[f]$ may also include some points $(\mathbf{x}, z)$ for which $\mathbf{x}$ is a relative boundary point of $K(f)$. This is illustrated by the function of one variable $f(x) = 1/x$ for $0 < x \leq 1$. (See Fig. 4.10a.) Here $K(f)$ has relative boundary consisting of the two points $x = 0$ and $x = 1$. The epigraph $[f]$ consists of all $(x, z)$ with $z \geq 1/x$,  $0 < x \leq 1$. Its relative boundary consists of all $(x, z)$ with $z = 1/x$,  $0 < x < 1$, as well as the points $(1, z)$ with $z \geq 1$. For this example, the relative boundary is the actual boundary. If the example is modified by considering $f$ to be the function of $x$ and $y$ with $K(f)$ equal to the set of points $(x, 0)$ for which $0 < x \leq 1$ and $f(x, y) = 1/x$ on this set, then $d = 1$

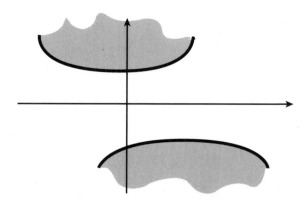

**Figure 4.9.** Epigraph of a convex function, hypograph of a concave function.

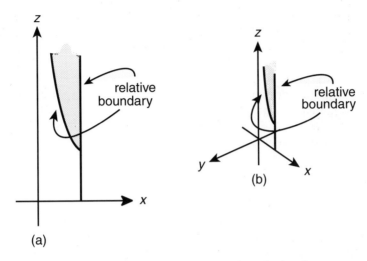

**Figure 4.10.** (a) and (b): Relative boundary of epigraph.

and $n = 2$, and the relative boundary is as described, regarded as a set in the plane $y = 0$. This is shown in Fig. 4.10b. In this example every point of the epigraph is a boundary point.

In general, if $f(\mathbf{x})$ is finite at a relative boundary point $\mathbf{x}$ of $K(f)$, then the relative boundary of $[f]$ includes all points $(\mathbf{x}, z)$ with $z \geq f(\mathbf{x})$. All these points are included in $[f]$, and none can be relative interior points of $[f]$ (Problem 4.17). The examples of the previous paragraph illustrate this rule.

### Supporting Hyperplanes of an Epigraph

Let $H$ be a hyperplane in $\mathcal{R}^{n+1}$. Its equation can always be written in one of the two forms:

$$\mathbf{y}^t\mathbf{x} - z = w, \tag{4.40a}$$

$$\mathbf{y}^t\mathbf{x} = w. \tag{4.40b}$$

In the first case we say that $H$ is a *nonvertical* hyperplane; in the second we say that $H$ is a *vertical* hyperplane.

We apply this definition to the supporting hyperplane of the epigraph $[f]$ at a point $(\mathbf{x}_0, f(\mathbf{x}_0))$ on the graph of $f$. From the preceding paragraphs, we see that such a point is a relative boundary point of $[f]$. By the results of Section 4.3, a supporting plane to $[f]$ exists at the point. We assert that *if* $\mathbf{x}_0 \in K(f)^I$, *then there is no vertical supporting hyperplane, so every supporting hyperplane to $[f]$ at the point is nonvertical.*

We must show that there is no vertical supporting hyperplane to $[f]$ at a point $(\mathbf{x}_0, f(\mathbf{x}_0))$ for which $\mathbf{x}_0$ is a relative interior point of $K(f)$. Suppose that $H$ is a vertical supporting plane at the point and thus $H$ has equation of form

(4.40b). Then we can assume that every point $(\mathbf{x}, z)$ of $[f]$ satisfies the inequality $\mathbf{y}'\mathbf{x} - w \leq 0$, with equality for $\mathbf{x} = \mathbf{x}_0$. Hence the inequality holds for all $\mathbf{x}$ in $K(f)$, and $[f]$ is not contained in $H$. Therefore the plane (4.40b) is the equation of a hyperplane in $\mathcal{R}^n$ which is a supporting plane to $K(f)$ at the relative interior point $\mathbf{x}_0$. But in Section 4.3 we saw that this was impossible; see the Remark at the end of the section. Therefore our assertion is proved. [The point is that the equation (4.40b) does not contain $z$ and hence can be interpreted either as the equation of a hyperplane in $\mathcal{R}^{n+1}$—a vertical one—or as that of a hyperplane in $\mathcal{R}^n$.]

REMARK ON THE 2-DIMENSIONAL CASE. A common application of convexity is to sets bounded by closed curves in the plane. Let such a curve $C$ be given by an equation $r = f(\theta)$ in polar coordinates, where $f(\theta)$ has continuous derivatives through the second order for all $\theta$, $f$ has period $2\pi$ in $\theta$, and $f(\theta) > 0$ for all $\theta$. The curve $C$ is then an oval enclosing the origin, and $C$ is the boundary of a bounded closed region $R$. At each point of $C$ there is a nonzero tangent vector $\mathbf{T}$ pointing in the direction of increasing $\theta$. We denote by $\phi$ the angle from the direction of the $x$-axis to $\mathbf{T}$. It is shown in elementary calculus that

$$\frac{d\phi}{d\theta} = \frac{-rr'' + r^2 + 2r'^2}{(r'^2 + r^2)^{3/2}}.$$

The absolute value of this quantity is the *curvature* of the oval at the corresponding point.

  *If $d\phi/d\theta > 0$ for all $\theta$, then the closed region $R$ bounded by $C$ is a convex region.* To show this, we first remark that $\phi$ is a monotone strictly increasing function of $\theta$, increasing by $2\pi$ as $\theta$ increases from 0 to $2\pi$ (see Chern, 1967). The value of $x$ has a minimum $a$ on $C$ and a maximum $b$, with $a < b$. For $x = b$ there is exactly one point on $C$ at which $\phi$ can be given the value $\pi/2$. (See Fig. 4.11.) As $\theta$ increases from the value at the point, $\phi$ can be chosen to vary continuously and increases to the value $3\pi/2$ when $x$ reaches $a$. As $\theta$ increases further, $\phi$ increases, to reach the value $5\pi/2$ when $x = b$ again and $C$ has been traced once. For each value of $x$ between $a$ and $b$, there is exactly one point on the upper arc of $C$, on which $\phi$ is between $\pi/2$ and $3\pi/2$; this follows from the mean value theorem, since if there were two such points, the tangent would have to be vertical for some intermediate point on the arc. Similarly, for $x$ between $a$ and $b$, there is exactly one point on C on the lower arc, for which $\phi$ is between $3\pi/2$ and $5\pi/2$. This shows that the region $R$ can be represented as the set of all points $(x, y)$ such that $F(x) \leq y \leq G(x)$; here the functions $F$ and $G$ are continuous for $a \leq x \leq b$ and, by the curvature assumption, $F''(x) > 0$ and $G''(x) < 0$ for $a < x < b$. Hence $F$ is a convex function, and $G$ is a concave function. Thus the region $R$ is the intersection of the epigraph $[F]$ of the convex function $F$ and the hypograph $[G]$ of the concave function $G$. As the intersection of two convex sets, $R$ is

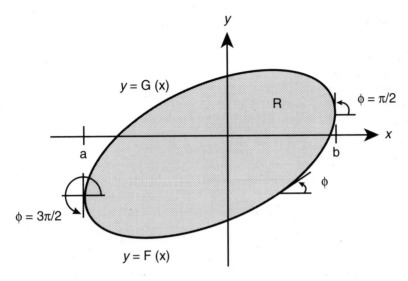

**Figure 4.11.** Convex region bounded by a convex oval in $\mathcal{R}^2$.

itself convex, as asserted. One also refers to the boundary curve $C$ as a convex oval.

## PROBLEMS

**4.16.** Let $f$ be the convex function in $\mathcal{R}^1$ such that $f(x) = x$ for $0 < x \leq 1$, $f = \infty$ otherwise. Find a supporting hyperplane to $[f]$ at each of the points given: $(\frac{1}{2}, \frac{1}{2})$, $(1, 1)$, $(1, 2)$, $(0, 0)$, $(0, 1)$, $(0, 2)$.

**4.17.** Prove: If $f(\mathbf{x})$ is finite at a relative boundary point $\mathbf{x}$ of $K(f)$, then the relative boundary of $[f]$ includes all points $(\mathbf{x}, z)$ with $z \geq f(\mathbf{x})$.

## 4.5  CONJUGATE OF CONVEX AND CONCAVE FUNCTIONS

### About sup and inf

In the calculus, one introduces the concept of the *least upper bound*, abbreviated lub, of a set of real numbers. This is the largest number in the set, if there is one, or else it is the smallest number which is greater than all numbers of the set; if the set is bounded above, then the lub always exists. For example, the set of values of the function $(x - 1)/x$ for $x > 0$ is bounded above by 1, which is also the lub of the set. It is also common to denote the lub by sup (for *supremum*), and we will use that notation, writing

$$\sup_{x}\left\{\frac{x - 1}{x} \mid 0 < x < \infty\right\} = 1.$$

When the set is not bounded above, we assign the value $+\infty$ to the lub or sup. Thus, if $N$ is the set of all positive integers,

$$\sup_{n}\left\{\frac{n!}{n^2}\,|\,n\in N\right\} = +\infty.$$

The *greatest lower bound* or glb or inf (for *infimum*) of a set is defined similarly as the smallest number in the set, if there is one, or else as the largest number which is smaller than all numbers of the set, or, when the set is not bounded below, as $-\infty$.

When we know that the set has a largest number, then we can write that number both as the sup of the set or as the max of the set:

$$\max_{x}\{\sin x\,|\,0\le x\le 2\pi\} = 1.$$

Similarly we can write min for the inf when there is a smallest number in the set.

Often the context indicates which values of the variables are allowed, and we can write, for example, $\sup_x f(\mathbf{x})$ or just $\sup f$; or we can write $\sup_E f$ if $\mathbf{x}$ is restricted to set $E$. The inf and sup are related by the rule

$$\sup_{E} f = -\inf_{E}(-f). \tag{4.50}$$

(See Problem 4.21.)

The problem of finding the sup or inf of a function can be regarded as a generalization of the problem of finding maxima and minima. We can extend the concept of duality to the general problems and say, for example, that the two programs:

$$\text{find}\quad \sup_{E} f,$$

$$\text{find}\quad \inf_{B} g$$

are dual if $\sup_E f = \inf_B g$.

On occasion we deal with a function $f(\mathbf{x}, \mathbf{y})$ defined for $\mathbf{x}\in E$ and $\mathbf{y}\in F$, where $E$ is a set in $\mathcal{R}^n$ and $F$ is a set in $\mathcal{R}^m$. Then we can ask for such quantities as

$$\sup_{\mathbf{x}}\{f(\mathbf{x}, \mathbf{y})\,|\,\mathbf{x}\in E, \mathbf{y}\in F\}.$$

Here the result depends on $\mathbf{y}$. For example, if $E$ is the interval $[0, 1]$ and $F$ is the interval $[-\pi, \pi]$, then

$$\sup_{x}\{x^2\sin y\,|\,x\in E,\ y\in F\} = \begin{cases} \sin y, & \text{if } 0\le y\le \pi, \\ 0, & \text{otherwise,} \end{cases}$$

$$\sup_{y}\{x^2\sin y\,|\,x\in E,\ y\in F\} = x^2.$$

In each case we could take the sup of the resulting values. There is a general rule that *the order does not matter*; that is,

$$\sup_x \sup_y \{f(\mathbf{x}, \mathbf{y}) \mid \mathbf{x} \in E,\ \mathbf{y} \in F\} = \sup_y \sup_x \{f(\mathbf{x}, \mathbf{y}) \mid \mathbf{x} \in E,\ \mathbf{y} \in F\}. \qquad (4.51)$$

For a proof, see Dieudonné (1960, p. 24; see also Problem 4.20). It should be observed that in each case one may have to allow $+\infty$ as a number in the set considered. When $+\infty$ does occur, both sides of the equation have the value $+\infty$. For example, if $E$ is the interval $0 \le x < 1$ and $F$ is the interval $0 < y < 1$ and $f(x, y) = \ln(1/(x + y))$, then

$$\sup_y \sup_x \{f(x, y) \mid x \in E,\ y \in F\} = \sup_x \sup_y \{f(x, y) \mid x \in E,\ y \in F\} = \infty$$

and here, for $x = 0$, $\sup_y f = \infty$. (See Problem 4.19(b).) There is a similar rule for successive inf operations.

We need two other rules, which we state for sup only:

$$\sup_E cf(\mathbf{x}) = c \sup_E f(\mathbf{x}), \qquad (4.52)$$

where $c$ is a positive constant; if $f(\mathbf{x})$ and $g(\mathbf{x})$ are both defined on set $E$; then

$$\sup_E \{f(\mathbf{x}) + g(\mathbf{x})\} \le \sup_E f + \sup_E g. \qquad (4.53)$$

(See Problem 4.21.)

By the *conjugate* of a convex function $f$ in $\mathcal{R}^n$ we mean the function $f^c$ defined by the equation

$$f^c(\mathbf{y}) = \sup_x (\mathbf{y}^t \mathbf{x} - f(\mathbf{x})) \qquad (4.54)$$

for all $\mathbf{y}$ in $\mathcal{R}^n$.

**Example 1.** $f(x) = 2x + 3$ in $\mathcal{R}^1$. Then

$$f^c(y) = \sup_x (yx - 2x - 3) = \sup_x ((y - 2)x - 3).$$

If $y = 2$, this yields the finite value $-3$. Otherwise, the sup is $+\infty$. So $f^c(y)$ is finite only at one point. (See Problem 4.23(a).)

**Example 2.** $f(x) = x^2$ in $\mathcal{R}^1$. Then

$$f^c(y) = \sup_x (yx - x^2).$$

The quadratic function has its maximum for $x = y/2$, and the maximum value is $y^2/4 = f^c(y)$. (See Problem 4.23(b).)

***Example 3.***   $f(\mathbf{x}) = 0$ for all $\mathbf{x} \in \mathcal{R}^n$. Then $f^c(\mathbf{0}) \doteq 0$; $f^c(\mathbf{y}) = +\infty$ otherwise. (See Problem 4.23(c).)

For each $\mathbf{y}$, $\mathbf{y}^t\mathbf{x} - f(\mathbf{x})$ is finite for $\mathbf{x}$ in $K(f)$, which is nonempty. Hence $f^c(\mathbf{y})$ *is finite or* $+\infty$ *for every* $\mathbf{y}$.

It is useful to think of $f^c(\mathbf{y})$ for a convex $f$ as the smallest $z$ such that $z \geq \mathbf{y}^t\mathbf{x} - f(\mathbf{x})$ for all $\mathbf{x}$ or $f(\mathbf{x}) \geq \mathbf{y}^t\mathbf{x} - z$ for all $\mathbf{x}$. (See Fig. 4.12.) Thus let $\mathbf{y}^t\mathbf{x} - z = w$ be a nonvertical hyperplane $E$ in $\mathcal{R}^{n+1}$. Then $E$ lies below the epigraph $[f]$ of the convex function $f$ precisely when $f^c(\mathbf{y}) \leq w$. Indeed,

$$z = \mathbf{y}^t\mathbf{x} - w \leq f(\mathbf{x})$$

for all $\mathbf{x}$ if and only if, for all $\mathbf{x}$, $\mathbf{y}^t\mathbf{x} - f(\mathbf{x}) \leq w$ and hence if and only if

$$\sup[\mathbf{y}^t\mathbf{x} - f(\mathbf{x})] = f^c(\mathbf{y}) \leq w.$$

We refer to this property as the *location test*.

The set $K(f^c)$ on which $f^c$ has finite values is nonempty. Indeed, $K(f)^I$, the relative interior of $K(f)$, is nonempty (Section 3.1); we choose a point $\mathbf{x}_0$ in $K(f)^I$. Then as in Section 4.4 there is a nonvertical supporting hyperplane (4.40a) for $[f]$ through the point $(\mathbf{x}_0, f(\mathbf{x}_0))$. Since the hyperplane passes through this point, for suitable $\mathbf{y}_0$ its equation can be written in the form:

$$z = f(\mathbf{x}_0) + \mathbf{y}_0^t(\mathbf{x} - \mathbf{x}_0).$$

Since the epigraph of $f$ lies above this hyperplane, we conclude that

$$f(\mathbf{x}) \geq f(\mathbf{x}_0) + \mathbf{y}_0^t(\mathbf{x} - \mathbf{x}_0) \tag{4.55}$$

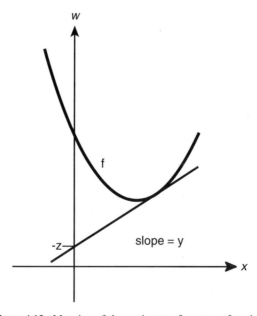

**Figure 4.12.** Meaning of the conjugate of a convex function.

for all $\mathbf{x}$. Hence for all $\mathbf{x}$

$$\mathbf{y}_0^t \mathbf{x} - f(\mathbf{x}) \le \mathbf{y}_0^t \mathbf{x}_0 - f(\mathbf{x}_0), \tag{4.56}$$

so $f^c(\mathbf{y}_0)$ is finite.

Furthermore, if $\mathbf{y}_1$ and $\mathbf{y}_2$ are in $K(f^c)$ and

$$\mathbf{y} = (1 - \alpha)\mathbf{y}_1 + \alpha\mathbf{y}_2$$

for some $\alpha$ on $(0, 1)$, then by the rules (4.52) and (4.53)

$$\begin{aligned}
f^c(\mathbf{y}) &= \sup_{\mathbf{x}}\{(1 - \alpha)\mathbf{y}_1^t \mathbf{x} + \alpha\mathbf{y}_2^t \mathbf{x} - f(\mathbf{x})\} \\
&= \sup_{\mathbf{x}}\{(1 - \alpha)(\mathbf{y}_1^t \mathbf{x} - f(\mathbf{x})) + \alpha(\mathbf{y}_2^t \mathbf{x} - f(\mathbf{x}))\} \\
&\le \sup_{\mathbf{x}}\{(1 - \alpha)(\mathbf{y}_1^t \mathbf{x} - f(\mathbf{x}))\} + \sup_{\mathbf{x}}\{\alpha(\mathbf{y}_2^t \mathbf{x} - f(\mathbf{x}))\} \\
&= (1 - \alpha)\sup_{\mathbf{x}}\{\mathbf{y}_1^t \mathbf{x} - f(\mathbf{x})\} + \alpha\sup_{\mathbf{x}}\{\mathbf{y}_2^t \mathbf{x} - f(\mathbf{x})\} \\
&= (1 - \alpha)f^c(\mathbf{y}_1) + \alpha f^c(\mathbf{y}_2).
\end{aligned}$$

This shows that $K(f^c)$ is a convex set in $\mathcal{R}^n$ and that $f^c$ is convex on $K(f^c)$. Thus *for every convex $f$, $f^c$ is a convex function.*

Similarly, if $g$ is concave in $\mathcal{R}^n$, one defines its conjugate to be the function

$$g^c(\mathbf{y}) = \inf_{\mathbf{x}}(\mathbf{y}^t \mathbf{x} - g(\mathbf{x})) \tag{4.57}$$

and verifies that $g^c(\mathbf{y})$ is a concave function in $\mathcal{R}^n$. If $f$ is convex, then $-f$ is concave. One verifies that

$$(-f)^c(\mathbf{y}) = -f^c(-\mathbf{y}). \tag{4.58}$$

(See Problem 4.24(a).)

### Further Properties of Conjugates

We mention several properties, with proofs left to the problems (Problem 4.24) except for a few difficult proofs, for which references are given. A convex function in $\mathcal{R}^n$ is said to be *closed* if its epigraph is a closed set in $\mathcal{R}^{n+1}$. Throughout the functions are assumed to be convex in $\mathcal{R}^n$.

1. If $k$ is a constant scalar, then $(f + k)^c = f^c - k$.
2. If $\mathbf{b}$ is a constant vector, then $f(\mathbf{x} - \mathbf{b})^c = f^c(\mathbf{y}) + \mathbf{b}^t \mathbf{y}$.
3. If $k$ is a positive constant, then $(kf)^c(\mathbf{y}) = kf^c((1/k)\mathbf{y})$.
4. If $\mathbf{c}$ is constant, then $(f + \mathbf{c}^t\mathbf{x})^c = f^c(\mathbf{y} - \mathbf{c})$.
5. If $f_1$ and $f_2$ are convex in $\mathcal{R}^n$ and $f_1 \ge f_2$, then $f_1^c \le f_2^c$.

6. If $f$ is convex in $\mathcal{R}^n$, then $f^{cc} = f$ if and only if $f$ is closed. (See Stoer and Witzgall, 1970, pp. 160–161.)

7. If $f$ is convex in $\mathcal{R}^n$, then $f^{ccc} = f^c$, so $f^c$ is closed. (See Stoer and Witzgall, 1970, pp. 160–161.)

8. If $f$ is strictly convex in $\mathcal{R}^n$ and $f$ has continuous partial derivatives through the second order in $\mathcal{R}^n$, then $f^c(\mathbf{y})$ has a finite value for each $\mathbf{y}$ such that $\mathbf{y} = \nabla f(\mathbf{x})$ for some $\mathbf{x}$. When this holds, the $\mathbf{x}$ is unique, so the inverse of this mapping exists, defining $\mathbf{x} = \mathbf{x}(\mathbf{y})$ and

$$f^c(\mathbf{y}) = \mathbf{y}^t \mathbf{x}(\mathbf{y}) - f(\mathbf{x}(\mathbf{y})).$$

9. Let $h(x)$ be convex in $\mathcal{R}^n$, with $K(h) = \mathcal{R}^n$, and let $h$ be *positively homogeneous* (Section 4.1). That is, for each nonnegative constant $\lambda$, let

$$h(\lambda \mathbf{x}) = \lambda h(\mathbf{x}) \quad \text{for all } \mathbf{x}.$$

Then $h^c(\mathbf{y}) = 0$ if $\mathbf{y}^t \mathbf{x} \le h(\mathbf{x})$ for all $\mathbf{x}$, and $h^c(\mathbf{y}) = \infty$ otherwise.

10. Let $f(\mathbf{x}) = 0$ on the convex set $S$ in $\mathcal{R}^n$; let $f(\mathbf{x}) = +\infty$ otherwise. Then $f$ is convex, and

$$f^c(\mathbf{y}) = \sup_{\mathbf{x}} \{ \mathbf{y}^t \mathbf{x} \mid \mathbf{x} \in S \}.$$

This function is known as the *support function of S*.

11. One defines the *infimum convolution* of two convex functions $f_1$, $f_2$ in $\mathcal{R}^n$ as

$$f_1 \sqcup f_2 = \inf_{\mathbf{x}} \{ f_1(\mathbf{x}) + f_2(\mathbf{z} - \mathbf{x}) \} = h(\mathbf{z}).$$

One has then the two rules:

$$(f_1 + f_2)^c = f_1^c \sqcup f_2^c,$$
$$(f_1 \sqcup f_2)^c = f_1^c + f_2^c \qquad \text{if } K(f_1)^I \cap K(f_2)^I \ne \varnothing.$$

(See Stoer and Witzgall, 1970, pp. 166–167, 181.)

12. Let $n = k + l$, and let a convex function $f$ on $\mathcal{R}^n$ be written as $f(\mathbf{x}_1, \mathbf{x}_2)$, where $\mathbf{x}_1$ varies in $\mathcal{R}^k$ and $\mathbf{x}_2$ varies in $\mathcal{R}^l$. One calls $f$ *separable* if one can write $f$ as $f_1(\mathbf{x}_1) + f_2(\mathbf{x}_2)$ in terms of convex functions $f_1$ in $\mathcal{R}^k$ and $f_2$ in $\mathcal{R}^l$. For such a separable function $f$ one has the rule:

$$f^c(\mathbf{y}_1, \mathbf{y}_2) = f_1^c(\mathbf{y}_1) + f_2^c(\mathbf{y}_2).$$

13. Let $A$ be a positive semidefinite square matrix of order $n$. Then for each constant vector $\mathbf{b}$ in $\mathcal{R}^n$, the function

$$f(\mathbf{x}) = \tfrac{1}{2} \mathbf{x}^t A \mathbf{x} + \mathbf{b}^t \mathbf{x}$$

is convex and

$$f^c(A\mathbf{x} + \mathbf{b}) = \tfrac{1}{2}\mathbf{x}^t A\mathbf{x},$$

$$K(f^c) = \{A\mathbf{x} + \mathbf{b} \mid \mathbf{x} \in \mathcal{R}^n\}.$$

If $A$ is positive definite, so $f$ is strictly convex, then

$$f^c(\mathbf{y}) = \tfrac{1}{2}(\mathbf{y} - \mathbf{b})^t A^{-1}(\mathbf{y} - \mathbf{b}).$$

14. Let $A$ be an $m \times n$ matrix and $\mathbf{b}$ a vector of $\mathcal{R}^m$. Let $f(\mathbf{x})$ be a linear function $\mathbf{c}^t\mathbf{x}$ for $A\mathbf{x} = \mathbf{b}$, and let $f(\mathbf{x}) = +\infty$ otherwise. Let the equation $A\mathbf{x} = \mathbf{b}$ be satisfied for some $\mathbf{x}$. Then $f$ is convex in $\mathcal{R}^n$, and

$$f^c(\mathbf{y}) = -\mathbf{v}^t\mathbf{b} \qquad \text{for } \mathbf{y} = \mathbf{c} - A^t\mathbf{v}$$

and $f^c(\mathbf{y}) = +\infty$ otherwise.

## PROBLEMS

**4.18.** For each set given, find the sup and state when it is a maximum:

(a) $\{x \mid \ln x < 0\}$.

(b) $\{\tan x \mid 0 \le x < \pi/2\}$.

(c) $\{1/(x + y) \mid 1 < x < 2, -1 < y < 0\}$.

**4.19.** Verify the rule (4.51) for the following cases:

(a) $f(x, y) = x^2/y$ for $1 \le x \le 2, 2 \le y \le 3$.

(b) $f(x, y) = \ln[1/(x + y)]$ for $0 \le x < 1$ and $0 < y < 1$.

**\*4.20.** Prove the rule (4.51) for the case of $f(x, y)$ with $E = \{x \mid 0 \le x \le 1\}$ and $F = \{y \mid 0 \le y \le 1\}$ by showing that the left side (and also the right side) equals

$$\sup\{f(x, y) \mid 0 \le x \le 1, 0 \le y \le 1\}.$$

**4.21.** Prove the rules:

(a) (4.50).

(b) (4.52).

(c) (4.53).

**4.22.** For the rule (4.53) give an example for which the equals sign is correct and one for which the $<$ sign is correct.

**4.23.** Verify the results of the following examples in Section 4.5:

(a) Example 1.

(b) Example 2.

**(c)** Example 3.

**4.24.** Prove the rules concerning conjugates stated in Section 4.5:

    **(a)** (4.58).

    **(b)** Rule 1.

    **(c)** Rule 2.

    **(d)** Rule 3.

    **(e)** Rule 4.

    **(f)** Rule 5.

    **(g)** Rule 8. [Hint: Seek the critical points of the strictly concave function $\mathbf{y}^t\mathbf{x} - f(\mathbf{x})$.]

    **(h)** Rule 9. [Hint: Observe that for each $\mathbf{y}$ the function $\mathbf{y}^t\mathbf{x} - h(\mathbf{x})$, as a function of $\mathbf{x}$, has value 0 for $\mathbf{x} = \mathbf{0}$, and hence the sup is at least equal to 0. Next let $c \geq 0$ be the maximum of this function for $\|\mathbf{x}\| \leq 1$. Show that by the positive homogeneity of $h$, if $c = 0$, then the sup is 0 for that $\mathbf{y}$, and if $c > 0$, then the sup is $+\infty$. Show that these two cases correspond to the two cases of the assertion.]

    **(i)** Rule 10.

    **(j)** Rule 12.

    **(k)** Rule 13. [Hint: Apply rule 8.]

    **(l)** Rule 14.

## 4.6  FENCHEL DUALITY THEOREM

In $\mathcal{R}^n$ let $f$ be convex; let $g$ be concave so that $-g$ is convex, and accordingly, if it takes on some finite values, $f - g$ is also convex in $\mathcal{R}^n$. We consider two programs:

  Program I:  Find $\inf_\mathbf{x}(f(\mathbf{x}) - g(\mathbf{x}))$ for $\mathbf{x} \in K(f) \cap K(g)$.

  Program II: Find $\sup_\mathbf{y}(g^c(\mathbf{y}) - f^c(\mathbf{y}))$ for $\mathbf{y} \in K(f^c) \cap K(g^c)$.

We will see that under proper conditions these two programs are dual. We first give conditions that ensure that the sup for Program II is less than or equal to the inf for Program I.

**Lemma A.**  *If $K(f) \cap K(g)$ and $K(f^c) \cap K(g^c)$ are nonempty, then*

$$-\infty < \sup_\mathbf{y}(g^c(\mathbf{y}) - f^c(\mathbf{y})) \leq \inf_\mathbf{x}(f(\mathbf{x}) - g(\mathbf{x})) < \infty. \qquad (4.60)$$

*Moreover* $f(\mathbf{x}_0) - g(\mathbf{x}_0) = g^c(\mathbf{y}_0) - f^c(\mathbf{y}_0)$ *holds for a pair of feasible solutions* $\mathbf{x}_0$ *in* $K(f) \cap K(g)$ *and* $\mathbf{y}_0$ *in* $K(f^c) \cap K(g^c)$ *of Programs I and II if and only if* $f^c(\mathbf{y}_0) = \mathbf{y}_0^t \mathbf{x}_0 - f(\mathbf{x}_0)$ *and* $g^c(\mathbf{y}_0) = \mathbf{y}_0^t \mathbf{x}_0 - g(\mathbf{x}_0)$.

*Proof.* For $\mathbf{y}$ in $K(f^c) \cap K(g^c)$, we have by the definition of conjugate functions: for all $\mathbf{x}$,

$$f^c(\mathbf{y}) \geq \mathbf{y}^t \mathbf{x} - f(\mathbf{x}) \quad \text{and} \quad g^c(\mathbf{y}) \leq \mathbf{y}^t \mathbf{x} - g(\mathbf{x}). \tag{4.61}$$

Hence

$$g^c(\mathbf{y}) - f^c(\mathbf{y}) \leq f(\mathbf{x}) - g(\mathbf{x})$$

for all $\mathbf{x}$. This holds in particular for each $\mathbf{x}$ in $K(f) \cap K(g)$, which is nonempty. It follows that the sup and inf are both finite and satisfy (4.60).

Next suppose that for a particular $\mathbf{x}, \mathbf{y}$ one has equality in the last inequality; that is, $g^c(\mathbf{y}) - f^c(\mathbf{y}) = f(\mathbf{x}) - g(\mathbf{x})$. If we add this equality to the first inequality in (4.61), we obtain an inequality that is the opposite of the second one in (4.61). Hence, for this particular $\mathbf{x}, \mathbf{y}$, the second inequality in (4.61) becomes an equation. Similar reasoning shows that the first inequality in (4.61) also becomes an equation (Problem 4.25). Thus Lemma A is proved. ∎

**Duality Theorem of Fenchel.** *If* $K(f)^I \cap K(g)^I$ *is nonempty and* $\inf_\mathbf{x}[(f(\mathbf{x}) - g(\mathbf{x})]$ *is finite, then Program II has an optimal solution* $\mathbf{y}_0$ *and*

$$\inf_\mathbf{x}[f(\mathbf{x}) - g(\mathbf{x})] = g^c(\mathbf{y}_0) - f^c(\mathbf{y}_0) = \max_\mathbf{y}[g^c(\mathbf{y}) - f^c(\mathbf{y})]. \tag{4.62}$$

*Proof.* By hypothesis, $\mu = \inf_\mathbf{x}[f(\mathbf{x}) - g(\mathbf{x})]$ is finite. Let $f_1(\mathbf{x}) = f(\mathbf{x}) - \mu$ so that $f_1^c(\mathbf{y}) = f^c(\mathbf{y}) + \mu$ (rule 1 in Section 4.5). Also

$$f_1(\mathbf{x}) - g(\mathbf{x}) \geq 0. \tag{4.63}$$

We claim that the epigraph $[f_1]$ and the hypograph $[g]$ have no relative interior points in common. Suppose that $(\mathbf{x}, z) \in [f_1] \cap [g]$. Then $\mathbf{x} \in K(f_1) \cap K(g) = K(f) \cap K(g)$ and $f_1(\mathbf{x}) \leq z \leq g(\mathbf{x})$. By (4.63), $f_1(\mathbf{x}) = z = g(\mathbf{x})$. Hence $(\mathbf{x}, z)$ is a relative boundary point of both $[f_1]$ and $[g]$.

Since $[f_1]$ and $[g]$ have no relative interior points in common, the Separation Theorem of Section 4.3 applies, and they can be properly separated by a hyperplane $E$. We claim that $E$ is nonvertical. If it were vertical, then the hyperplane $E_0 : \{\mathbf{x} \mid (\mathbf{x}, z) \in E\}$, which is the orthogonal projection of $E$ on the $n$-dimensional space of coordinates $(x_1, \ldots, x_n)$, would separate $K(f_1) = K(f)$ and $K(g)$ without containing both. This contradicts the converse of the Separation Theorem (see the Remark at the end of Section 4.3), since $K(f), K(g)$ have relative interior points in common.

Thus $E$ is nonvertical and can be represented as $\{(\mathbf{x}, z) \mid \mathbf{y}_0^t \mathbf{x} - z = w_0\}$. Then $(\mathbf{y}_0, w_0)$ is a point of the intersection $[f_1^c] \cap [g^c]$. Indeed, $f_1(\mathbf{x}) \geq \mathbf{y}_0^t \mathbf{x} - w_0$ and

$g(\mathbf{x}) \leq \mathbf{y}_0^t \mathbf{x} - w_0$ for all $\mathbf{x}$, or $w_0 \geq \mathbf{y}_0^t \mathbf{x} - f_1(\mathbf{x})$ for all $\mathbf{x}$ and $w_0 \leq \mathbf{y}_0^t \mathbf{x} - g(\mathbf{x})$ for all $\mathbf{x}$. Hence $f_1^c(\mathbf{y}_0) \leq w_0 \leq g^c(\mathbf{y}_0)$, and therefore

$$g^c(\mathbf{y}_0) - f_1^c(\mathbf{y}_0) = g^c(\mathbf{y}_0) - \mu - f^c(\mathbf{y}_0) \geq 0$$

or

$$g^c(\mathbf{y}_0) - f^c(\mathbf{y}_0) \geq \mu = \inf_{\mathbf{x}}[f(\mathbf{x}) - g(\mathbf{x})]. \tag{4.64}$$

We now apply Lemma A and conclude from (4.60) that (4.62) is valid, and so the theorem is proved. ∎

REMARK 1. If Program II has feasible solutions—that is, if $K(f^c)$ meets $K(g^c)$—then Lemma A shows that $\inf_{\mathbf{x}}(f(\mathbf{x}) - g(\mathbf{x}))$ is finite, so this hypothesis in the Fenchel Duality Theorem is satisfied.

**Corollary.** *Let $f$ be convex and $g$ concave in $\mathcal{R}^n$. If $[f]$ and $[g]$ have no relative interior points in common, but $K(f)$ and $K(g)$ have relative interior points in common, then there exists a nonvertical hyperplane $E = \{(\mathbf{x}, z) \mid \mathbf{y}^t \mathbf{x} - z = w\}$ which properly separates $[f]$ and $[g]$.*

*Proof.* One applies the same argument as in the proof of the Fenchel theorem, with $f$ instead of $f_1$. Thus as in that proof $[f]$ and $[g]$ can be properly separated by a hyperplane $E$, and by the same argument as above, $E$ is nonvertical. ∎

*Example 1.* We seek the minimum of the function in $\mathcal{R}^1$ equal to $x^2$ for $-1 \leq x \leq 2$. We apply the Fenchel theorem to seek a dual. To that end we take $f(x) = x^2$ in $\mathcal{R}^1$ and $g(x) = 0$ on the interval $[-1, 2]$; $g(x) = -\infty$ otherwise. We find that $f^c(y) = y^2/4$ in $\mathcal{R}^1$ and that $g^c(y) = 2y$ for $y \leq 0$, $g^c(y) = -y$ for $y \geq 0$. Thus $g^c(y) - f^c(y)$ is the concave function equal to $2y - y^2/4$ for $y \leq 0$ and to $-y - y^2/4$ for $y \geq 0$. The hypotheses of the Fenchel theorem are satisfied, and hence (4.62) holds; we verify that here the maximum value of $g^c(y) - f^c(y)$ is 0 and that this is the minimum of $f - g$. (See Problem 4.27.)

*Example 2.* We consider a standard linear programming problem (Section 3.6): to minimize the linear function $\mathbf{c}^t \mathbf{x}$ for $A\mathbf{x} = \mathbf{b}$ and $\mathbf{x} \geq 0$. As usual, $A$ is an $m \times n$ matrix, and $\mathbf{b}$ and $\mathbf{c}$ are constant vectors. We take $f(\mathbf{x}) = \mathbf{c}^t \mathbf{x}$ for $A\mathbf{x} = \mathbf{b}$ and $f(\mathbf{x}) = +\infty$ otherwise. We take $g(\mathbf{x}) = 0$ for $\mathbf{x} \geq 0$ and $g(\mathbf{x}) = -\infty$ otherwise. By rule 14 of Section 4.5, $f^c(\mathbf{y}) = -\mathbf{b}^t \mathbf{v}$ if $\mathbf{y}$ can be represented as $\mathbf{c} - A^t \mathbf{v}$ for some $\mathbf{v}$ and $f^c(\mathbf{y}) = +\infty$ otherwise. We find that $g^c(\mathbf{y}) = 0$ for $\mathbf{y} \geq 0$ and that $g^c(\mathbf{y}) = -\infty$ otherwise. If the set $\{\mathbf{x} \mid A\mathbf{x} = \mathbf{b}\}$ contains points interior to the set $\{\mathbf{x} \mid \mathbf{x} \geq 0\}$, and the function $f - g$ is bounded below, then the Fenchel theorem applies, and we obtain a dual problem: to maximize the linear function $\mathbf{b}^t \mathbf{v}$ for $\mathbf{y} = \mathbf{c} - A^t \mathbf{v} \geq 0$; that is, for $A^t \mathbf{v} \leq \mathbf{c}$. Observe that the dual variable $\mathbf{v}$ is now a vector in $\mathcal{R}^m$. Verification of details is left to Problem 4.28. We remark that from the theory of linear programming, when $f - g$ is bounded below, it has a minimizer.

***Example 3.*** We consider the problem of minimizing a quadratic function subject to equality restraints:

$$\text{minimize} \quad \phi(\mathbf{x}) = \tfrac{1}{2}\mathbf{x}^t A \mathbf{x} - \mathbf{r}^t \mathbf{x}$$

for $B\mathbf{x} = \mathbf{g}$, as in Section 3.4. We assume that $A$ is positive definite, that $B$ is an $m \times n$ matrix, and that feasible points exist.

We take $f(\mathbf{x}) = \phi(\mathbf{x})$ in $\mathcal{R}^n$, and $g(\mathbf{x}) = 0$ for $B\mathbf{x} = \mathbf{g}$ and $g(\mathbf{x}) = -\infty$ otherwise. Then by rule 13 of Section 4.5,

$$f^c(\mathbf{y}) = \tfrac{1}{2}(\mathbf{y} + \mathbf{r})^t A^{-1}(\mathbf{y} + \mathbf{r})$$

in $\mathcal{R}^n$, and by rule 14, $g^c(\mathbf{y}) = -\mathbf{v}^t \mathbf{g}$ for $\mathbf{y} = -B^t\mathbf{v}$ for some $\mathbf{v}$. Accordingly

$$g^c(\mathbf{y}) - f^c(\mathbf{y}) = -\mathbf{g}^t\mathbf{v} - \tfrac{1}{2}(\mathbf{r} - B^t\mathbf{v})^t A^{-1}(\mathbf{r} - B^t\mathbf{v})$$

if $\mathbf{y} = -B^t\mathbf{v}$ for some $\mathbf{v}$ in $\mathcal{R}^m$ and $g^c - f^c = -\infty$ otherwise. The Fenchel theorem applies, and the dual problem is that of maximizing the function on the right side of the last displayed equation in $\mathcal{R}^m$. Since $A$ is positive definite, the inf of $f - g$ is a minimum. The result agrees with that obtained in Section 3.4 (cf. (3.43)).

REMARK 2. Even though the Fenchel theorem is formulated in terms of convex and concave functions, it is striking how few properties of convexity are involved in the proofs. In particular, Lemma A does not really involve convexity at all. In Problems 4.31 and 4.32 we pursue this topic.

## PROBLEMS

**4.25.** Write out in detail the proof of Lemma A in Section 4.6.

**4.26.** Write out in detail the proof of the Corollary in Section 4.6.

**4.27.** Verify all the steps in Example 1 in Section 4.6.

**4.28.** Verify all the steps in Example 2 in Section 4.6.

**4.29.** Apply the Fenchel theorem to find the minimum of the function $x_1^2 + x_2$ for $x_1^2 + x_2^2 \leq 1$.

**4.30.** Verify all the steps in Example 3 in Section 4.6.

**4.31.** Let $f$ be a function defined in $\mathcal{R}^n$ with finite values on the set $K(f)$ and value $+\infty$ otherwise; let $g$ be a function defined in $\mathcal{R}^n$ with finite values on the set $K(g)$ and value $-\infty$ otherwise. Let the conjugates $f^c$ and $g^c$ be defined by (4.54) and (4.57). Show that Lemma A of Section 4.6 is valid, with the same proof. (Thus convexity is not involved.)

**4.32.** Apply the definitions of Problem 4.31 to formulate Fenchel's theorem for the functions $f$ and $g$ in $\mathcal{R}^1$, where $f(x) = x^3$ for $-1 \leq x \leq 1$, $f(x) = \infty$ otherwise and $g(x) = 0$ for $-1 \leq x \leq 1$, $g(x) = -\infty$ otherwise. Show that the conclusion of the theorem is valid for these choices of $f$ and $g$. (Thus in some cases one obtains a duality theorem for nonconvex functions.)

## 4.7  ROCKAFELLAR DUALITY THEOREM

The Rockafellar theorem is a generalization of the Fenchel theorem. Linear terms are added and a linear substitution is introduced. One introduces two functions:

$$\psi(\mathbf{x}) = f(\mathbf{x}) - g(A\mathbf{x} - \mathbf{b}) + \mathbf{c}^t\mathbf{x}, \tag{4.70}$$

$$\phi(\mathbf{y}) = g^c(\mathbf{y}) - f^c(A^t\mathbf{y} - \mathbf{c}) + \mathbf{b}^t\mathbf{y}. \tag{4.71}$$

Here $f$ is convex in $\mathcal{R}^n$, $g$ is concave in $\mathcal{R}^m$, $A$ is an $m$ by $n$ matrix, $\mathbf{b} \in \mathcal{R}^m$, and $\mathbf{c} \in \mathcal{R}^n$.

If both functions $\psi$ and $\phi$ have some finite values, then $\psi$ is convex in $\mathcal{R}^n$ and $\phi$ is concave in $\mathcal{R}^m$. For $A$ the identity matrix and $\mathbf{b} = \mathbf{c} = \mathbf{0}$, these reduce to the functions $f - g$, $g^c - f^c$ of the Fenchel theorem.

We consider two new programs:

Program I:  Find $\inf_x \psi(\mathbf{x})$.
Program II: Find $\sup_y \phi(\mathbf{y})$.

The Rockafellar theorem asserts that under certain conditions, these programs are dual.

In order to motivate the choice of the two objective functions, we consider a 1-dimensional case: $m = n = 1$. We let $A$ be a constant $a \neq 0$. The objective functions become

$$\psi(x) = f(x) - g(ax - b) + cx, \qquad \phi(y) = g^c(y) - f^c(ay - c) + by.$$

We apply the Fenchel theorem, with the function $f$ of that theorem replaced by $f_1(x) = f(x) + cx$ and the function $g$ of that theorem replaced by $g_1(x) =$

$g(ax - b)$. We assume that the hypotheses of the Fenchel theorem are satisfied by $f_1$ and $g_1$. The conclusion then states that the inf of $\psi(x)$ equals the max of the function $g_1^c(y) - f_1^c(y)$. We now verify that this is the same as the maximum of the function $\phi(y)$. Rule 4 of Section 4.5 gives $f_1^c(y) = f^c(y - c)$. We assert that

$$g(ax - b)^c = g^c\left(\frac{y}{a}\right) + \frac{b}{a}y$$

and leave the proof as an exercise (Problem 4.33). Accordingly

$$\inf_x \psi(x) = \max_y \left[ g^c\left(\frac{y}{a}\right) + \frac{b}{a}y - f^c(y - c) \right].$$

The value of the right-hand side is unchanged if we replace $y$ by $ay$ inside the brackets. Thus $\inf_x \psi(x) = \max_y \phi(y)$, as claimed above.

For the general case of the Rockafellar theorem, a more involved proof is required, involving geometry in $\mathcal{R}^n$. As preparation for the theorem, we consider a convex function $f$ in $\mathcal{R}^n$ and a linear variety $M$ in $\mathcal{R}^n$ which is represented parametrically as

$$\{\mathbf{x} \mid \mathbf{x} = A\mathbf{u} - \mathbf{b}\}, \tag{4.72}$$

where $A$ is an $n$ by $m$ matrix, $\mathbf{b}$ a vector of $\mathcal{R}^n$. We will assume that $M$ contains a point $\mathbf{x}_0$ of $K(f)$. We denote by $M^o$ the vertical linear variety consisting of all points $(\mathbf{x}, z)$ of $\mathcal{R}^{n+1}$ for which $\mathbf{x}$ is in $M$.

We also consider a linear variety $N$ in $\mathcal{R}^{n+1}$ which projects on $M$; that is, $N$ is the intersection of $M^o$ with a nonvertical hyperplane $E = \{(\mathbf{x}, z) \mid \mathbf{y}^t\mathbf{x} - z = w\}$ in $\mathcal{R}^{n+1}$. See Fig. 4.13. The figure illustrates the case when $n = 2$ and $N$ is a 1-dimensional linear variety in the $x_1z$-plane of 3-dimensional space. Thus $M$ is the $x_1$-axis, $M^o$ is the $x_1z$-plane, $E$ has the equation

$$y_1x_1 + y_2x_2 - z = w.$$

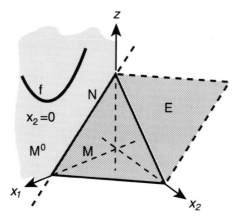

**Figure 4.13.** Geometry for Rockafellar theorem.

The parametric equation (4.72) of $M$ reduces to $x_1 = au - b,\ x_2 = 0$.)

From (4.72), we can represent $N$ as follows:

$$N = \{(A\mathbf{u} - \mathbf{b},\ z)\,|\,\mathbf{v}^t\mathbf{u} - z = w',\ \mathbf{u} \in \mathcal{R}^m\}, \tag{4.73}$$

where

$$\mathbf{v} = A^t\mathbf{y}, \quad w' = w + \mathbf{y}^t\mathbf{b}. \tag{4.74}$$

(See Problem 4.34.) Equation (4.73) represents $N$ as the range of the mapping of the hyperplane $E' : \mathbf{v}^t\mathbf{u} - z = w'$ in $\mathcal{R}^{m+1}$ into $\mathcal{R}^{n+1}$ by the equations $\mathbf{x} = A\mathbf{u} - \mathbf{b},\ z = z$. The hyperplane $E'$ is uniquely determined by $N$. For a given choice of $\mathbf{v}$ and $w'$, and hence of the hyperplane $E'$, each choice of $\mathbf{y}$ and $w$ satisfying (4.74) leads to a representation of $N$ as above; that is, as the intersection of the vertical hyperplane $M^o$ with the nonvertical hyperplane $E$.

We remark that $N$ lies below $[f]$ if and only if

$$w' \geq \sup_{\mathbf{u}}(\mathbf{v}^t\mathbf{u} - f(A\mathbf{u} - \mathbf{b})), \tag{4.75}$$

for $N$ lies below $[f]$ if and only if $f(A\mathbf{u} - \mathbf{b}) \geq \mathbf{v}^t\mathbf{u} - w'$ for all $\mathbf{u}$ in $\mathcal{R}^m$ or $w' \geq \mathbf{v}^t\mathbf{u} - f(A\mathbf{u} - \mathbf{b})$ for all such $\mathbf{u}$.

We write

$$\hat{f}(\mathbf{u}) = f(A\mathbf{u} - \mathbf{b}). \tag{4.76}$$

This is a convex function in $\mathcal{R}^m$ (as pointed out in Section 3.1); see Problem 4.35. [For the special case of Fig. 4.13, $\hat{f}(u) = f(au - b,\ 0)$, and its graph is essentially that shown in the figure. Its graph is precisely that shown when $a = 1$ and $b = 0$; otherwise, there is a change of scale and shift of origin on the $x_1$-axis.] Thus we conclude:

**Lemma B.**    *The linear variety $N$ lies below $[f]$ if and only if $\hat{f}^c(\mathbf{v}) \leq w'$.*

We remark that Lemma B can be interpreted as an application of the location test of Section 4.5 to the hyperplane $E'$ and the function $\hat{f}(\mathbf{u})$.

## Stable Points

We call the point $\mathbf{x}_0$ of $K(f)$ a stable point for $f$, or say that $f$ is $\mathbf{x}_0$-*stable*, if, for each choice of $N$ as above lying below $[f]$, there is a nonvertical hyperplane $E_1$ in $\mathcal{R}^{n+1}$ that separates $N$ from $[f]$. There is a similar definition for concave functions, or one can simply define $\mathbf{x}_0$ to be stable for concave $g$ if it is stable for the convex function $f = -g$.

We claim that every *relative interior* point $\mathbf{x}_0$ of $K(f)$ is stable for $f$. To show this, we apply the corollary of the Fenchel Duality Theorem (Section 4.6) to the function $f$, and the function $g$ which has the value $z$ if $(\mathbf{x},\ z)$ is in $N$ and the value $-\infty$ otherwise. The graph of $g$ is the linear variety $N$, so $g$ is simply a linear function on $M$. Hence $g$ is concave. Since $N$ is below $[f]$, $[f]$ and $[g]$ have no relative interior points in common. However, $\mathbf{x}_0$ is a relative interior point of $K(f)$ and of $K(g) = M$. Thus the corollary applies and the nonvertical separating hyperplane $E_1$ exists. It lies below $[f]$ and above $N$.

Certain boundary points of $K(f)$ may also be stable for $f$. For example, let $f(\mathbf{x}) = 0$ for $\mathbf{x} \geq \mathbf{0}$ and $f(\mathbf{x}) = \infty$ otherwise. Then the boundary points of $K(f) = \{\mathbf{x} \mid \mathbf{x} \geq \mathbf{0}\}$ are stable for $f$. We verify this for $\mathbf{x}_0 = \mathbf{0}$. If now $M$ and $N$ are as above and $N$ is below $[f]$, then

$$z = \mathbf{v}^t \mathbf{u} - w' \leq 0 \qquad \text{for } \mathbf{x} = A\mathbf{u} \geq \mathbf{0}.$$

Here the term $\mathbf{v}^t \mathbf{u}$ must itself be nonpositive, for $\mathbf{x} = A\mathbf{u}$ nonnegative; otherwise, we could make this term arbitrarily large positive for some nonnegative $\mathbf{x} = A\mathbf{u}$ (Problem 4.36). Thus

$$-\mathbf{v}^t \mathbf{u} \geq 0 \qquad \text{for } \mathbf{x} = A\mathbf{u} \geq \mathbf{0}.$$

We can now apply the Farkas lemma (Section 2.8) and conclude that there is a vector $\mathbf{q} \geq \mathbf{0}$ such that $-\mathbf{v} = A^t \mathbf{q}$. Hence we can write $N$ as

$$z = -\mathbf{q}^t A\mathbf{u} - w' \qquad \text{for } \mathbf{x} = A\mathbf{u}.$$

This shows that $N$ lies in the hyperplane $E_1 : z = -\mathbf{q}^t \mathbf{x} - w'$. Here both terms on the right, in the expression for $z$, are nonpositive for $\mathbf{x} \geq \mathbf{0}$, since $\mathbf{q} \geq \mathbf{0}$ and since $N$ lies below $[f]$ at $\mathbf{x} = \mathbf{0}$. Therefore $\mathbf{x}_0 = \mathbf{0}$ is a stable point for $[f]$.

By similar arguments we can verify that each boundary point of $K(f)$ is stable for $f$ if $f(\mathbf{x}) = \mathbf{c}^t \mathbf{x}$ when $\mathbf{x} \geq \mathbf{0}$, $f(\mathbf{x}) = +\infty$ otherwise. (See Problem 4.37.)

**Example 1.** As in Fig. 4.13, let $n = 2$ and $m = 1$; let $f$ have the value $x_2$ for $x_2 > 0$ and the value 1 for $x_2 = 0$, the value $+\infty$ otherwise. Then $f$ is convex and each boundary point of $K(f)$ is *unstable* for $f$. If we take $N$ to be the line $x_1 = u$, $x_2 = 0$, $z = \frac{1}{2}$, then each boundary point of $K(f)$ lies in $M$, the projection of $N$ onto the plane $z = 0$ and $N$ lies below $[f]$, but there is no nonvertical plane $y_1 x_1 + y_2 x_2 - z = w'$ separating $N$ from $[f]$. (See Problem 4.38.)

**Lemma C.** *Let $g(\mathbf{u})$ be concave in $\mathcal{R}^m$, let $M$ be the linear variety $\mathbf{u} = A\mathbf{x} - \mathbf{b}$ in $\mathcal{R}^m$, let $\mathbf{u}_0 = A\mathbf{x}_0 - \mathbf{b}$ be a point of $M \cap K(g)$ at which $g$ is stable, and let $\hat{g}(\mathbf{x}) = g(A\mathbf{x} - \mathbf{b})$. Then for every $\mathbf{y}$ in $K(\hat{g}^c)$,*

$$\hat{g}^c(\mathbf{y}) = \max_{\mathbf{w}} \{g^c(\mathbf{w}) + \mathbf{w}^t \mathbf{b} \mid \mathbf{y} = A^t \mathbf{w}\}. \tag{4.77}$$

This is proved in Section 4.8.

We now return to the functions $\psi(\mathbf{x})$ and $\phi(\mathbf{y})$ defined by (4.70) and (4.71), under the hypotheses stated, and the corresponding Programs I and II. As for the programs of Section 4.6, we show that when feasible solutions of these programs exist, the sup of the second program is at most equal to the inf of the first program and that both sup and inf are finite:

**Lemma D.**  *If* $\mathbf{x}$ *and* $\mathbf{y}$ *are feasible solutions of Programs I and II, respectively, then*

$$\infty > \psi(\mathbf{x}) \geq \phi(\mathbf{y}) > -\infty.$$

*Moreover* $\psi(\mathbf{x}_0) = \phi(\mathbf{y}_0)$ *holds for feasible* $\mathbf{x}_0$, $\mathbf{y}_0$ *if and only if*

$$f^c(A^t\mathbf{y}_0 - \mathbf{c}) = \mathbf{y}_0^t A\mathbf{x}_0 - f(\mathbf{x}_0) - \mathbf{c}^t\mathbf{x}_0$$

*and*

$$g^c(\mathbf{y}_0) = \mathbf{y}_0^t A\mathbf{x}_0 - g(A\mathbf{x}_0 - \mathbf{b}) - \mathbf{y}_0^t\mathbf{b}.$$

*Proof.*  We reason as in the proof of Lemma A in Section 4.6. By definition of $f^c$, $f^c(\mathbf{z}) \geq \mathbf{z}^t\mathbf{x} - f(\mathbf{x})$ for all $\mathbf{x}$ and thus

$$f^c(A^t\mathbf{y} - \mathbf{c}) \geq (A^t\mathbf{y} - \mathbf{c})^t\mathbf{x} - f(\mathbf{x}) = \mathbf{y}^t A\mathbf{x} - f(\mathbf{x}) - \mathbf{c}^t\mathbf{x}.$$

Similarly $g^c(\mathbf{y}) \leq \mathbf{y}^t\mathbf{z} - g(\mathbf{z})$ for all $\mathbf{z}$ in $\mathcal{R}^m$ so that, with $\mathbf{z} = A\mathbf{x} - \mathbf{b}$,

$$g^c(\mathbf{y}) \leq \mathbf{y}^t(A\mathbf{x} - \mathbf{b}) - g(A\mathbf{x} - \mathbf{b}) = \mathbf{y}^t A\mathbf{x} - g(A\mathbf{x} - \mathbf{b}) - \mathbf{y}^t\mathbf{b}.$$

These hold for all $\mathbf{x}$ and $\mathbf{y}$. By combining the two inequalities, we get $\psi(\mathbf{x}) \geq \phi(\mathbf{y})$ as claimed. As in the proof of Lemma A, if equality holds for one pair $\mathbf{x}$, $\mathbf{y}$, then we have equality in the two previous inequalities. (See Problem 4.39.) ∎

We now state and prove a theorem which is a special case of the Rockafellar Duality theorem.

**Theorem R.**  *Consider the Programs I and II where, as above, $f$ is convex in $\mathcal{R}^n$, $g$ is concave in $\mathcal{R}^m$, $A$ is an $m$ by $n$ matrix, $\mathbf{b} \in \mathcal{R}^m$, $\mathbf{c} \in \mathcal{R}^n$, and the functions $\psi$ and $\phi$ are defined by (4.70) and (4.71). Let $f_1(\mathbf{x}) = f(\mathbf{x}) + \mathbf{c}^t\mathbf{x}$; let $g_1(\mathbf{x}) = g(A\mathbf{x} - \mathbf{b})$. We make the following hypotheses:*

1. $K(f_1)^I \cap K(g_1)^I$ *is nonempty.*
2. *A point* $\mathbf{x}_0$ *in* $K(f_1) \cap K(g_1)$ *exists such that $g$ is stable at the point* $\mathbf{u}_0 = A\mathbf{x}_0 - \mathbf{b}$.
3. *The quantity* $\inf_{\mathbf{x}} \psi(\mathbf{x})$ *is finite.*

*Then Program II has an optimal solution* $\mathbf{y}_0$ *and*

$$\inf_{\mathbf{x}} \psi(\mathbf{x}) = \phi(\mathbf{y}_0) = \max_{\mathbf{y}} \phi(\mathbf{y}).$$

*Proof.* By the Fenchel Duality theorem, applied to $f_1$ and $g_1$,

$$\inf_{\mathbf{x}} \psi(\mathbf{x}) = \max_{\mathbf{y}} \phi_1(\mathbf{y}),$$

where $\phi_1(\mathbf{y}) = g_1^c(\mathbf{y}) - f_1^c(\mathbf{y})$. In particular, $g_1^c(\mathbf{y})$ is finite for some $\mathbf{y}$.

Now by rule 4 of Section 4.5, $f_1^c(\mathbf{y}) = (f(\mathbf{x}) + \mathbf{c}^t\mathbf{x})^c = f^c(\mathbf{y} - \mathbf{c})$. By hypothesis 2, we can now apply Lemma C and write for $\mathbf{y}$ in $K(g_1^c)$,

$$g_1^c(\mathbf{y}) = (g(A\mathbf{x} - \mathbf{b}))^c(\mathbf{y}) = \hat{g}^c(\mathbf{y}) = \max_{\mathbf{w}}\{g^c(\mathbf{w}) + \mathbf{w}^t\mathbf{b} \mid \mathbf{y} = A^t\mathbf{w}\}.$$

Hence

$$\phi_1(\mathbf{y}) = \max_{\mathbf{w}}\{g^c(\mathbf{w}) + \mathbf{w}^t\mathbf{b} \mid \mathbf{y} = A^t\mathbf{w}\} - f^c(\mathbf{y} - \mathbf{c})$$

$$= \max_{\mathbf{w}}\{g^c(\mathbf{w}) + \mathbf{w}^t\mathbf{b} - f^c(\mathbf{y} - \mathbf{c}) \mid \mathbf{y} = A^t\mathbf{w}\}.$$

Therefore, since the order of the two max operations can be reversed as in (4.51),

$$\max_{\mathbf{y}} \phi_1(\mathbf{y}) = \max_{\mathbf{w}} \max_{\mathbf{y}}\{g^c(\mathbf{w}) + \mathbf{w}^t\mathbf{b} - f^c(\mathbf{y} - \mathbf{c}) \mid \mathbf{y} = A^t\mathbf{w}\}$$

$$= \max_{\mathbf{w}}(g^c(\mathbf{w}) + \mathbf{w}^t\mathbf{b} - f^c(A^t\mathbf{w} - \mathbf{c})) = \max_{\mathbf{y}} \phi(\mathbf{y})$$

and the theorem is proved.                                                        ∎

*Example 2.* We consider a typical problem of linear programming: to minimize $\psi(\mathbf{x}) = \mathbf{c}^t\mathbf{x}$ for $\mathbf{x} \geq 0$ and $A\mathbf{x} \geq \mathbf{b}$. We let $f(\mathbf{x}) = 0$ for $\mathbf{x} \geq 0$ and $f(\mathbf{x}) = \infty$ otherwise, and let $g(\mathbf{u}) = 0$ for $\mathbf{u} \geq 0$ and $g(\mathbf{u}) = -\infty$ otherwise. Accordingly $\psi(\mathbf{x}) = f(\mathbf{x}) - g(A\mathbf{x} - \mathbf{b}) + \mathbf{c}^t\mathbf{x}$, as desired. Next we find that $f_1(\mathbf{x}) = \mathbf{c}^t\mathbf{x}$ for $\mathbf{x} \geq 0$ and $g_1(\mathbf{x}) = 0$ for $A\mathbf{x} - \mathbf{b} \geq 0$ and that $K(f_1)^I \cap K(g_1)^I$ is the set $G = \{\mathbf{x} \mid \mathbf{x} > 0, \ A\mathbf{x} - \mathbf{b} > 0\}$, which we assume to be nonempty. In particular, $G$ contains a point $\mathbf{x}_0 > 0$ such that $\mathbf{u}_0 = A\mathbf{x}_0 - \mathbf{b} > 0$ so that $\mathbf{u}_0$ is in the relative interior of $K(g)$ and is hence is a point at which $g$ is stable. Now $f^c(\mathbf{w}) = 0$ for $\mathbf{w} \leq 0$ and $f^c(\mathbf{w}) = \infty$ otherwise, while $g^c(\mathbf{y}) = 0$ for $\mathbf{y} \geq 0$, $g^c(\mathbf{y}) = -\infty$ otherwise. Hence $\phi(\mathbf{y}) = \mathbf{b}^t\mathbf{y}$ for $A^t\mathbf{y} \leq \mathbf{c}$ and $\mathbf{y} \geq 0$. If now $\psi(\mathbf{x})$ is bounded below, then Theorem R applies and $\inf \psi = \max \phi$. This is of course a familiar duality theorem for linear programming (see Section 4.6). From the theory of linear programming, we know that if $\psi$ is bounded below, then $\psi$ takes on its minimum. In most cases $K(\psi)$ is a bounded closed set, so $\psi$ must be bounded below and take on its minimum. (Also, as in Remark 1 in Section 4.6, whenever Program II has a feasible solution, $\psi$ is bounded below by Lemma D.) (See Problem 4.40.)

## Discussion

As noted, Theorem R is a special case of a theorem of Rockafellar. In Stoer and Witzgall (1970), a stronger theorem is proved. To obtain this, the Fenchel Duality Theorem is first generalized by replacing the hypothesis that $K(f)$ and $K(g)$ have common relative interior points by the hypothesis that there are points $\mathbf{x}_f$ and $\mathbf{x}_g$ of $K(f) \cap K(g)$ such that $f$ is $\mathbf{x}_f$-stable and $g$ is $\mathbf{x}_g$-stable. This is then used to show that in Theorem R the conclusion is valid if the hypotheses 1 and 2 are replaced by the hypothesis that there are feasible points $\mathbf{x}_f$ and $\mathbf{x}_g$ of Program I such that $f$ is stable at $\mathbf{x}_g$ and $g$ is stable at $\mathbf{u}_g = A\mathbf{x}_g - \mathbf{b}$. Further generalizations are also given in Stoer and Witzgall (1970). In particular, Stoer and Witzgall term a convex or concave function $f$ *stable* if every point of $K(f)$ is a point of stability for $f$. It is then shown that many functions appearing in applications are stable.

From these results we can now verify that for the linear programming example considered above, the duality theorem is valid if we replace the condition that the set $G$ be nonempty by the requirement that the set

$$K(f_1) \cap K(g_1) = \{\mathbf{x} \mid \mathbf{x} \geq 0, \ A\mathbf{x} - \mathbf{b} \geq 0\}$$

is nonempty.

## 4.8   PROOF OF LEMMA C

We prove an analogous lemma for *convex* functions and then convert it to the concave case of Lemma C. Thus we first prove the following:

**Lemma C'.**   *Let $f(\mathbf{x})$ be convex in $\mathcal{R}^n$, let $M$ be the linear variety $\mathbf{x} = A\mathbf{u} - \mathbf{b}$ in $\mathcal{R}^n$, and let $\mathbf{x}_0$ be a point of $M \cap K(f)$ at which $f$ is stable. Let $\hat{f}(\mathbf{u}) = f(A\mathbf{u} - \mathbf{b})$. Then for every $\mathbf{v}$ in $K(\hat{f}^c)$,*

$$\hat{f}^c(\mathbf{v}) = \min_{\mathbf{y}}\{f^c(\mathbf{y}) + \mathbf{y}^t\mathbf{b} \mid \mathbf{v} = A^t\mathbf{y}\}. \tag{4.80}$$

*Proof.*   Let $N$ be a linear variety as in Section 4.7, lying below $[f]$ and represented as the intersection of $M^o$ with a nonvertical hyperplane $E: \mathbf{y}^t\mathbf{x} - z = w$. Since $\mathbf{x}_0$ is stable for $f$, a hyperplane $E_1$ exists, separating $N$ from $[f]$. This hyperplane meets the linear variety $M^o$ in a linear variety $N_1$ of the same dimension as $N$; these two linear varieties must be parallel (or coincident), since $E_1$ lies above $N$. (See Problem 4.41.) Further $N_1$ must lie above $N$. Thus we can construct a hyperplane $E_0$ parallel to $E_1$ that *passes through* $N$, and $E_0$

also lies below $[f]$. Therefore $N$ can be represented as above as the intersection of $M^o$ with a nonvertical hyperplane $E_0 : \mathbf{y}_0^t \mathbf{x} - z = w_0$, which *lies below* $[f]$:

$$N = \{(\mathbf{x}, z) \mid \mathbf{y}_0^t \mathbf{x} - z = w_0\} \cap \{(A\mathbf{u} - \mathbf{b}, z) \mid \mathbf{u} \in \mathcal{R}^m\}. \tag{4.81}$$

By the location test of Section 4.5, $f^c(\mathbf{y}_0) \le w_0$.

As in Section 4.7 we can also represent $N$ as the range of the mapping $\mathbf{x} = A\mathbf{u} - \mathbf{b}$ of the linear variety $E' : \mathbf{v}^t \mathbf{u} - z = w'$ in $\mathcal{R}^m$ into $\mathcal{R}^n$, where

$$\mathbf{v} = A^t \mathbf{y}_0 \quad \text{and} \quad w' = w_0 + \mathbf{y}_0^t \mathbf{b} \tag{4.82}$$

and by Lemma B, $\hat{f}^c(\mathbf{v}) \le w'$. Thus

$$w' \ge f^c(\mathbf{y}_0) + \mathbf{y}_0^t \mathbf{b}. \tag{4.83}$$

We now reason as follows: We select a $\mathbf{v}$ representable as $A^t\mathbf{y}$, for some $\mathbf{y}$ for which $\hat{f}^c(\mathbf{v})$ is finite, and choose $w' \ge \hat{f}^c(\mathbf{v})$. Then there is a corresponding linear variety $N$ with this $\mathbf{v}$ and $w'$; in particular, $N$ can be represented as the intersection of $M$ with $E_0$, as above, so (4.82) and (4.83) hold. It follows that

$$w' \ge \inf_{\mathbf{y}}\{f^c(\mathbf{y}) + \mathbf{y}^t b \mid \mathbf{v} = A^t\mathbf{y}\}. \tag{4.84}$$

This holds for every $w' \ge \hat{f}^c(\mathbf{v})$. It follows that

$$\hat{f}^c(\mathbf{v}) \ge \inf_{\mathbf{y}}\{f^c(\mathbf{y}) + \mathbf{y}^t b \mid \mathbf{v} = A^t\mathbf{y}\}. \tag{4.85}$$

We remark that when $\mathbf{v}$ is *not* representable as $A^t\mathbf{y}$, then $\hat{f}^c(\mathbf{v}) = +\infty$. (See Problem 4.42.)

We also have an inequality opposite to (4.85):

$$\hat{f}^c(\mathbf{v}) \le \inf_{\mathbf{y}}\{f^c(\mathbf{y}) + \mathbf{y}^t \mathbf{b} \mid \mathbf{v} = A^t\mathbf{y}\}. \tag{4.86}$$

To show this, we note that

$$f^c(\mathbf{y}) = \sup_{\mathbf{x}}\{\mathbf{y}^t \mathbf{x} - f(\mathbf{x}) \mid \mathbf{x} \in \mathcal{R}^n\}.$$

This is clearly greater than or equal to the sup for $\mathbf{x} \in M \subset \mathcal{R}^n$. Attaching the restriction $\mathbf{v} = A^t\mathbf{y}$, we cannot increase the sup and we conclude that

$$\inf_{\mathbf{y}}\{f^c(\mathbf{y}) + \mathbf{y}^t \mathbf{b} \mid \mathbf{v} = A^t\mathbf{y}\} \ge \inf_{\mathbf{y}}\left\{ \sup_{\mathbf{u}}\{\mathbf{y}^t(A\mathbf{u} - \mathbf{b}) + \mathbf{y}^t\mathbf{b} - f(A\mathbf{u} - \mathbf{b}) \mid \mathbf{v} = A^t\mathbf{y}\} \right\}$$

$$\ge \inf_{\mathbf{y}}\left( \sup_{\mathbf{u}}(\mathbf{v}^t\mathbf{u} - f(A\mathbf{u} - \mathbf{b})) \right) = \hat{f}^c(\mathbf{v}) \quad \text{(no dependence on } \mathbf{y}\text{)}.$$

Combining (4.85) and (4.86), we obtain equality

$$\hat{f}^c(\mathbf{v}) = \inf_{\mathbf{y}}\{f^c(\mathbf{y}) + \mathbf{y}^t\mathbf{b} \mid \mathbf{v} = A^t\mathbf{y}\}. \tag{4.87}$$

Here the inf can be replaced by min, since $\mathbf{v} \in K(\hat{f}^c)$. Indeed, as above, the linear variety

$$N = \{(A\mathbf{u} - \mathbf{b},\, z) \mid \mathbf{v}^t\mathbf{u} - z = \hat{f}^c(\mathbf{v})\} \tag{4.88}$$

lies below $[f]$ by Lemma B. Hence as above there exists a nonvertical plane

$$E_0 = \{(\mathbf{x},\, z) \mid \mathbf{y}_0^t\mathbf{x} - z = w_0\}$$

that contains $N$ and lies below $[f]$. In other words, $\mathbf{v} = A^t\mathbf{y}_0$, $\hat{f}^c(\mathbf{v}) = w_0 + \mathbf{y}_0^t\mathbf{b}$, and $w_0 \geq f^c(\mathbf{y}_0)$ for some $\mathbf{y}_0$, $w_0$. Then $\hat{f}^c(\mathbf{v}) \geq f^c(\mathbf{y}_0) + \mathbf{y}_0^t\mathbf{b}$. By (4.87), equality must hold here. Accordingly inf can be replaced by min in (4.87).

Thus Lemma C' is proved. To obtain the desired result for a concave function $g$, we let $f(\mathbf{x}) = -g(\mathbf{x})$ so that $f$ is convex and is stable at $\mathbf{x}_0$, a point of $M \cap K(f)$. Next $\hat{f}(\mathbf{u}) = -g(A\mathbf{u} - \mathbf{b}) = -\hat{g}(\mathbf{u})$, and by the rule (4.58), $f^c(\mathbf{y}) = -g^c(-\mathbf{y})$ and $\hat{f}^c(\mathbf{v}) = -\hat{g}^c(-\mathbf{v})$. If $\mathbf{v}$ is in $K(\hat{f}^c)$, then $-\mathbf{v}$ is in $K(\hat{g}^c)$, and we conclude from Lemma C' that

$$-\hat{g}^c(-\mathbf{v}) = \min_{\mathbf{y}}\{-g^c(-\mathbf{y}) + \mathbf{y}^t\mathbf{b} \mid \mathbf{v} = A^t\mathbf{y}\}.$$

Here we can replace $\mathbf{v}$ by $-\mathbf{v}$ on both sides and then replace $\mathbf{y}$ by $-\mathbf{y}$ inside the bracket on the right. Then the rule that the minimum of the negative of a function equals the negative of the maximum of the function yields the result:

$$\hat{g}^c(\mathbf{v}) = \max_{\mathbf{y}}\{g^c(\mathbf{y}) + \mathbf{y}^t\mathbf{b} \mid \mathbf{v} = A^t\mathbf{y}\}. \tag{4.89}$$

This is the same as (4.77) except for some changes in notation (and interchange of $m$ and $n$). ∎

***Example 1.*** Let $f(\mathbf{x})$ be the linear function $\mathbf{c}^t\mathbf{x}$ in $\mathcal{R}^n$. Then we verify that $f^c(\mathbf{y}) = 0$ for $\mathbf{y} = \mathbf{c}$ and $f^c(\mathbf{y}) = \infty$ otherwise, that $\hat{f}^c(\mathbf{v})$ is $+\infty$ except for the value $\mathbf{c}^t\mathbf{b}$ for $\mathbf{v} = A^t\mathbf{c}$, and that (4.80) is satisfied. (See Problem 4.43.)

***Example 2.*** Let $f(x_1, x_2) = x_1^2 + x_2^2$, where $x_1 = 3u$, $x_2 = 4u$, so $\hat{f}(u) = 25u^2$. Here $A = [3, 4]^t$. One verifies that (4.80) is satisfied, with both sides equal to $v^2/100$. (See Problem 4.45.)

## PROBLEMS

**4.33.** Let $a \neq 0$ and $b$ be constants, and let $g$ be concave in $\mathcal{R}^1$.

(a) Show that the function $g(ax - b)$ is concave in $\mathcal{R}^1$.

(b) Show that the conjugate of $g(ax - b)$ is the concave function $g^c(y/a) + (b/a)y$.

(c) Illustrate the result of (b) for the case of $g(x) = 1 - |x|$, $a = 2$, $b = 3$.

**4.34.** Let $M$ be the linear variety (4.72), and let $M^o$ be the corresponding vertical linear variety in $\mathcal{R}^{n+1}$, as described following (4.72). Let $N$ be a linear variety in $\mathcal{R}^{n+1}$ that is the intersection of $M^o$ with a nonvertical hyperplane $E$ in $\mathcal{R}^{n+1}$ whose equation is $\mathbf{y}^t\mathbf{x} - z = w$. Verify that $N$ can be represented as in equations (4.73), (4.74), and that $N$ is the range of the mapping of the hyperplane $E' : \mathbf{v}^t\mathbf{u} - z = w'$ in $\mathcal{R}^{m+1}$ into $\mathcal{R}^{n+1}$ which takes $(\mathbf{u}, z)$ to $(A\mathbf{u} - \mathbf{b}, z)$.

**4.35.** Let $A$ be an $n \times m$ matrix, $\mathbf{b}$ a vector of $\mathcal{R}^n$. Let $f$ be convex in $\mathcal{R}^n$. Show that $f(A\mathbf{u} - \mathbf{b})$ is convex in $\mathcal{R}^m$ provided that this function is finite for at least one value of $\mathbf{u}$. [Here it is understood that convex functions are defined as in Section 4.4.]

**4.36.** Let $A$ be an $n \times m$ matrix, let $\mathbf{v}$ be a given vector in $\mathcal{R}^m$, and let $w'$ be a constant. Let $z = \mathbf{v}^t\mathbf{u} - w'$ be a function of $\mathbf{u}$ in $\mathcal{R}^m$ such that $z \leq 0$ whenever $A\mathbf{u} \geq 0$. Show that also $\mathbf{v}^t\mathbf{u} \leq 0$ whenever $A\mathbf{u} \geq 0$.

**4.37.** Let $f$ be the convex function in $\mathcal{R}^n$ with values $\mathbf{c}^t\mathbf{x}$ for $\mathbf{x} \geq 0$, where $\mathbf{c}$ is a constant vector, and value $+\infty$ otherwise.

(a) Show that $\mathbf{x}_0 = 0$ is stable for $f$. [Hint: Imitate the procedure used before Example 1 in Section 4.7, for the case $\mathbf{c} = 0$.]

(b) Show that every boundary point $\mathbf{x}_0$ of $K(f)$ is stable for $f$. [Hint: Given $N : z = \mathbf{v}^t\mathbf{u} - w'$, $\mathbf{x} = A\mathbf{u} + \mathbf{x}_0$ below [$f$] so that $z = \mathbf{v}^t\mathbf{u} - w' \leq \mathbf{c}^t\mathbf{x}$ for $\mathbf{x} = A\mathbf{u} + \mathbf{x}_0 \geq 0$, let $k = \sup(z - \mathbf{c}^t\mathbf{x})$ for $(\mathbf{x}, z)$ in $N$ and $\mathbf{x} \geq 0$, show that $k \leq 0$ and that the hyperplane $E_1 : z = \mathbf{c}^t\mathbf{x} + k$ separates $N$ from [$f$]. Note that this simple proof covers case (a) and the case $\mathbf{c} = 0$.]

**4.38.** Verify the results in Example 1 of Section 4.7.

**4.39.** Consider Lemma D in Section 4.7. In the text, a proof is given for the first statement. Imitate the proof of Lemma A in Section 4.6 to prove the remaining assertion of the Lemma.

**4.40.** Verify the results in Example 2 in Section 4.7.

**4.41.** As in the proof of Lemma C', let $N$ and $N_1$ be linear varieties in $\mathcal{R}^{n+1}$ that are the intersections of a vertical linear variety $M^o$ with respective nonvertical hyperplanes $E$ and $E_1$. Show that if $N$ lies below $E_1$, then $N$ and $N_1$ are parallel (or coincident).

**4.42.** Let $\hat{f}(\mathbf{u})$ be defined as in Lemma C' in Section 4.8. Let $\mathbf{v}$ be a vector of $\mathcal{R}^m$ that is *not* representable as $A^t\mathbf{y}$ for suitable $\mathbf{y}$. Show that $\hat{f}^c(\mathbf{v}) = +\infty$. [Hint: Let $\hat{f}(\mathbf{u}_0) = \mathbf{x}_0 \in K(f)$. By linear algebra (see Appendix A.17) the hypothesis implies that $\mathbf{v}$ is not in the row space of matrix $A$ so that $\mathbf{v} = \mathbf{p} + \mathbf{q}$ where $\mathbf{p} \neq \mathbf{0}$ is in the null space and $\mathbf{q}$ is in the row space of $A$, and hence $\mathbf{p}$ is orthogonal to the row space. Now show that $\mathbf{v}^t\mathbf{u} - f(A\mathbf{u} - \mathbf{b})$ is unbounded above for $\mathbf{u}$ of the form $\mathbf{u}_0 + \lambda\mathbf{p}$, with $\lambda > 0$.]

**4.43.** Verify the results in Example 1 in Section 4.8.

**4.44.** With reference to Problem 4.42, show by example that it can occur that $\hat{f}^c(\mathbf{v}) = +\infty$ even though $\mathbf{v}$ is representable as $A^t\mathbf{y}$ for some $\mathbf{y}$.

**4.45.** Verify the results in Example 2 in Section 4.8.


## 4.9   NORMS, DUAL NORMS, MINKOWSKI NORMS

As in Section 3.8 we define a norm in $\mathcal{R}^n$ to be a function $\rho(\mathbf{x})$ with the properties:

  **(i)**   $\rho(\mathbf{x}) > 0$ except that $\rho(\mathbf{0}) = 0$,

  **(ii)**  $\rho(c\mathbf{x}) = |c|\rho(\mathbf{x})$,

  **(iii)** $\rho(\mathbf{x} + \mathbf{y}) \leq \rho(\mathbf{x}) + \rho(\mathbf{y})$.

From each norm $\rho$ in $\mathcal{R}^n$ one obtains a *distance function* by defining the distance from $\mathbf{x}$ to $\mathbf{y}$ to be $\rho(\mathbf{y} - \mathbf{x})$. One then verifies that this function has the familiar properties associated with distance (Problem 4.46).

Examples of norms in $\mathcal{R}^n$ are the $A$-norms introduced in Section 1.14 and the $p$-norms introduced in Section 3.8 (of which the Euclidean norm is the special case $p = 2$); the $p$-norms are considered further at the end of this section.

*Each norm $\rho$ in $\mathcal{R}^n$ is a convex function, with finite values for all $\mathbf{x}$.* (See Problem 1.93.) Hence it is continuous in $\mathcal{R}^n$. One can also prove the continuity directly (Problem 4.47). The important $p$-norm is easily seen to be continuously differentiable for $\mathbf{x} \neq \mathbf{0}$ provided that $1 < p < \infty$ and its gradient is never $\mathbf{0}$ (see Problem 4.48).


### Equivalence of Norms

Although there are many norms in $\mathcal{R}^n$, we can regard them as roughly the same, up to scale factors. We justify this statement by showing that *for each two norms $\rho_1(\mathbf{x})$, $\rho_2(\mathbf{x})$ in $\mathcal{R}^n$, there is a positive constant $k$ such that $\rho_1(\mathbf{x}) \leq k\rho_2(\mathbf{x})$ for all $\mathbf{x}$.* To prove this, we remark that the function $\rho_1(\mathbf{x})/\rho_2(\mathbf{x})$ is continuous and positive on the bounded closed set $\|\mathbf{x}\|_2 = 1$ and hence has a positive maximum $k$. The conclusion now follows from property (ii).

In Section 1.9 we considered the level sets and sublevel sets of a convex function. For a norm $\rho$, each level set has a very simple character. By the property (ii), on each ray $\mathbf{x} = t\mathbf{b}$, $t \geq 0$, through the origin, where $\mathbf{b} \neq \mathbf{0}$, the norm is a linear function of $t$ and increases from 0 to $+\infty$ as $t$ varies from 0 to $+\infty$. Hence $\rho$ takes on the value $c > 0$ at exactly one point on each ray. If we fix $c > 0$ and vary $\mathbf{b}$ on the unit sphere, $\|\mathbf{b}\|_2 = 1$, then the value of $\rho$ at the point on the ray for which $\rho = c$ becomes a continuous function of $\mathbf{b}$. Equivalently, if we use spherical coordinates as in Section 1.9, then the level set $\rho = c$ is given by an equation $r = F(\phi, \theta_1, \ldots, \theta_{n-2})$ in appropriate spherical coordinates, where $F$ is continuous. The continuity follows from the results of Section 4.2. Thus, as in Section 1.9, the level sets form a center configuration at the origin, at which $\rho$ has its minimum value, 0. The corresponding o-sublevel set $\rho(\mathbf{x}) < c$ is a convex set, as pointed out in Section 1.9. Furthermore it is an open region in $\mathcal{R}^n$ (Problem 4.49). The boundary of this o-sublevel set is the level set $\rho(\mathbf{x}) = c$, as follows from the discussion given above. If we adjoin the boundary to the o-sublevel set, we obtain the closed region $\rho(\mathbf{x}) \leq c$, which is also convex, as pointed out in Section 1.9. Furthermore it is a bounded set (Problem 4.50). Such a convex set, forming a bounded closed region in $\mathcal{R}^n$, is termed a *convex body*. There is a large literature about convex bodies; an important reference is Bonnesen and Fenchel (1987). We denote the convex body $\{\mathbf{x} \mid \rho(\mathbf{x}) \leq 1\}$ by $B_\rho$ and its boundary, the level set $\{\mathbf{x} \mid \rho(\mathbf{x}) = 1\}$, by $S_\rho$; the set $S_\rho$ is a bounded closed set in $\mathcal{R}^n$.

## Dual Norm

Associated with each norm $\rho$ in $\mathcal{R}^n$, there is a second norm $\sigma$, called the dual norm of $\rho$, and we will write $\sigma = \rho^D$ to indicate the relationship. The dual $\sigma$ is defined by the equation

$$\sigma(\mathbf{y}) = \max_{\mathbf{x}} \{\mathbf{x}^t \mathbf{y} \mid \rho(\mathbf{x}) = 1\}. \qquad (4.90)$$

Since $\rho$ is continuous in $\mathcal{R}^n$ and $S_\rho$ is a bounded closed set, the maximum exists, so $\sigma$ is a well-defined function in $\mathcal{R}^n$. We will see that it is also a norm.

From this definition, one obtains a general "Hölder inequality":

$$\mathbf{x}^t \mathbf{y} \leq \rho(\mathbf{x})\sigma(\mathbf{y}). \qquad (4.91)$$

This is true for $\rho(x) = 1$ from (4.90). By the property (ii), it is valid for all $\mathbf{x}$ (Problem 4.51). We observe that the dual of the $p$-norm is the $q$-norm, where $(1/p) + (1/q) = 1$, as in Section 3.8, and the Hölder inequality (4.91) implies the inequality (3.84); for a proof, see the end of this section.

We now verify that $\sigma$ has the basic properties (i)–(iii) of a norm. For $\mathbf{y} = \mathbf{0}$, clearly $\sigma(\mathbf{y}) = 0$, by (4.90). For $\mathbf{y} \neq \mathbf{0}$, the ray $\mathbf{x} = t\mathbf{y}$ meets the level set $\rho = 1$ at a unique point at which $t > 0$, so $\mathbf{x}^t\mathbf{y}$ has the positive value $t\|\mathbf{y}\|_2^2$ at the point. Accordingly $\sigma(\mathbf{y}) > 0$. Thus the first property is proved. Proof of the second one is left as an exercise (Problem 4.52). For the third property we have

$$\sigma(\mathbf{y} + \mathbf{z}) = \max_{\mathbf{x}}\{\mathbf{x}^t(\mathbf{y} + \mathbf{z}) \mid \rho(\mathbf{x}) = 1\} = \max_{\mathbf{x}}\{\mathbf{x}^t\mathbf{y} + \mathbf{x}^t\mathbf{z} \mid \rho(\mathbf{x}) = 1\}$$

$$\leq \max_{\mathbf{x}}\{\mathbf{x}^t\mathbf{y} \mid \rho(\mathbf{x}) = 1\} + \max_{\mathbf{x}}\{\mathbf{x}^t\mathbf{z} \mid \rho(\mathbf{x}) = 1\} = \sigma(\mathbf{y}) + \sigma(\mathbf{z}).$$

If $\rho$ is continuously differentiable for $\mathbf{x} \neq \mathbf{0}$, then we have the further rules:

**(iv)** $\nabla\rho(c\mathbf{x}) = \nabla\rho(\mathbf{x})$ for $c > 0$ and $\mathbf{x} \neq \mathbf{0}$.
**(v)** $\mathbf{x}^t\nabla\rho(\mathbf{x}) = \rho(\mathbf{x})$ for $\mathbf{x} \neq \mathbf{0}$.
**(vi)** If $\mathbf{y} = \lambda\nabla\rho(\mathbf{x})$ and $\mathbf{x} \neq \mathbf{0}$, $\lambda > 0$, then $\lambda = \sigma(\mathbf{y})$.

*Proofs.* By (ii) for $c > 0$, $\rho(c\mathbf{x}) = c\rho(\mathbf{x})$. For $\mathbf{x} \neq \mathbf{0}$, we differentiate with respect to $x_i$ to obtain

$$c\frac{\partial\rho}{\partial x_i}(c\mathbf{x}) = c\frac{\partial\rho}{\partial x_i}(\mathbf{x}).$$

Rule (iv) follows by division by $c$. Rule (v) follows from Euler's theorem on homogeneous functions (Kaplan, 1991, p. 101). Here $\rho(\mathbf{x})$ is homogeneous of degree 1 and its gradient is homogeneous of degree 0.

For rule (vi) we consider the maximum problem (4.90) and apply Lagrange multipliers to the function $z = \mathbf{x}^t\mathbf{y}$ with side condition $\rho(\mathbf{x}) = 1$. We obtain the equations $\mathbf{y} - \lambda\nabla\rho = \mathbf{0}$, $\rho(\mathbf{x}) = 1$. Hence at the maximum, with the aid of (v),

$$z = \lambda\mathbf{x}^t\nabla\rho = \lambda\rho(\mathbf{x}) = \lambda.$$

Thus $\lambda = \sigma(\mathbf{y}) > 0$. Now, if $\mathbf{y} = \lambda\nabla\rho(\mathbf{x})$ and $\mathbf{x} \neq \mathbf{0}$, then by (iv) the same equation holds with $\mathbf{x}$ replaced by $\mathbf{x}_1 = (1/\rho(\mathbf{x}))\mathbf{x}$ so that $\rho(\mathbf{x}_1) = 1$. Since also $\lambda > 0$, these are the same as the equations obtained by Lagrange multipliers so that again $\sigma(\mathbf{y}) = \lambda$.                    ∎

We observe that (4.90) is a maximum problem for a linear function. We can replace the side condition here by $\rho(\mathbf{x}) \leq 1$, for as pointed out at the end of Section 4.1, a linear function takes on its maximum only at (relative) boundary points, and here the set $S_\rho = \{\mathbf{x} \mid \rho(\mathbf{x}) = 1\}$ is the boundary (and relative boundary) of the set $B_\rho = \{\mathbf{x} \mid \rho(\mathbf{x}) \leq 1\}$. This shows that the dual norm $\sigma$ is the conjugate of a convex function, as in rule 10. at the end of Section 4.5; it is the conjugate of the convex function $f$ equal to 0 on $B_\rho$ and equal to $+\infty$ elsewhere, and $\sigma$ is hence the *support function* for the convex body $B_\rho$.

## Minkowski Norms

To each convex body $E$ containing the origin in its interior, we can associate a convex function $\rho(\mathbf{x})$ with most of the properties of a norm; we call it a Minkowski norm. We set $\rho(\mathbf{0}) = 0$, and to each $\mathbf{x}$ in $\mathcal{R}^n$ other than $\mathbf{0}$, we assign the value $\rho(\mathbf{x}) = \|\mathbf{x}\|_2 / \|\mathbf{x}_1\|_2$, where $\mathbf{x}_1$ is the unique point at which the ray from the origin to $\mathbf{x}$ meets the boundary of $E$; we are here applying the line-intersection rule of Section 4.2. Thus $E$ becomes $B_\rho$ and its boundary becomes $S_\rho$. (See Fig. 4.14). We verify that $\rho(\mathbf{x})$ satisfies the rules (i) and (iii) and, instead of (ii), the following rule:

**(ii')** $\rho(c\mathbf{x}) = c\rho(\mathbf{x})$ if $c \geq 0$ (positive homogeneity, as in Section 4.1).

The convexity of $\rho(\mathbf{x})$ follows from the properties (i), (ii'), and (iii) (Problem 4.53). The proofs of (i) and (ii') are left as exercises (Problem 4.54). For (iii) we first observe that, since $B_\rho$ is convex,

$$\rho((1 - \alpha)\mathbf{x} + \alpha\mathbf{y}) \leq 1 \qquad \text{for } 0 \leq \alpha \leq 1, \ \mathbf{x}, \ \mathbf{y} \in B_\rho. \tag{4.92}$$

Now for arbitrary $\mathbf{x}, \mathbf{y}$ (not $\mathbf{0}$), by (i) and (ii'),

$$\rho\left(\frac{\mathbf{x}}{\rho(\mathbf{x})}\right) = 1 \quad \text{and} \quad \rho\left(\frac{\mathbf{y}}{\rho(\mathbf{y})}\right) = 1.$$

Therefore, by (4.92),

$$\rho\left((1 - \alpha)\frac{\mathbf{x}}{\rho(\mathbf{x})} + \alpha\frac{\mathbf{y}}{\rho(\mathbf{y})}\right) \leq 1.$$

Rule (iii) now follows if we take $\alpha = \rho(\mathbf{y})/k$, where $k = \rho(\mathbf{x}) + \rho(\mathbf{y})$ so that $1 - \alpha = \rho(\mathbf{x})/k$. (The rule is immediate if one of $\mathbf{x}, \mathbf{y}$ is $\mathbf{0}$.) If $E$ happens to have

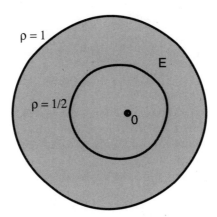

**Figure 4.14.** Minkowski norm.

symmetry with respect to the origin, so $E$ is unaffected by replacing each point $\mathbf{x}$ of $E$ by $-\mathbf{x}$, then (ii′) is equivalent to (ii), and the Minkowski norm is a norm as we have defined it. If $E$ fails to have this symmetry, then the Minkowski norm is not a norm in our sense. In most applications one has the symmetry required. However, there are cases in which it is natural to use an "unsymmetric norm." (See Section 4.11 below.)

If a Minkowski norm is differentiable for $\mathbf{x} \neq \mathbf{0}$, then it also has the properties (iv), (v), and (vi), with the same proofs as before. One defines the dual of a Minkowski norm by (4.90) and verifies that it is again a Minkowski norm. Its properties are as above. For a full discussion of Minkowski norms, see Stoer and Witzgall (1970, pp. 81–87, 131–133) and Bonnesen and Fenchel (1987, pp. 23–25).

### Polar Set

For an arbitrary nonempty set $E$ in $\mathcal{R}^n$, the set

$$E^\pi = \{\mathbf{y} \in \mathcal{R}^n \mid \mathbf{y}^t \mathbf{x} \leq 1 \text{ for all } \mathbf{x} \in E\}$$

is called the polar set of $E$. The polar set contains the vector $\mathbf{0}$ and is hence nonempty. Further we verify that $E^\pi$ *is a closed convex set* (Problem 4.55).

We are here interested in the case when $E$ is a convex body containing the origin as interior point. Then $E^\pi$ *is also a convex body having the origin as interior point* (Problem 4.55).

***Example 1.*** In the $xy$-plane we consider two square convex bodies: $A$ is the square with vertices $(\pm 1, 0)$, $(0, \pm 1)$; $B$ is the square with vertices $(\pm 1, \pm 1)$. We verify that $A^\pi = B$ and $B^\pi = A$ (Fig. 4.15). (See Problem 4.56.)

The example suggests a general rule: *If $E_1$ and $E_2$ are convex bodies in $\mathcal{R}^n$ having the origin as interior point and $E_1^\pi = E_2$, then $E_2^\pi = E_1$, or equivalently,*

$$E^{\pi\pi} = E \tag{4.93}$$

for every convex body in $\mathcal{R}^n$ having the origin as interior point.

We prove that the rule (4.93) is valid. To that end, we show first that $E$ is contained in $E^{\pi\pi}$. We observe that $E^{\pi\pi}$ consists of all $\mathbf{z}$ such that $\mathbf{z}^t\mathbf{y} \leq 1$ for all $\mathbf{y}$ in $E^\pi$, that is, for all $\mathbf{y}$ such that $\mathbf{y}^t\mathbf{x} \leq 1$ for all $\mathbf{x}$ in $E$. But for each $\mathbf{x}$ in $E$, $\mathbf{x}^t\mathbf{y} \leq 1$ for all $\mathbf{y}$ in $E^\pi$. Therefore, $\mathbf{x} \in E^{\pi\pi}$. Accordingly $E \subset E^{\pi\pi}$.

Next we show that for each $\mathbf{z}$ not in $E$, $\mathbf{z} \notin E^{\pi\pi}$. Let $\mathbf{x}^*$ be the point of $E$ closest (in Euclidean distance) to $\mathbf{z}$. Then the vector $\mathbf{x}^* - \mathbf{z}$ is nonzero and is normal to a supporting hyperplane of $E$ at $\mathbf{x}^*$. We construct a hyperplane parallel to this hyperplane through the midpoint of the segment joining $\mathbf{z}$ to $\mathbf{x}^*$. The equation of this hyperplane can be written in the form $\mathbf{y}^t\mathbf{x} = w$; the hyper-

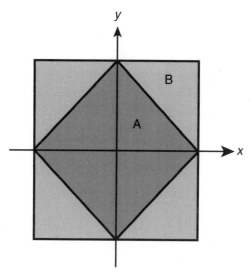

**Figure 4.15.** Examples of polar sets.

plane separates $\mathbf{z}$ from $E$ (see Fig. 4.16), and since $\mathbf{0}$ is in $E$, $w \neq 0$. After multiplying by a scalar, we can write the equation in the form $\mathbf{y}^t \mathbf{x} = 1$. Since $\mathbf{0}$ is in $E$, we must have $\mathbf{y}^t \mathbf{x} < 1$ for all $\mathbf{x} \in E$ and $\mathbf{y}^t \mathbf{z} > 1$. This shows that $\mathbf{y} \in E^\pi$ and $\mathbf{z} \notin E^{\pi\pi}$.

Combining the two results, we conclude that the rule (4.93) is valid as asserted. (The proof we have given shows that the rule is valid for every convex body containing $\mathbf{0}$.) ∎

### Application to Dual Norms

Let $\rho$ be a norm in $\mathcal{R}^n$, let $\sigma$ be its dual. From (4.90) and the properties of $\rho$, we can write

$$\sigma(\mathbf{y}) = \max_{\mathbf{x}} \{\mathbf{x}^t \mathbf{y} \mid \rho(\mathbf{x}) \leq 1\}.$$

It follows that $\sigma(\mathbf{y}) \leq 1$ precisely when $\mathbf{x}^t \mathbf{y} \leq 1$ for all $\mathbf{x}$ in $B_\rho$. Therefore $B_\sigma$ is *the polar set of* $B_\rho$.

Now $B_\rho$ is a convex body containing $\mathbf{0}$ as interior point and hence the rule (4.93) applies. Therefore

$$B_\sigma^\pi = B_\rho^{\pi\pi} = B_\rho.$$

Accordingly, if $\tau$ is the dual norm of $\sigma$, then $B_\tau = B_\rho$. Furthermore, since $\tau$ is a norm, $\tau(\mathbf{0}) = 0$. It follows that the norms $\tau$ and $\rho$ coincide. They agree, with

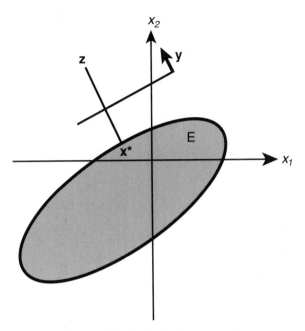

**Figure 4.16.** Proof of rule on polar sets.

value 1, on the boundary $S_\tau = S_\rho$ of $B_\tau = B_\rho$, and hence, by the rule (ii), they agree for all $\mathbf{x}$ in $\mathcal{R}^n$. We conclude that if $\rho^D = \sigma$, then $\sigma^D = \rho$, or

$$\rho^{DD} = \rho. \tag{4.94}$$

This rule also applies to Minkowski norms, since the proof just given can be repeated, with (ii′) replacing (ii).

The rule (4.94) is illustrated by the $p$-norms. The $p$-norm and $q$-norm (with $(1/p) + (1/q) = 1$) are duals of each other; in particular, the 1-norm and $\infty$-norm are duals of each other and the 2-norm is dual to itself.

**More on the $p$-Norm**

We show here how the basic properties of this norm can be derived from the theory of convex sets. We first remark that the $p$-norm can be interpreted as a Minkowski norm. To that end, we note that for $1 \le p < \infty$, the function $|x|^p$ of $x$ is a convex function in $\mathcal{R}^1$ (Problem 4.57). Hence the function

$$\phi(\mathbf{x}) = \sum_{i=1}^{n} |x_i|^p \tag{4.95}$$

is a convex function in $\mathcal{R}^n$. The sublevel set $\phi(\mathbf{x}) \le 1$ is hence an $n$-dimensional convex set $B$. The set $B$ is bounded, since it is contained in the set $\{\mathbf{x} \,|\, \|\mathbf{x}\| \le \sqrt{n}\}$

(Problem 4.58). It is closed, since $\phi$ is continuous in $\mathcal{R}^n$. Therefore $B$ is a convex body, and the origin $\mathbf{0}$ is an interior point of $B$. On each ray starting at the origin: $\mathbf{x} = t\mathbf{a}$, the function $\phi$ is a strictly increasing function of $t$, going from 0 to $\infty$ as $t$ goes from 0 to $\infty$. It follows that the boundary of $B$ is the level set: $\phi(\mathbf{x}) = 1$.

The corresponding Minkowski norm has the value 1 on the boundary of $B$, has the value 0 at the origin, and varies linearly on each ray from the origin; these properties uniquely determine the norm (Problem 4.58(b)). But the properties are all satisfied by the $p$-norm (Problem 4.59). Therefore *the $p$-norm is the Minkowski norm for the convex body $B$*. In this case the convex body $B$ is symmetric, so the $p$-norm is also symmetric and hence has all the properties (i), (ii), (iii).

Next we verify that the dual of the $p$-norm is the $q$-norm, where $q = \infty$ for $p = 1$, and $p^{-1} + q^{-1} = 1$ for $p > 1$. The proof for the case $p = 1$ is considered below. For $p > 1$ the $p$-norm $\rho(\mathbf{x})$ is continuously differentiable in $\mathcal{R}^n$ except at the origin (Problem 4.48), and we can use Lagrange multipliers to solve the maximum problem (4.90). For $\mathbf{y} = \mathbf{0}$ the problem is trivial, and we find, as expected, that $\sigma(\mathbf{0}) = 0$. We turn to the case of $\mathbf{y} \neq \mathbf{0}$. Since the $p$-norm depends only on the absolute values of the components of $\mathbf{x}$ and is hence unaffected by multiplying some components of $\mathbf{x}$ by $-1$, we find that the dual norm $\sigma(\mathbf{y})$ has the analogous property: It is a function of the absolute values of the components of $\mathbf{y}$ (see Problem 4.61). Hence we need to solve the maximum problem only for the case when all $y_i$ are nonnegative. A maximizer $\mathbf{x}$ then exists with all $x_i$ nonnegative, for if any of the $x_i$ are negative, then the value of the inner product $\mathbf{x}^t\mathbf{y}$ is not decreased by replacing each of these $x_i$ by $|x_i|$. At the maximizer we are then led to the equation in (vi), with $\rho(\mathbf{x}) = 1$ so that $\mathbf{x} \neq \mathbf{0}$. Calculating the gradient, we find

$$y_i = \lambda x_i^{p-1}, \qquad i = 1, \ldots, n. \tag{4.96}$$

Here $\lambda$ must be positive, since $\mathbf{y} \neq \mathbf{0}$. Thus by (vi), $\lambda = \sigma(\mathbf{y})$. Now $\mathbf{x}^t\mathbf{y} = \sigma(\mathbf{y})$, where the components of $\mathbf{x}$ are given by (4.96). By combining these equations and noting that $p/(p-1) = q$, we obtain the expected formula:

$$\sigma(\mathbf{y}) = \left( \sum_{i=1}^{n} |y_i|^q \right)^{1/q}. \tag{4.97}$$

The $q$-norm is the dual of the $p$-norm, where $p^{-1} + q^{-1} = 1$. Accordingly, the Hölder formula is valid:

$$|\mathbf{x}^t\mathbf{y}| \leq \|\mathbf{x}\|_p \|\mathbf{y}\|_q. \tag{4.98}$$

We use absolute values on the left here, because the norms are symmetric.

It remains to take care of the case $p = 1$. Here the remark about dependence on absolute values and existence of a maximizer with nonnegative $x_i$ for $\mathbf{y} \geq 0$ remains valid. Thus we are led to the *linear program*:

$$maximize \quad \mathbf{y}^t\mathbf{x} \text{ for } x_1 + \cdots + x_n = 1, \qquad \mathbf{x} \geq \mathbf{0},$$

where $\mathbf{y}$ is a given vector with nonnegative components. As in Section 3.7, the maximizer is a basic feasible solution of the equation $x_1 + \cdots + x_n = 1$; that is, a vector with all components 0 except for one which is 1. We conclude that for $\mathbf{y} \geq 0$, $\sigma(\mathbf{y}) = \max(y_1, \ldots, y_n)$, and hence for arbitrary $\mathbf{y}$,

$$\sigma(\mathbf{y}) = \max(|y_1|, \ldots, |y_n|) = \|\mathbf{y}\|_\infty.$$

Since we know that the dual of a norm is a norm, we have also proved that the $\infty$-norm is a norm and proved the validity of the Hölder inequality (4.98) for $p = 1$ and $q = \infty$.


## PROBLEMS

**4.46.** Let $\rho(\mathbf{x})$ be a norm in $\mathcal{R}^n$ and hence have the properties (i)–(iii). Define the distance between $\mathbf{x}$ and $\mathbf{y}$ in $\mathcal{R}^n$ to be $d(\mathbf{x}, \mathbf{y}) = \rho(\mathbf{y} - \mathbf{x})$. Show that

(a) $d(\mathbf{x}, \mathbf{y}) > 0$ if $\mathbf{x} \neq \mathbf{y}$.

(b) $d(\mathbf{x}, \mathbf{x}) = 0$.

(c) $d(\mathbf{x}, \mathbf{y}) = d(\mathbf{y}, \mathbf{x})$.

(d) $d(\mathbf{x}, \mathbf{z}) \leq d(\mathbf{x}, \mathbf{y}) + d(\mathbf{y}, \mathbf{z})$ (triangle inequality).

**4.47.** Prove the continuity of a norm $\rho$ in $\mathcal{R}^n$ by showing from the properties (i)–(iii) that for given $\mathbf{x}$ in $\mathcal{R}^n$ and $\varepsilon > 0$, one can choose $\delta > 0$ so that $\|\mathbf{x} - \mathbf{y}\|_2 < \delta$ implies $|\rho(\mathbf{x}) - \rho(\mathbf{y})| < \varepsilon$. [Hint: Write

$$\rho(\mathbf{y}) = \rho(\mathbf{x} + \mathbf{y} - \mathbf{x}) \leq \rho(\mathbf{x}) + \rho(\mathbf{y} - \mathbf{x})$$

and the similar relations with $\mathbf{x}$ and $\mathbf{y}$ interchanged to conclude that $|\rho(\mathbf{x}) - \rho(\mathbf{y})| \leq \rho(\mathbf{x} - \mathbf{y})$, and then use equivalence of norms to write $\rho(\mathbf{x} - \mathbf{y}) \leq k\|\mathbf{x} - \mathbf{y}\|_2$ for a positive constant $k$.]

**4.48.** Let $\rho(\mathbf{x})$ be the $p$-norm $(\sum |x_i|^p)^{1/p}$ in $\mathcal{R}^n$, where $1 < p < \infty$.

(a) Show that $\rho(\mathbf{x})$ has continuous partial derivatives for all $\mathbf{x} \neq \mathbf{0}$ in $\mathcal{R}^n$.

(b) Show that $\nabla \rho \neq \mathbf{0}$ for $\mathbf{x} \neq \mathbf{0}$.

**4.49.** Let $\rho(\mathbf{x})$ be a norm in $\mathcal{R}^n$. Show that for $c > 0$ the o-sublevel set $\rho(\mathbf{x}) < c$ is an open region. [Hint: Show by continuity that every point of the set has a neighborhood in the set. Show that by convexity every two points of the set can be joined by a line segment in the set. Observe that the origin is in the set, so it is not empty.]

**4.50.** Let $\rho(\mathbf{x})$ be a norm in $\mathcal{R}^n$. Show that for $c > 0$ the sublevel set $\rho(\mathbf{x}) \le c$ is a bounded set. [Hint: Use equivalence of the norm $\rho$ with the Euclidean norm.]

**4.51.** Complete the proof of the Hölder inequality (4.91) for dual norms.

**4.52.** Show that $\sigma(\mathbf{y})$ as defined by (4.90) in terms of the norm $\rho$ has the property (ii).

**4.53.** Let $\rho(\mathbf{x})$ be a function defined in $\mathcal{R}^n$ and having the properties (i) and (iii) and the property (ii'). Show that $\rho(\mathbf{x})$ is convex in $\mathcal{R}^n$.

**4.54.** Let $E$ be a convex body having the origin as interior point. Let $\rho(\mathbf{x})$ be the corresponding Minkowski norm, as defined in Section 4.9.

(a) Show that $\rho(\mathbf{x})$ has the property (ii').

(b) Let $E$ be the circular region of radius 2 and center $(1, 0)$ in the $xy$-plane. Show that except for points $(0, y)$, $\rho(\mathbf{x}) \ne \rho(-\mathbf{x})$, so the rule (ii) is not valid for this Minkowski norm.

(c) Let $E$ be a convex body in the $xy$-plane for which the origin is an interior point and whose boundary is given in polar coordinates by the equation $r = \phi(\theta)$. Show that the corresponding Minkowski norm is given by the equation $\rho = r/\phi(\theta)$ for $r > 0$ (and the value 0 at the origin). Generalize this result to $\mathcal{R}^3$.

**4.55.** Let $E$ be a nonempty set in $\mathcal{R}^n$ and $E^\pi$ its polar set as defined in Section 4.9.

(a) Show that $E^\pi$ is a closed convex set.

(b) Show that if $E$ is a convex body having the origin as interior point, then so also is $E^\pi$. [Hint: In view of (a), one must show that under the stated hypothesis, $E^\pi$ is bounded and has the origin as interior point. Show that since $\|\mathbf{x}\|_2 < M$ in $E$ for some constant $M$, all $\mathbf{y}$ with $\|\mathbf{y}\|_2 < 1/M$ are in $E^\pi$. Further show that since for some positive $\varepsilon$ all $\mathbf{x}$ with $\|\mathbf{x}\|_2 < \varepsilon$ are in $E$, every $\mathbf{y}$ in $E^\pi$ satisfies $\|\mathbf{y}\|_2 < 1/\varepsilon$.]

**4.56.** Verify the assertions in Example 1 in Section 4.9.

**4.57.** Verify that, for $1 \le p < \infty$, the function $f(x) = |x|^p$ is convex on $\mathcal{R}^1$. [Hint: For $p = 1$ the result is immediate. For $p > 1$ use the result of Problem 1.15(e). In particular, show that $f'(x)$ is continuous for all $x$ and that $f'(x)$ is monotone strictly increasing for all $x$, so that $f$ is strictly convex.]

**4.58.** Let $\phi(\mathbf{x})$ be defined by (4.95), and let $B$ be the set $\phi(\mathbf{x}) \le 1$.

(a) Show that $B$ is contained in the set $\|\mathbf{x}\|_2 \le \sqrt{n}$.

**(b)** Show that the Minkowski norm associated with $B$ is the unique function that is linear on each ray from the origin, with value 0 at the origin and value 1 on the boundary of $B$, which is the level set $\phi(\mathbf{x}) = 1$.

**(c)** Show that the $p$-norm in $\mathcal{R}^n$ has the value 0 at the origin and the value 1 on the boundary of $B$, also that it is a linear function of $t$ on each ray $\mathbf{x} = t\mathbf{a}$ from the origin.

**4.59.** Show from the definition (4.90) of the dual norm of a norm $\rho$ that if the norm $\rho(\mathbf{x})$ is unchanged when any component $x_i$ of $\mathbf{x}$ is replaced by its negative, then $\sigma(\mathbf{y})$ has the analogous property: It is unchanged when any component $y_i$ of $\mathbf{y}$ is replaced by its negative.

**\*4.60.** Generalize the result of Problem 4.59 by showing that if the norm $\rho$ has the property that $\rho(A\mathbf{x}) = \rho(\mathbf{x})$ for all $\mathbf{x}$, where $A$ is an orthogonal matrix of order $n$, then $\sigma$ has the analogous property: $\sigma(A\mathbf{y}) = \sigma(\mathbf{y})$ for all $\mathbf{y}$.

**4.61.** Verify all the steps leading to the formula (4.97) for the dual of the $p$-norm for $1 < p < \infty$.

## 4.10 GENERALIZED FERMAT–WEBER PROBLEM

In Sections 3.10 and 3.11 we have obtained duals for the Fermat–Weber problem and certain generalizations of it. In this section we extend these results further. A key step was provided by Witzgall (see Witzgall, 1964), who showed that the Rockafellar theory could be applied to obtain a dual for a Fermat–Weber problem with a general Minkowski norm, not necessarily symmetric. We prove a general duality theorem that includes that of Witzgall and has even broader application; in particular, it applies to the multifacility location problem.

It will be convenient to write $f'(\mathbf{x}) = f(-\mathbf{x})$ for each function $f$ defined in $\mathcal{R}^n$. We let $\rho$ be a Minkowski norm in $\mathcal{R}^n$, $\sigma$ its dual; both norms are convex functions in $\mathcal{R}^n$. As in Section 4.9, $\rho^D = \sigma$, $\sigma^D = \rho$, and we can refer to these as a pair of dual norms. Since the norms are allowed to be unsymmetric, $\rho(-\mathbf{x})$ is not necessarily equal to $\rho(\mathbf{x})$. We verify that $\rho'(\mathbf{x})$ is also a Minkowski norm (Problem 4.62).

We need the conjugates of these norms, which are also convex functions in $\mathcal{R}^n$: *The conjugate of a Minkowski norm $\rho(\mathbf{x})$ with dual $\sigma$ is the function $\rho^c(\mathbf{y})$ equal to 0 for $\sigma(\mathbf{y}) \leq 1$ and to $+\infty$ for $\sigma(\mathbf{y}) > 1$.* (See Problem 4.63(a).)

There is a similar statement with $\sigma$ and $\rho$ interchanged. By the rule (4.58), it now follows that the conjugate of the *concave* function $-\sigma(\mathbf{y})$ is the function $q(\mathbf{x})$ equal to 0 for $\rho'(\mathbf{x}) \leq 1$ and to $-\infty$ otherwise. (See Problem 4.63(b)).

We also observe that by the Hölder inequality (4.91), one has the rule

$$-\mathbf{x}^t\mathbf{y} \leq \rho'(\mathbf{x})\sigma(\mathbf{y}). \tag{4.100}$$

We also verify the rules

$$(w\rho)^D = \frac{1}{w}\rho^D, \quad (\rho')^D = (\rho^D)', \quad (\rho')^c = (\rho^c)', \tag{4.101}$$

where $w$ is a positive constant (Problem 4.64).

We can now state our duality theorem:

**Theorem D.** *Let $\rho$ and $\sigma$ be dual Minkowski norms in $\mathcal{R}^m$. Let $A$ be a constant $m \times n$ matrix of rank $m < n$. Let $\mathbf{b} \in \mathcal{R}^m$ and $\mathbf{c} \in \mathcal{R}^n$ be constant vectors. Let the equation $A^t\mathbf{y} = \mathbf{c}$ have a solution $\mathbf{y}_0$ with $\rho'(\mathbf{y}_0) < 1$. Then the following programs are dual:*

$$minimize \quad \psi(\mathbf{x}) = \sigma(A\mathbf{x} - \mathbf{b}) + \mathbf{c}^t\mathbf{x} \quad in \; \mathcal{R}^n; \tag{4.102}$$

$$maximize \quad \phi(\mathbf{y}) = \mathbf{b}^t\mathbf{y} \quad for \quad A^t\mathbf{y} = \mathbf{c} \quad and \quad \rho'(\mathbf{y}) \le 1. \tag{4.103}$$

REMARK. For a symmetric norm $\rho$, the prime ($'$) can be dropped.

*Proof of the Theorem.* We apply the Rockafellar Theorem R of Section 4.7. We take the function $f$ to be identically 0 and $g(\mathbf{u}) = -\sigma(\mathbf{u})$. Thus the function $\psi(\mathbf{x})$ of (4.70) reduces to the function $\psi$ in (4.102). As in Example 3 in Section 4.5, the conjugate of $f$ is the function $f^c(\mathbf{v})$ which equals 0 for $\mathbf{v} = \mathbf{0}$ and otherwise equals $+\infty$. Hence $f^c(A^t\mathbf{y} - \mathbf{c})$ has the value 0 for $A^t\mathbf{y} = \mathbf{c}$ and otherwise equals $+\infty$. The function $g^c(\mathbf{y})$ has the value 0 for $\rho'(\mathbf{y}) \le 1$ and is otherwise equal to $-\infty$. Accordingly $\phi(\mathbf{y})$ reduces to the linear function $\mathbf{b}^t\mathbf{y}$ for $A^t\mathbf{y} = \mathbf{c}$ and $\rho'(\mathbf{y}) \le 1$; otherwise $\phi$ has the value $-\infty$.

For Theorem R we form the functions $f_1(\mathbf{x}) = f(\mathbf{x}) + \mathbf{c}^t\mathbf{x}$ and $g_1(\mathbf{x}) = g(A\mathbf{x} - \mathbf{b})$. We see at once that $K(f_1)$ and $K(g_1)$ reduce to $\mathcal{R}^n$, so hypothesis 1 of Theorem R is satisfied. Hypothesis 2 is fulfilled, since $g$ is stable at every relatively interior point of $K(g) = \mathcal{R}^m$.

We next remark that $\tau(\mathbf{x}) = \sigma(A\mathbf{x})$ is a Minkowski norm in $\mathcal{R}^n$. (See Problem 4.65.) Let $E$ denote the constraint set for the program (4.103). The point $\mathbf{y}_0$ is in $E$, so $E$ is nonempty. We write $2d = 1 - \rho'(\mathbf{y}_0)$, a positive constant.

For $\mathbf{y} \in E$ and $\mathbf{x} \in \mathcal{R}^n$,

$$\phi(\mathbf{y}) = \mathbf{b}^t\mathbf{y} = \mathbf{y}^t\mathbf{b} = -\mathbf{y}^t(A\mathbf{x} - \mathbf{b}) + \mathbf{y}^t A\mathbf{x} = -\mathbf{y}^t(A\mathbf{x} - \mathbf{b}) + \mathbf{c}^t\mathbf{x} \tag{4.104}$$

so that by the rule (4.100),

$$\phi(\mathbf{y}) \le \rho'(\mathbf{y})\sigma(A\mathbf{x} - \mathbf{b}) + \mathbf{c}^t\mathbf{x} \le \psi(\mathbf{x}).$$

Therefore $\psi$ is bounded below and hypothesis 3 of Theorem R is satisfied.

Now for each $\mathbf{x} \in \mathcal{R}^n$,

$$\psi(\mathbf{x}) = \sigma(A\mathbf{x} - \mathbf{b}) + (A^t\mathbf{y}_0)^t\mathbf{x} \ge \sigma(A\mathbf{x}) - \sigma(\mathbf{b}) - \rho'(\mathbf{y}_0)\sigma(A\mathbf{x})$$

by (4.100) and rule (iii) of Section 4.9. Therefore

$$\psi(\mathbf{x}) \geq \tau(\mathbf{x})(1 - \rho'(\mathbf{y}_0)) - \sigma(\mathbf{b}) = 2d\tau(\mathbf{x}) - \sigma(\mathbf{b}). \qquad (4.105)$$

Accordingly $\psi$ has limit $+\infty$ as $\tau(\mathbf{x}) \to \infty$. Choose $\tau_0$ so that $\psi(\mathbf{x}) \geq \psi(\mathbf{0}) = \sigma(-\mathbf{b})$ for $\tau(\mathbf{x}) \geq \tau_0$. Then, by continuity, $\psi$ has a minimum at most equal to $\psi(\mathbf{0})$ on the bounded closed set $\{\mathbf{x} \mid \tau(\mathbf{x}) \leq \tau_0\}$, and this is the minimum of $\psi$ in $\mathcal{R}^n$. Thus $\psi$ has a minimizer $\mathbf{x}^*$. [*Remark*. The inequality (4.105) shows that

$$\psi(\mathbf{x}) > \psi(\mathbf{0}) = \sigma(-\mathbf{b}) \quad \text{for} \quad \tau(\mathbf{x}) > \tau_1 = \frac{\sigma(\mathbf{b}) + \sigma(-\mathbf{b})}{1 - \rho'(\mathbf{x}_0)}$$

and hence the minimizer must lie in the set $\tau(\mathbf{x}) \leq \tau_1$.]

By Theorem R, the minimum $\psi(\mathbf{x}^*)$ equals the maximum of $\phi$. Therefore Theorem D is proved.                                                                                       ∎

We observe that in general the minimizer $\mathbf{x}^*$ is not unique. The maximizer $\mathbf{y}^*$ of the linear function $\phi$ is normally unique. It can fail to be unique if the relative boundary of $E$ includes a line segment.

## 4.11   APPLICATION TO FACILITY LOCATION

We here show the result of Witzgall (1964). Given $k$ distinct fixed points $\mathbf{b}_1$, $\mathbf{b}_2, \ldots, \mathbf{b}_k$ in $\mathcal{R}^n$, we choose a Minkowski norm $\tau$ in $\mathcal{R}^n$. As above we denote the norm $\tau(-\mathbf{x})$ by $\tau'(\mathbf{x})$. We select $2k$ positive scalars $w_1, w_1', \ldots, w_k, w_k'$ and set

$$\psi(\mathbf{x}) = \sum_{i=1}^{k} [w_i \tau(\mathbf{x} - \mathbf{b}_i) + w_i' \tau'(\mathbf{x} - \mathbf{b}_i)].$$

The program "minimize $\psi$ in $\mathcal{R}^n$" is clearly a generalization of the Fermat–Weber problem; we recover the original problem by setting $n = 2$, requiring $\tau$ to be the Euclidean norm, and choosing all $w_i$ and all $w_i'$ to be $\frac{1}{2}$. Witzgall found the dual of the generalized problem to be the following:

$$\textit{maximize} \quad \phi(\mathbf{y}) = -\sum_{i=1}^{k} \mathbf{b}_i^t(\mathbf{y}_i + \mathbf{y}_i')$$

in $\mathcal{R}^m$, where $m = 2kn$ and $\mathbf{y} \in \mathcal{R}^m$ is written as

$$(\mathbf{y}_1^t, \mathbf{y}_1'^t, \ldots, \mathbf{y}_k^t, \mathbf{y}_k'^t)^t,$$

with each $\mathbf{y}_i$ and $\mathbf{y}'_i$ in $\mathcal{R}^n$, subject to the constraints

$$\sum_{i=1}^{k}(\mathbf{y}_i + \mathbf{y}'_i) = \mathbf{0}$$

and

$$\eta(\mathbf{y}_i) \leq w_i \quad \text{and} \quad \eta'(\mathbf{y}'_i) \leq w'_i, \qquad i = 1, \ldots, k,$$

where $\eta$ is the dual norm of $\tau$.

In the case of a symmetric norm $\tau$, Witzgall found the dual problem to be that of maximizing $\phi(\mathbf{y})$ in $\mathcal{R}^m$ where $m = kn$, the vectors $\mathbf{y}'_i$ are deleted from the expressions for $\phi$, for $\mathbf{y}$ and in the sum of the first constraint, and of the remaining constraints only those concerning the $\mathbf{y}_i$ are retained. In particular, for the original Fermat–Weber problem in the plane considered in Section 3.10, the same dual is obtained as given in Section 3.10. (See Problem 4.66.)

We verify that Witzgall's duality theorem is a special case of our Theorem D. To that end, we first observe that the function

$$\sigma(\mathbf{y}) = \sum_{i=1}^{k}(w_i \tau(\mathbf{y}_i) + w'_i \tau'(\mathbf{y}'_i)) \tag{4.110}$$

is a norm in $\mathcal{R}^m$. (See Problem 4.67.) If we denote the vector

$$(\mathbf{b}^t_1, \ \mathbf{b}^t_1, \ \ldots, \ \mathbf{b}^t_k, \ \mathbf{b}^t_k)^t$$

of $\mathcal{R}^m$ by $\mathbf{b}$ and let $A$ be the $m \times n$ matrix $(I_n, \ldots, I_n)^t$ in block notation, then we find that

$$\psi(\mathbf{x}) = \sigma(A\mathbf{x} - \mathbf{b}).$$

(See Problem 4.68.)

We now apply Theorem D to obtain the dual program. The function $\psi$ is given by (4.102) with $\mathbf{c} = \mathbf{0}$. Hence the dual function is given by (4.103) with $\mathbf{c} = \mathbf{0}$ and $\rho = \sigma^D$; thus

$$\phi(\mathbf{y}) = \mathbf{b}^t\mathbf{y} \quad \text{for } A^t\mathbf{y} = \mathbf{0} \quad \text{and} \quad \rho'(\mathbf{y}) \leq 1. \tag{4.111}$$

The equation $A^t\mathbf{y} = \mathbf{0}$ is satisfied by $\mathbf{y} = \mathbf{y}_0 = \mathbf{0}$, which satsfies $\rho'(\mathbf{y}_0) < 1$. The equation $A^t\mathbf{y} = \mathbf{0}$ becomes

$$\sum_{i=1}^{k}(\mathbf{y}_i + \mathbf{y}'_i) = \mathbf{0}.$$

The condition $\rho'(\mathbf{y}) \leq 1$ we analyze as follows. We first apply the rule 12 of Section 4.5 to the function $-\sigma$. The rule is stated for a separable convex function formed of two terms; by applying it repeatedly, we can apply it to the separable concave function $-\sigma$ formed of $2k$ terms. We conclude that

$$(-\sigma)^c = \sum_{i=1}^{k} ((-w_i\tau)^c + (-w_i'\tau')^c). \tag{4.112}$$

As noted at the beginning of Section 4.10, the conjugate of $-\sigma$ is the function equal to 0 for $\rho'(\mathbf{y}) \leq 1$ and to $-\infty$ otherwise. Thus the condition $\rho'(\mathbf{y}) \leq 1$ is *equivalent* to the condition $(-\sigma)^c(\mathbf{y}) = 0$. But in (4.112) a similar consideration applies to each term on the right. Since each term has only the values 0 and $-\infty$, the sum can reduce to 0 when and only when every term on the right is 0. By (4.101) the dual of $w_i\tau$ is $(1/w_i)\eta$, and the dual of $w_i'\tau'$ is $(1/w_i')\eta'$. Hence the condition $\rho'(\mathbf{y}) \leq 1$ is equivalent to the condition that for $i = 1, \ldots, k$, $\eta'(\mathbf{y}_i) \leq w_i$ and $\eta(\mathbf{y}_i') \leq w_i'$.

Thus application of Theorem D leads to the dual program

$$maximize \quad \phi(\mathbf{y}) = \mathbf{b}^t\mathbf{y},$$

where

$$\sum_{i=1}^{k} (\mathbf{y}_i + \mathbf{y}_i') = \mathbf{0},$$

and for $i = 1, \ldots, k$,

$$\eta'(\mathbf{y}_i) \leq w_i \quad and \quad \eta(\mathbf{y}_i') \leq w_i'.$$

If we now replace each $\mathbf{y}_i$ by its negative, this program becomes the same as that found by Witzgall.

## Applications Involving an Unsymmetric Norm

We wish to find the best location for a center of distribution of objects by airplane, and to that end we seek to choose the location that minimizes total travel time to the $k$ distribution points. If we take the speed $w$ of the prevailing wind into account, then the time for the displacement vector $\mathbf{x}$ becomes

$$\tau(\mathbf{x}) = \frac{\|\mathbf{x}\|}{\sqrt{v_0^2 - w^2 \sin^2 \theta} + w \cos \theta}, \tag{4.113}$$

where $v_0$ is airspeed of the plane and $\theta$ is the angle between the prevailing wind direction and the vector $\mathbf{x}$. For simplicity, we consider the case where $w$ is small compared to $v_0$, so that the term in $w^2$ under the square root sign can be replaced by 0, and we have

$$\tau(\mathbf{x}) = \frac{\|\mathbf{x}\|}{v_0 + w \cos \theta}. \tag{4.114}$$

We now verify that $\tau$ is a Minkowski norm if $v_0 \geq 2w$ and that for $w > 0$ this norm is unsymmetric (Problem 4.69). Thus we are led to a Fermat–Weber problem with unsymmetric norm. There are other physical effects leading to such lack of symmetry, such as travel on a river.

## Multifacility Location

We take a very simple case of this problem in order to illustrate the ideas. We assume two given points $(p_1, q_1)$, $(p_2, q_2)$ in the $xy$-plane and try to place two new points $(x_1, y_1)$ and $(x_2, y_2)$ in the plane so as to minimize the distance between the new points plus the sum of the distances of the new points from the old ones. Thus we seek to minimize

$$
\begin{aligned}
\psi(x_1, y_1, x_2, y_2) = {} & \|(x_1 - p_1, \; y_1 - q_1)^t\| + \|(x_1 - p_2, \; y_1 - q_2)^t\| \\
& + \|(x_2 - p_1, \; y_2 - q_1)^t\| + \|(x_2 - p_2, \; y_2 - q_2)^t\| \\
& + \|(x_2 - x_1, \; y_2 - y_1)^t\|.
\end{aligned}
$$

Here the norm is the Euclidean norm in $\mathcal{R}^2$.

We observe that the each term on the right can be written in the form

$$
\|A_i(x_1, \; y_1, \; x_2, \; y_2)^t - (r_i, \; s_i)^t\|.
$$

For example, for the first term on the right

$$
A_i = \begin{bmatrix} 1 & 0 & 0 & 0 \\ 0 & 1 & 0 & 0 \end{bmatrix},
$$

and $r_i = p_1$, $s_i = q_1$. For the last term,

$$
A_i = \begin{bmatrix} -1 & 0 & 1 & 0 \\ 0 & -1 & 0 & 1 \end{bmatrix}
$$

and $r_i = s_i = 0$. Thus the right side can be written as

$$
\sum_{i=1}^{5} \|A_i(x_1, \; y_1, \; x_2, \; y_2)^t - (r_i, \; s_i)^t\|.
$$

If we now consider $\mathcal{R}^{10}$ as formed from five successive copies of $\mathcal{R}^2$, as above, then we obtain a norm $\sigma$ in $\mathcal{R}^{10}$ from the Euclidean norms in the 2-dimensional subspaces. If we denote by $\mathbf{y}$ a vector of this $\mathcal{R}^{10}$, and write

$$
\mathbf{y} = (\mathbf{y}_1^t, \ldots, \mathbf{y}_5^t)^t
$$

in terms of the 2-dimensional vectors $\mathbf{y}_i$, then

$$\sigma(\mathbf{y}) = \sum_{i=1}^{5} \|\mathbf{y}_i\|$$

and we can write

$$\psi(\mathbf{x}) = \sigma(A\mathbf{x} - \mathbf{b}),$$

where $A = (A_1^t, \ldots, A_5^t)^t$, in block notation, $\mathbf{x}$ is the vector $(x_1, y_1, x_2, y_2)^t$ of $\mathcal{R}^4$, and $\mathbf{b}$ is the vector $(r_1, s_1, \ldots, r_5, s_5)^t$ of $\mathcal{R}^{10}$.

Thus the multifacility optimization problem considered here is a special case of (4.102), and its dual is obtained from Theorem D.

The procedure followed in this example clearly generalizes to a multifacility location problem in $\mathcal{R}^n$, with any number of fixed points and any number (at least 2) of new location points in $\mathcal{R}^n$ and with an arbitrary norm $\tau$ in $\mathcal{R}^n$. One could also introduce positive numerical coefficients and allow for a nonsymmetric norm as in the problem considered by Witzgall. In each case what is obtained is a form of (4.102) with $\mathbf{c} = \mathbf{0}$ and an appropriate norm $\sigma$ and matrix $A$.

In particular, for our 2-dimensional example, we can apply Theorem D, as for the facility location problem of Witzgall, to obtain the dual:

$$maximize \quad \phi(\mathbf{y}) = \mathbf{b}^t \mathbf{y},$$

where

$$\mathbf{b} = (p_1, q_1, p_2, q_2, p_1, q_1, p_2, q_2, 0, 0)^t,$$

subject to the constraints

$$\|\mathbf{y}_i\| \le 1, \qquad i = 1, \ldots, 5,$$

and

$$y_{11} + y_{21} - y_{51} = 0, \qquad y_{12} + y_{22} - y_{52} = 0,$$

$$y_{31} + y_{41} + y_{51} = 0, \qquad y_{32} + y_{42} + y_{52} = 0,$$

where $y_{ij}$, $j = 1, 2$, are the components of $\mathbf{y}_i$. For this example, simple geometric reasoning locates a minimizer of $\psi$, and the duality is easily verified (Problem 4.70).

### Differentiable Case

In Theorem D, let the norm $\sigma$ be differentiable at $\mathbf{u}^* = A\mathbf{x}^* - \mathbf{b}$, where $\mathbf{x}^*$ is a minimizer of $\psi$. Then the gradient of $\psi$ reduces to $\mathbf{0}$ at $\mathbf{x}^*$. If we let

$$\mathbf{y}^* = -\nabla\sigma(\mathbf{u}), \qquad \mathbf{u} = A\mathbf{x}^* - \mathbf{b}, \tag{4.115}$$

then we verify that $\phi(\mathbf{y}^*) = \psi(\mathbf{x}^*)$ so that $\mathbf{y}^*$ is a maximizer of $\phi$. (See Problem 4.71.) Thus the equation (4.115) allows one to recover a maximizer of $\phi$ from a minimizer of $\psi$.

Similarly let a maximizer $\mathbf{y}^*$ of $\phi$ be a relative boundary point of the constraint set $E$ at which the norm $\rho'$ is differentiable. Then $\mathbf{y}^*$ is a maximizer of $\phi$ subject to the side conditions $\rho' \leq 1$ and $A^t\mathbf{y} = \mathbf{c}$. We verify that the Karush–Kuhn–Tucker conditions are applicable so that there exist a vector $\mu$ and a positive scalar $\lambda$ such that

$$\mathbf{b} - A\mu = \lambda\nabla\rho'(\mathbf{y}^*). \tag{4.116}$$

We verify that with $\mathbf{x}^* = \mu$, $\psi(\mathbf{x}^*) = \phi(\mathbf{y}^*)$. Thus the equation (4.116) allows us to recover a minimizer of $\psi$. (See Problem 4.72.)

## PROBLEMS

**4.62.** Let $\rho$ be a Minkowski norm in $\mathcal{R}^n$. Write $\rho'(\mathbf{x}) = \rho(-\mathbf{x})$. Verify that $\rho'$ is also a Minkowski norm in $\mathcal{R}^n$.

**4.63.** Let $\rho, \sigma$ be dual Minkowski norms in $\mathcal{R}^n$.

   **(a)** Show that $\rho^c(\mathbf{y}) = 0$ for $\sigma(\mathbf{y}) \leq 1$ and $\rho^c(\mathbf{y}) = \infty$ otherwise.

   **(b)** Apply the result of **(a)** and the rule (4.58) to show that the conjugate of the *concave* function $-\sigma(\mathbf{y})$ is the function $q(\mathbf{x})$ equal to 0 for $\rho'(\mathbf{x}) \leq 1$ and to $-\infty$ otherwise.

**4.64.** Let $w$ be a positive constant and let $\rho$ be a Minkowski norm in $\mathcal{R}^n$. Verify the rules

   **(a)** $(w\rho)^D = \rho^D/w$.

   **(b)** $(\rho')^D = (\rho^D)'$.

   **(c)** $(\rho')^c = (\rho^c)'$.

**4.65.** Let $A$ be an $m \times n$ matrix of rank $m$, and let $\sigma$ be a Minkowski norm in $\mathcal{R}^m$. Show that $\tau(\mathbf{x}) = \sigma(A\mathbf{x})$ is a Minkowski norm in $\mathcal{R}^n$.

**4.66.** Show that the dual found by Witzgall of the Fermat–Weber problem considered by him for the case of a symmetric norm $\tau$ includes as a special case the dual given in Section 3.10.

**4.67. (a)** Let $p = m + n$, and let each vector of $\mathcal{R}^p$ be written as

$$\mathbf{z} = (x_1, \ldots, x_m, y_1, \ldots, y_n)^t$$

in terms of vectors $\mathbf{x}$ in $\mathcal{R}^m$ and $\mathbf{y}$ in $\mathcal{R}^n$. Let $\rho$ be a Minkowski norm in $\mathcal{R}^m$, $\sigma$ a Minkowski norm in $\mathcal{R}^n$; let $a$ and $b$ be positive constants. Show that the function that assigns to each $\mathbf{z}$, as above, in $\mathcal{R}^p$ the number $a\rho(\mathbf{x}) + b\sigma(\mathbf{y})$ is a Minkowski norm in $\mathcal{R}^p$.

**(b)** Apply the result of part **(a)** repeatedly to show that the function

$$\sigma(\mathbf{y}) = \sum_{i=1}^{k}(w_i\tau(\mathbf{y}_i) + w_i'\tau'(\mathbf{y}_i'))$$

introduced in (4.110) is a norm in $\mathcal{R}^m$.

**4.68.** Let $\psi(\mathbf{x})$ be as defined in the first displayed equation of Section 4.11. As in the subsequent discussion, let

$$\mathbf{b} = (\mathbf{b}_1^t, \mathbf{b}_1^t, \ldots, \mathbf{b}_k^t, \mathbf{b}_k^t)^t$$

and let $A$ be the $m \times n$ matrix $(I_n, \ldots, I_n)^t$ in block notation. Verify that

$$\psi(\mathbf{x}) = \sigma(A\mathbf{x} - \mathbf{b}).$$

**4.69. (a)** Let $v_0 > 0$ and $w \geq 0$ be constants. For each vector $\mathbf{x}$ in $\mathcal{R}^2$ let $\tau(\mathbf{x})$ be defined as in (4.114), where $\theta$ is the polar angle of $\mathbf{x}$. Show that $\tau$ is a Minkowski norm if $v_0 \geq 2w$ and that for $w > 0$, this norm is unsymmetric. [Hint: The sublevel set $\tau(\mathbf{x}) \leq 1$ is bounded by the oval $r = v_0 + w\cos\theta$ in polar coordinates. Apply the results given at the end of Section 4.4 to show that this set is a bounded closed convex region, and then show that $\tau$ is the corresponding Minkowski norm. The lack of symmetry can be verified directly.]

**(b)** Consider the problem of delivery of articles by airplane moving at airspeed $v_0$ with wind speed $w$, as discussed in Section 4.11. Justify the formula (4.113) for the time for the plane to make the displacement $\mathbf{x}$, where $\theta$ is the angle between the prevailing wind direction and $\mathbf{x}$.

**4.70.** Consider the multifacility location problem for two given points and two new facilities given in Section 4.11.

**(a)** Show by geometric reasoning that $\psi$ is minimized if the two new facilities are located at the same point on the line segment joining the two given points $(p_i, q_i)$, $i = 1, 2$, and that for each such location $\psi$ has its minimum value equal to twice the distance between the two given points.

**(b)** Verify the duality theorem given for this problem for the special case $p_1 = -1$, $p_2 = 1$, $q_1 = q_2 = 0$ by showing that $\psi_{\min} = \phi_{\max} = 4$.

**4.71.** In Theorem D, let the norm $\sigma$ be differentiable at $\mathbf{u}^* = A\mathbf{x}^* - \mathbf{b}$, where $\mathbf{x}^*$ is a minimizer of $\psi$. Define $\mathbf{y}^*$ by (4.115), and verify that $\phi(\mathbf{y}^*) = \psi(\mathbf{x}^*)$.

**4.72.** In Theorem D, let a maximizer $\mathbf{y}^*$ of $\phi$ be a relative boundary point of the constraint set $E$ at which the norm $\rho'$ is differentiable so that $\mathbf{y}^*$ is a maximizer of $\phi$ subject to the side conditions $\rho' \leq 1$ and $A^t\mathbf{y} = \mathbf{c}$.

**(a)** Verify that the Karush–Kuhn–Tucker conditions are applicable so that there exist a vector $\mu$ and a positive scalar $\lambda$ such that (4.116) holds. [Hint: Show that the linear independence assumption of Section 2.6 is satisfied by showing that if it fails, then the linear variety $A^t\mathbf{y} = \mathbf{c}$ would fail to meet the set $\{\mathbf{y} \mid \rho'(\mathbf{y}) < 1\}$.]

**(b)** Verify that, with $\mathbf{x}^* = \mu$, $\psi(\mathbf{x}^*) = \phi(\mathbf{y}^*)$ so that $\mathbf{x}^*$ is a minimizer of $\psi$.

**4.73.** Show that the duality theorem proved in Section 3.8 is a special case of Theorem D.

**4.74.** Show that the duality theorem proved in Section 3.11 is a special case of Theorem D.

## REFERENCES

Bonnesen, T., and Fenchel, W. (1987). *Theory of Convex Bodies*, BSC Associates, Moscow, ID.

Chern, S. S. (1967). "Curves and surfaces in Euclidean space," in *Studies in Global Geometry and Analysis*, 16–56, Math. Assn. of America, Englewood Cliffs, NJ.

Dieudonné, J. (1960). *Foundations of Modern Analysis*, Academic Press, New York.

Kaplan, W. (1991). *Advanced Calculus*, 4th ed., Addison-Wesley, Reading, MA.

Kaplan, W., and Yang, W. H. (1997). "Duality theorem for a generalized Fermat–Weber problem", in *Mathematical Programming*, 76, 285–297.

Stoer, J., and Witzgall, C. (1970). *Convexity and Optimization in Finite Dimensions*, Springer, Berlin.

Webster, R. (1994). *Convexity*, Oxford University Press, Oxford.

Witzgall, C. (1964). "Optimal location of a central facility, mathematical models and concepts," National Bureau of Standards Report 8388, Washington, DC.

# Appendix

## Linear Algebra

In this appendix we give a concise review of aspects of linear algebra needed for this book. For a full discussion one is referred to Davis (1965), Kaplan (1991, ch. 1), Perlis (1991), and Strang (1976).

### A.1  MATRICES

A (real) *matrix* is a rectangular array of quantities having real values such as

$$\begin{bmatrix} 1 & 5 \\ 2 & -7 \end{bmatrix}, \quad \begin{bmatrix} a & b & c \\ x & x^2 & e^x \end{bmatrix},$$

where $a$, $b$, $c$, $x$ have real values. An $m \times n$ matrix has $m$ rows and $n$ columns; one says that the matrix has *size* $(m, n)$. The matrices shown have sizes $(2, 2)$ and $(2, 3)$, respectively. One often denotes a matrix by a single letter, such as $A$, and writes

$$A = \begin{bmatrix} 1 & 5 \\ 2 & -7 \end{bmatrix}.$$

One also writes $A = (a_{ij})$, where the ranges of the indices $i$ and $j$ are from 1 to $m$ and 1 to $n$, respectively, and $a_{ij}$ is the *ij*-entry of the matrix, the quantity in the $i$th row and $j$th column.

### A.2  SPECIAL MATRICES

A $1 \times n$ matrix is called a *row vector*; an $m \times 1$ matrix is called a *column vector*. When $m = n$, the $m \times n$ matrix $A$ is said to be a *square* matrix of *order* $n$; it has a determinant, denoted by det $A$. An important square matrix of order $n$ is the *identity matrix* $I = I_n$ for which $a_{ii} = 1$ for $i = 1, \ldots, n$ and all other $a_{ij} = 0$. The context generally indicates the order $n$. The *zero matrix* $O$ of size $(m, n)$ has all entries 0; the size is generally indicated by the context. One denotes by $\text{diag}(c_1, \ldots, c_n)$ the square matrix $A = (a_{ij})$ of order $n$ for which $a_{ii} = c_i$ for $i = 1, \ldots, n$ and $a_{ij} = 0$ otherwise; $A$ is said to be a *diagonal matrix*. In matrix theory, a number is called a *scalar*.

## A.3   TRANSPOSE

The *transpose* of matrix $A$ is $A^t = (b_{ij})$ where $b_{ij} = a_{ji}$ for all $i$ and $j$. Thus the transpose of a row vector is a column vector. We prefer to work with column vectors and often use the transpose to convert a row vector to a column vector. We commonly denote column vectors by lowercase boldface letters and write, for example,

$$\mathbf{a} = \begin{bmatrix} 3 \\ 5 \\ 2 \end{bmatrix} = (3, \ 5, \ 2)^t.$$

## A.4   OPERATIONS ON MATRICES

If $k$ is a scalar and $A = (a_{ij})$, $B = (b_{ij})$, then

$$kA = (ka_{ij}),$$

$$A + B = (a_{ij} + b_{ij}),$$

provided that $A$ and $B$ have the same size;

$$-B = (-1)B \quad \text{and} \quad A - B = A + (-B),$$

$$AB = C = (c_{ij}) \qquad \text{for } c_{ij} = a_{i1}b_{1j} + \cdots + a_{ir}b_{rj},$$

where $A$ has size $(m, r)$ and $B$ has size $(r, n)$;

$$A^2 = AA, \quad A^3 = A(A^2), \ldots,$$

provided that $A$ is square.

## A.5  RULES

The operations obey the following rules, which apply whenever the sizes of the matrices concerned make the statement meaningful:

$$AI = A, \quad IA = A, \quad A + O = A, \quad O + A = A;$$

$$AO = O, \quad OA = O, \quad kO = O, \quad A + B = O \quad \text{if and only if } B = -A;$$

$$A + B = B + A, \quad A + (B + C) = (A + B) + C, \quad (k + l)A = kA + lA;$$

$$A(BC) = (AB)C, \quad k(lA) = (kl)A, \quad A^{m+n} = A^m A^n, \quad (A^m)^n = A^{mn};$$

$$(A^t)^t = A, \quad (A + B)^t = A^t + B^t, \quad (AB)^t = B^t A^t.$$

## A.6  INVERSE MATRIX

Let matrix $A$ be square. Matrix $B$ is said to be an inverse of $A$ and $A$ is said to be invertible (or nonsingular) if $AB = I$. When such a matrix $B$ exists, one has also $BA = I$, and $B$ is the unique matrix such that $AB = I$. $B$ is also the unique matrix such that $BA = I$. One writes $B = A^{-1}$. If $A$ and $C$ are invertible and have the same order, then $AC$ is invertible, and its inverse is $C^{-1}A^{-1}$. If $A$ is invertible, then so is $A^t$ and its inverse is $(A^{-1})^t$. $A$ is invertible if and only if $\det A \neq 0$, and when it is, $\det A^{-1} = 1/\det A$.

## A.7  LINEAR EQUATIONS

Let $A$ have size $(m, n)$, and let $\mathbf{x} = (x_1, \ldots, x_n)^t$ and $\mathbf{b} = (b_1, \ldots, b_n)^t$ be column vectors. The equation $A\mathbf{x} = \mathbf{b}$ is then equivalent to $m$ linear equations in the $n$ unknowns $x_1, \ldots, x_n$. When $m = n$ and $A$ is invertible, the equations have the unique solution $\mathbf{x} = A^{-1}\mathbf{b}$. (See also A.19.)

## A.8  LINEAR INDEPENDENCE

For fixed positive integer $n$, all $n \times 1$ column vectors $\mathbf{x}$, $\mathbf{y}$, ... are said to form the *vector space* $\mathcal{R}^n$. Vectors $\mathbf{x}_1, \ldots, \mathbf{x}_m$ of $\mathcal{R}^n$ are said to be *linearly independent* if the only *linear combination*

$$c_1\mathbf{x}_1 + \cdots + c_m\mathbf{x}_m$$

of these vectors equal to $\mathbf{0} = (0, \ldots, 0)^t$ is $0\mathbf{x}_1 + \cdots + 0\mathbf{x}_m$; hence the vectors are linearly dependent (not linearly independent) if and only if one of the $m$ vectors can be expressed as a linear combination of the other vectors. There are $n$ linearly independent vectors in $\mathcal{R}^n$: namely the vectors $\mathbf{e}_1 = (1, 0, \ldots, 0)^t, \ldots, \mathbf{e_n} = (0, 0, \ldots, 1)^t$. There are no more than $n$ linearly independent vectors in $\mathcal{R}^n$. In general,

$$\mathbf{x} = (x_1, \ldots, x_n)^t = x_1\mathbf{e_1} + \cdots + x_n\mathbf{e_n}.$$

## A.9  SUBSPACES

A subset $W$ of $\mathcal{R}^n$ is said to be a subspace of $\mathcal{R}^n$ if it contains the vector $\mathbf{0} = (0, \ldots, 0)^t$ and if, for each scalar $k$ and each pair of vectors $\mathbf{x}$, $\mathbf{y}$ in $W$, the vectors $k\mathbf{x}$ and $\mathbf{x} + \mathbf{y}$ are in $W$. The *dimension* of $W$ is the integer $m \geq 0$ such that $W$ contains $m$, and no more than $m$, linearly independent vectors. If $m > 0$ and $\mathbf{v}_1, \ldots, \mathbf{v}_m$ are linearly independent vectors in $W$, then every vector $\mathbf{v}$ of $W$ can be expressed uniquely as a linear combination of $\mathbf{v}_1, \ldots, \mathbf{v}_m$; these $m$ vectors are said to form a *basis* of $W$. In particular, $\mathcal{R}^n$ has dimension $n$, and $\mathbf{e}_1, \ldots, \mathbf{e}_n$ form a basis of $\mathcal{R}^n$ called the *standard basis*. If $\mathbf{w}_1, \ldots, \mathbf{w}_k$ are vectors of $\mathcal{R}^n$, then all their linear combinations form a subspace $W$ of $\mathcal{R}^n$; if they are linearly independent, then $W$ has dimension $k$ and $\mathbf{w}_1, \ldots, \mathbf{w}_k$ form a basis of $W$. If they are linearly dependent and at least one of the vectors is not $\mathbf{0}$, then for some $m < k$, $m$ of the vectors are linearly independent and form a basis of $W$.

## A.10  RANK

For each $m \times n$ matrix $A = (a_{ij})$ the columns of $A$ are column vectors in $\mathcal{R}^m$ and the rows are the transposes of column vectors in $\mathcal{R}^n$. The linear combinations of the columns form a subspace of $\mathcal{R}^m$ called the *column space* of $A$; the linear combinations of the transposes of the rows form a subspace of $\mathcal{R}^n$ called the *row space* of $A$. These two subspaces have the same dimension, called the *rank* of $A$. The rank is at most equal to $\min(m, n)$ and when it equals $\min(m, n)$, $A$ is said to have *maximum rank*. When $A$ is square, so $m = n$, $A$ has maximum rank precisely when $A$ is invertible.

## A.11  INNER PRODUCT AND NORM

In $\mathcal{R}^n$ each pair of vectors $\mathbf{u}$, $\mathbf{v}$ has an inner product

$$(\mathbf{u}, \mathbf{v}) = \mathbf{u}^t\mathbf{v} = u_1v_1 + \cdots + u_nv_n.$$

Hence  $(\mathbf{u}, \mathbf{v}) = (\mathbf{v}, \mathbf{u})$,  $(\mathbf{u}, \mathbf{v} + \mathbf{w}) = (\mathbf{u}, \mathbf{v}) + (\mathbf{u}, \mathbf{w})$,  $(\mathbf{u}, k\mathbf{v}) = k(\mathbf{u}, \mathbf{v})$.  The vectors $\mathbf{u}$, $\mathbf{v}$ are said to be *orthogonal* if $(\mathbf{u}, \mathbf{v}) = 0$. The *Euclidean norm* of $\mathbf{u}$ is

$$\|\mathbf{u}\|_2 = (\mathbf{u}, \mathbf{u})^{1/2}.$$

Often the subscript 2 is omitted, when the context makes clear that the Euclidean norm is intended. This norm can be interpreted as distance from the vector **0** and it has the usual properties of distance in geometry. In particular, the Pythagorean theorem is valid: if **u**, **v** are orthogonal, then

$$\|\mathbf{u} + \mathbf{v}\|^2 = \|\mathbf{u}\|^2 + \|\mathbf{v}\|^2.$$

A vector of norm 1 is called a *unit vector.*

## A.12   ORTHOGONAL SYSTEMS

In $\mathcal{R}^n$ an orthogonal system of vectors is a set of nonzero vectors $\mathbf{u}_1, \ldots, \mathbf{u}_k$ such that $(\mathbf{u}_i, \mathbf{u}_j) = 0$ for $i \neq j$. If the vectors are all unit vectors, the orthogonal system is termed *orthonormal*. The standard basis of $\mathcal{R}^n$ is an orthonormal system. Each subspace $W$ of $\mathcal{R}^n$ containing more than one vector has an orthonormal basis; this can be obtained from a known basis $\{\mathbf{v}_i\}$ of $W$ by choosing each $\mathbf{u}_i$ as an appropriate linear combination of $\mathbf{v}_1, \ldots, \mathbf{v}_i$ (Gram–Schmidt process).

## A.13   ORTHOGONAL MATRICES

A square matrix $A = (a_{ij})$ of order $n$ is said to be orthogonal if its columns form an orthonormal system in $\mathcal{R}^n$; the transposes of its rows then also form an orthonormal system in $\mathcal{R}^n$. $A$ is orthogonal precisely when $A$ is invertible and $A^{-1} = A^t$.

## A.14   EIGENVALUES AND EIGENVECTORS

For each square matrix of order $n$, a nonzero vector **v** of $\mathcal{R}^n$ is an eigenvector of $A$ if $A\mathbf{v} = \lambda\mathbf{v}$ for some scalar $\lambda$; the scalar $\lambda$ is termed an eigenvalue of $A$, associated with the eigenvector **v**. Each eigenvalue of $A$ is a root of the *characteristic equation* of $A$:

$$\det(A - \lambda I) = 0,$$

an algebraic equation of degree $n$. This equation may have complex roots, which are termed complex eigenvalues of $A$.

## A.15   SYMMETRIC MATRICES

A square matrix $A$ is said to be symmetric if $A^t = A$. Let $A$ be a symmetric matrix. Then $A$ has only real eigenvalues and has a set of eigenvectors forming an orthonormal system. Furthermore there is an orthogonal matrix $Q$ such that $AQ = Q \operatorname{diag}(\lambda_1, \ldots, \lambda_n)$, where $\lambda_1, \ldots, \lambda_n$ are the eigenvalues of $A$, repeated according to multiplicity.

## A.16  SINGULAR VALUES

For each $m \times n$ matrix $A$ of rank $r > 0$, there are orthogonal matrices $Q_1$, $Q_2$ such that

$$A = Q_1 \Sigma Q_2,$$

where $\Sigma = (\sigma_{ij})$ and

$$\sigma_{11} \geq \sigma_{22} \geq \cdots \geq \sigma_{rr} > 0,$$

with $\sigma_{ij} = 0$ otherwise. The nonzero $\sigma_{ii}$ are called singular values of $A$.

## A.17  NULL SPACE

The null space of the $m \times n$ matrix $A$ is the set $N$ of all $\mathbf{x}$ in $\mathcal{R}^n$ such that $A\mathbf{x} = \mathbf{0}$. It is a subspace of $\mathcal{R}^n$, of dimension $n - r$, where $r$ is the rank of $A$. The null space of $A$ and the row space $R$ of $A$ are orthogonal complements of each other: that is, $\mathbf{p}^t \mathbf{q} = 0$ for $\mathbf{p}$ in $N$ and $\mathbf{q}$ in $R$, and each $\mathbf{x}$ in $\mathcal{R}^n$ can be expressed uniquely as $\mathbf{p} + \mathbf{q}$, where $\mathbf{p}$ is in $N$ and $\mathbf{q}$ is in $R$.

## A.18  RANGE

The column space $C$ of the $m \times n$ matrix $A$ coincides with the range of $A$: that is, with the set of all $\mathbf{y}$ in $\mathcal{R}^m$ such that $\mathbf{y} = A\mathbf{x}$ for some $\mathbf{x}$. $C$ and the null space of $A^t$ are orthogonal complements of each other. If $A$ has maximum rank, then $P = A(A^t A)^{-1} A^t$ is the orthogonal projection of $\mathcal{R}^m$ onto the range of $A$.

## A.19  MORE ON LINEAR EQUATIONS

Let $A$ be an $m \times n$ matrix. As in Section A.7 the matrix equation $A\mathbf{x} = \mathbf{b}$ is equivalent to $m$ linear equations in $n$ unkowns. The matrix equation also states that the vector $\mathbf{b}$ of $\mathcal{R}^m$ can be expressed as a linear combination of the column vectors $\mathbf{v}_1, \ldots, \mathbf{v}_n$ of $A$. Thus the matrix equation has a solution precisely when $\mathbf{b}$ is in the column space of $A$. This statement is equivalent to the statement that the matrices $A$ and $[A \mid \mathbf{b}]$ have the same rank. Here $[A \mid \mathbf{b}]$ is the *augmented* matrix of $A$: It is the $m \times (n + 1)$ matrix obtained by adjoining to $A$ an $(n + 1)$-th column equal to $\mathbf{b}$. When $\mathbf{b} = \mathbf{0}$ (case of *homogeneous* linear equations), there is always at least one solution: namely $\mathbf{x} = \mathbf{0}$; and the solutions form a subspace of $\mathcal{R}^n$ of dimension $n - r$, where $r$ is the rank of $A$. When $\mathbf{b} \neq \mathbf{0}$, the solutions (if there are any) are given by $\mathbf{x} = \mathbf{x}_0 + \mathbf{u}$, where $\mathbf{x}_0$ is one particular solution and $\mathbf{u}$ is an arbitrary solution of the corresponding homogeneous equation: $A\mathbf{x} = \mathbf{0}$.

## A.20 BLOCK MATRICES OR PARTITIONED MATRICES

A matrix can be partitioned by horizontal and vertical lines to form a rectangular array of matrices. Thus

$$
A = \left[
\begin{array}{cc|cc|ccc}
1 & 2 & 0 & 1 & 3 & 5 & 6 \\
3 & 4 & 0 & 2 & 4 & 1 & 7 \\
\hline
5 & 6 & 2 & 7 & 5 & 1 & 2
\end{array}
\right] = (A_{ij}),
$$

where

$$
A_{11} = \begin{bmatrix} 1 & 2 \\ 3 & 4 \end{bmatrix}, \qquad A_{21} = [5 \quad 6], \ldots
$$

Block matrices are used in this book only as a notation. However, they can be manipulated as ordinary matrices are; see, for example, Perlis (1991, sec. 1-6).

## REFERENCES

Davis, P. J. (1965). *The Mathematics of Matrices*, Blaisdell, Waltham, MA.

Kaplan, W. (1991). *Advanced Calculus*, 4th ed., Addison-Wesley, Reading, MA.

Perlis, S. (1991). *Theory of Matrices*, Dover, New York.

Strang, G. (1976). *Linear Algebra and Its Applications*, Academic Press, New York.

# Answers to Selected Problems

## CHAPTER 1

### Section 1.1

**1.1.** (a) No global max, global min is 1.

(b) Global min is 1, no global max.

(c) Local and global min of $-2$ at $x = 1$, no global max.

(e) Global max of $\frac{1}{2}$ at $n = 1$, global min of $10/101$ at $n = 10$.

(f) Global max of $9/8$ at $x = 3$, no min.

(g) Global max is $\sqrt{10}$, no min.

(h) Global min is $25\pi/8$.

**1.2.** (a) Critical points are $x = 1$, local min, and $x = -5$, local max.

(b) Local min at $x = 0$, local max at $x = -2$.

(c) Local min at $x = 2.0344$ and $5.1760$, local max at $x = 0.4636$ and $3.6052$.

(d) Local min at $x = 0$.

### Section 1.3

**1.6.** (a) and (b), strictly convex.

(c) and (d) not strictly convex.

### Section 1.4

**1.20.** Only (a), (c), and (e) are convex.

**1.24.** Inside and on the edges of the triangle with vertices $(0, 0), (1, 0), (0, 1)$.

## Section 1.5

**1.33.** $c = \sqrt{2}$ for **(a)** and $1/\sqrt{2}$ for **(b)**.

## Section 1.8

**1.37.** **(a)** Area is 42.5 square units.

**(b)** $\mathbf{x} = (3, 5, 1, 7)^t + t(2, 7, 1, 5)^t$, $\quad \mathbf{x} = (4, 4, 9, 1)^t + t(4, 1, 9, -5)^t$, not parallel.

**(c)** $(7/3, 8/3, 2/3, 16/3), (5/3, 1/3, 1/3, 11/3)$.

**(d)** Dimension is 4.

**(e)** Use standard basis for $\mathcal{R}^4$.

**1.39.** **(a)** Coincide when $\mathbf{a}_2 - \mathbf{a}_1$, $\mathbf{b}_1$, and $\mathbf{b}_2$ are all collinear.

**(b)** Intersect in a point when $\mathbf{b}_1$, $\mathbf{b}_2$ are not collinear and $\mathbf{a}_2 - \mathbf{a}_1$ is dependent on $\mathbf{b}_1$, $\mathbf{b}_2$.

**(c)** Parallel when $\mathbf{b}_1$, $\mathbf{b}_2$ are collinear but **(a)** fails.

**(d)** Skew when vectors of **(a)** are linearly independent.

**1.40.** **(a)** Rank of $A$ is 3, so $k = 2$.

**(b)** $(x_1, x_2, x_3)^t = (0.47, 0.93, 0.73)^t + D(t_1, t_2)^t$ and $x_4 = t_1, x_5 = t_2$, where

$$D = \begin{bmatrix} 0.33 & -0.73 & -0.47 \\ -0.33 & 0.53 & 0.07 \\ -0.33 & 0.13 & 0.27 \end{bmatrix}$$

**1.42.** **(a)** With $u = 3x_1 - 2x_2$, matrix is

$$H = \begin{bmatrix} 2\cos u - 12x_1 \sin u - 9x_1^2 \cos u & 4x_1 \sin u + 6x_1^2 \cos u \\ 4x_1 \sin u + 6x_1^2 \cos u & -4x_1^2 \cos u \end{bmatrix}$$

**(b)** Matrix is

$$H = \begin{bmatrix} 6x_1 + 8x_2 + 12x_3 & 8x_1 + 10x_2 & 12x_1 + 2x_3 \\ 8x_1 + 10x_2 & 10x_1 + 6x_2 & 0 \\ 12x_1 + 2x_3 & 0 & x_1 + 6x_3 \end{bmatrix}$$

## Section 1.11

**1.65.** With arbitrary constants $c_1, c_2$,

$$\int (c_1 - V(x))^{-1/2}\, dx = \pm \left(\frac{2}{m}\right)^{1/2} t + c_2.$$

**1.66. (a)** With arbitrary constants $c_1, c_2$,

$$\int \left( c_1 - \left( \frac{2g}{L} \right)(1 - \cos \theta) \right)^{-1/2} d\theta = \pm t + c_2.$$

**(b)** $0.0984, 0.1872, 0.2580$.

**(c)** $0.3208$.

**(d)** $2.016$ secs.

## Section 1.13

**1.74. (a)** $0$ for $x = 0$.

**(b)** $-1$ for $x = 0$.

**(c)** $1$ at $(0, \pm 1)$.

**(d)** $e^2 + e^{-1}$ at $(1, 1)$.

**1.80. (a)** $4$.

**(b)** $4.40$.

**(c)** $1.08$.

**(d)** $1.9601$.

**1.83. (a)** $b_{ij} = i^{4-j}$ for $i = 1, \ldots, 10, j = 1, \ldots, 4$.

**(c)** $\mathbf{u} = (-0.061932, 0.564462, 2.086833)^t$.

**(d)** $\mathbf{u} = (0.000139098, 9.675568)^t$.

## Section 1.15

**1.85. (a)** $\sqrt{58}$.

**(b)** $0.143$ rad.

**(c)** $2.74$ rad.

**1.86. (a)** $b_{11} = b_{12} = b_{22} = 1, b_{21} = 0$

**(b)** Diagonal of $B$ is $1$, $1$, $1.4142$, $1.8708$, $2.1712$, $2.4058$, $2.6129$, $2.8024$, $2.9787$, $3.1444$, $3.3013$. Entries $b_{i,i+1}$ are $1, 1, 0.7071, 0.5345$, $0.4157$, $0.3827$, $0.3568$, $0.3357$, $0.3180$; other entries are $0$.

## Section 1.16

**1.96.** $-0.0691$.

## Section 1.17

**1.97.** **(a)** $\mathbf{u} = (0.9963,\ 1.1026,\ 0.5367)^t$.

**(b)** $\mathbf{u} = (0,\ -0.2381,\ 0,\ 1.1111)^t$

# CHAPTER 2

## Section 2.1

**2.1.** **(a)** 1.3, exact value 1.3262.

**(b)** 1.1, exact value 1.0819.

**2.2.** **(a)** $7x + 4y - z = 9$.

**(b)** $0.8x - 0.96y - z = -1.28$.

**(c)** $2x + 2y + z = 9$.

**(d)** $x + 2y + z = 6$.

**2.5** Matrices are

**(a)** $\begin{bmatrix} 2 & -1 & 1 \\ 1 & 2 & -1 \\ 1 & -1 & 3 \end{bmatrix}$.   **(b)** $\begin{bmatrix} 2x_1 & -2x_2 \\ 2x_2 & 2x_1 \end{bmatrix}$.

**2.7.** $\begin{bmatrix} 4 & 8 \\ 4 & -8 \end{bmatrix}$.

**2.8.** **(a)** For example, $z_u = (vz^2 + vzw - uv^2z)/J$,
$w_u = (-vw^2 - uv^2w - vzw)/J$, where $J = u^2v^2 - uvz + uvw$

**2.10.** **(a)** $19/\sqrt{34}$.

**(b)** 0.

**(c)** $(2e^2/3)(\sin 2 + 3\cos 2)$.

## Section 2.2

**2.14.** **(a)** $(-2/\sqrt{5},\ -1/\sqrt{5})^t$, global min, $(2/\sqrt{5},\ 1/\sqrt{5})^t$, global max.

**(b)** Global min at $(\pm 1)(0.1222,\ 0.9925)^t$, global max at $(\pm 1)(0.9925,\ -0.1222)^t$.

**(c)** Global min for $x = 2^{-1/3}$, $y = \pm 2^{1/6}$.

**(d)** Local min for $x = 2^{-2/9}$, $y = 2^{1/9}$.

**2.15.** **(a)** Local max at $(1,\ 0,\ 0)^t$, $(0,\ 1,\ 0)^t$, $(0,\ 0,\ 1)^t$, and local min at $(\tfrac{1}{3})(-1,\ 2,\ 2)^t$, $(\tfrac{1}{3})(2,\ -1,\ 2)^t$, $(\tfrac{1}{3})(2,\ 2,\ -1)^t$.

(b) With $h = 1/\sqrt{2}$, at points $(\pm h, \pm h, \pm h, \pm h)^t$ have local max where $f$ is positive, local min where $f$ is negative; saddle points at $(\pm 1, 0, \pm 1, 0)$ etc.

## Section 2.4

**2.27.** (a) Solution leads to $(0, 0, 1)^t$.

## Section 2.6

**2.34.** (a) Max is $a$ at $x = 1$, min is $-a$ at $x = -1$.

(b) Max is 3 at $(1, 0)^t$, min is $-1$ at $(0, 1)^t$.

(c) Max is 2 at $(1, 1)^t$, min is $-2$ at $(-1, -1)^t$.

(d) $2/3$ at $(2/3, 1/3)^t$.

(e) $(2e)^{-1/2}$ at $(1/\sqrt{2}, 1/2)^t$.

(f) Max of $2\sqrt{3}/9$ at $x = z = \sqrt{2/3}$, $y = \sqrt{1/3}$, min of 0 at $(0, 1, 0)^t$ and $(1, 0, 1)^t$.

(g) 3 at $(0, 0, 1)^t$.

## Section 2.8

**2.41.** Min is $1/13$ at $(0, 2/13, 3/13)^t$.

## CHAPTER 3

## Section 3.1

**3.1.** (a) $(1, 0, 0), (0, 1, 0), (0, 0, 1)$.

(b) $(0, 1), (0, -1), (1, 2), (1, -2)$.

(c) $(\pm 1, \mp 1, 0), (0, \pm 1, \mp 1), (\pm 1, 0, \mp 1)$.

**3.2.** (a) Dimension is 2, boundary is the given set, interior is empty, relative boundary is formed of the edges of the triangle, and relative interior is the rest of the given set.

(b) Dimension is 2, boundary and relative boundary formed of 4 edges of a trapezoid, and interior and relative interior formed of rest of trapezoid.

(c) Dimension is 2, boundary is the given set, interior is empty, relative boundary formed of six edges of a hexagon, and relative interior formed of rest of hexagon.

**(d)** Dimension is 3, boundary and the relative boundary are formed of points with $2x^2 + y^2 + 3z^2 = 1$ and $z \geq 0$ and the points with $2x^2 + y^2 = 1$ and $z = 0$, and the interior and relative interior are the rest of the set.

**(e)** Dimension is 2, boundary is the given set, interior is empty, relative boundary is the circle $x_1^2 + x_2^2 + x_3^2 = 1$, $x_1 + x_2 + x_3 = 0$, and relative interior is the set $x_1^2 + x_2^2 + x_3^2 < 1$, $x_1 + x_2 + x_3 = 0$.

**3.11.** If $C$ has rank less than $m$, no conclusion in the case of strict convexity.

**3.16. (a)** The line segment joining the two points.

**(b)** Triangular surface with the three points as vertices.

**(c)** Solid tetrahedron with the four points as vertices.

**(d)** Triangular surface with the segment as one edge and the point as one vertex.

**(e)** Solid cone with the given point as apex and the given set as base.

## Section 3.3

**3.24. (a)** $(0, 0)$, global min.

**(b)** $(0, 0)$, saddle point.

**(c)** No critical point.

**(d)** Points of line $x + y = 0$, each a global minimizer.

**(e)** $(0, 0)$, global min.

**(f)** $(0, 0)$, saddle point.

**(g)** No critical point.

**(h)** $(-1, 3)$, global min.

**(i)** $(1, -2)$, global min.

**(j)** Points of line $x - y - 1 = 0$, each a global minimizer.

## Section 3.5

**3.30. (a)** minimizer $(0.1667, 0.5)^t$, dual is to maximize $\phi(y) = -1.5y^2 + 4y - 2.25$.

**(b)** minimizer $(0.6, 0, 0.8)^t$; dual is to maximize $\phi(\mathbf{y}) = -(\frac{1}{2})\mathbf{y}^t E \mathbf{y} + \mathbf{d}^t \mathbf{y} + k$ in $\mathcal{R}^2$, where

$$E = \begin{bmatrix} 1.6970 & 3.2727 \\ 3.2727 & 8.4545 \end{bmatrix},$$

$\mathbf{d} = (-0.9697, -2.7273)^t$, and $k = 6.6515$.

**3.37. (b)** $\mathbf{x}^{(1)} = (4, 1)^t$, $\mathbf{x}^{(2)} = \mathbf{x}^*$. **(c)** $\mathbf{x}^{(1)} = (11/14, 22/7)^t$, $\mathbf{x}^{(2)} = \mathbf{x}^*$.

## Section 3.7

**3.39. (a)** Max of 7/3 at $(1/3, 0)^t$.

    **(b)** Min of $-1.5$ at $(0, 1.5)^t$.

    **(c)** Min of 2 at $(1, 1)^t$.

    **(d)** No feasible points.

    **(e)** Max of 1 for all $\mathbf{x}$ with $x_1 + x_2 + x_3 = 1$ and all $x_i \geq 0$.

    **(f)** Min of 0 at $(0, 0, 1)^t$.

## Section 3.9

**3.66. (a)** Minimizer is $(-1.5, -0.5, 0)^t$.

    **(b)** $d_0 = (-4, -6, -2)^t$,

        $d_1 = (-2.1641, -0.5910, 2.0559)^t$,

        $d_2 = (-0.2298, 0.1806, -0.1333)^t$.

## CHAPTER 4

### Section 4.4

**4.16.** At $(\frac{1}{2}, \frac{1}{2})$, $z = x$; at $(1, 1)$, $x = 1$ or $z = 2x - 1$, for example; at $(1, 2)$, $x = 1$; at $(0, 0)$, $z = 0$ or $z = -x$, for example; at $(0, 1)$ and at $(0, 2)$, $x = 0$.

### Section 4.6

**4.18. (a)** Sup is 1, no max.

    **(b)** Sup is $\infty$, no max.

    **(c)** Sup is $\infty$, no max.

# Index

# WILEY-INTERSCIENCE
## SERIES IN DISCRETE MATHEMATICS AND OPTIMIZATION

### ADVISORY EDITORS

RONALD L. GRAHAM

*AT & T Laboratories, Florham Park, New Jersey, U.S.A.*

JAN KAREL LENSTRA

*Department of Mathematics and Computer Science,*
*Eindhoven University of Technology, Eindhoven, The Netherlands*

ROBERT E. TARJAN

*Princeton University, New Jersey, and*
*NEC Research Institute, Princeton, New Jersey, U.S.A.*